The Limits of Att

The Limits of Attention

Temporal Constraints in Human Information Processing

Edited by

KIMRON SHAPIRO

Centre for Cognitive Neuroscience
The University of Wales, Bangor

OXFORD
UNIVERSITY PRESS

OXFORD
UNIVERSITY PRESS

Great Clarendon Street, Oxford OX2 6DP

Oxford University Press is a department of the University of Oxford.
It furthers the University's objective of excellence in research, scholarship,
and education by publishing worldwide in

Oxford New York

Athens Auckland Bangkok Bogotá Bombay Buenos Aires Calcutta
Cape Town Dar es Salaam Delhi Florence Hong Kong Istanbul
Karachi Kuala Lumpur Madrid Melbourne Mexico City Mumbai
Nairobi Paris São Paulo Shanghai Singapore Taipei Tokyo Toronto Warsaw

and associated companies in Berlin Ibadan

Oxford is a registered trade mark of Oxford University Press
in the UK and in certain other countries

Published in the United States
by Oxford University Press Inc., New York

© Oxford University Press 2001

British Library Cataloguing in Publication Data

Data available

Library of Congress Cataloguing in Publication Data

ISBN 0 19 850516 7 Hbk
ISBN 0 19 850515 9 Pbk

10 9 8 7 6 5 4 3 2 1

Typeset by EXPO Holdings, Malaysia
Printed in Great Britain
on acid-free paper by T.J. International Ltd, Padstow, Cornwall

Preface

The question of how we distribute attention over time, as opposed to over space, has been of considerable interest to those studying dual-task interference in the past ten years. *The Limits of Attention: Temporal Constraints on Human Information Processing*, is the results of one attempt to answer this question. Twelve sets of authors with wide-ranging perspectives on this issue have come together in the present volume to share their views, based on both published and as yet unpublished data. The topic of 'dual-task interference' has been around for a long time and is of interest not only to cognitive scientists, but indeed to the 'person on the street', as all humans endeavour to address the age-old question of why we cannot walk and at the same time chew gum (alternatively, pat our heads and rub our stomachs). The topics and related paradigms offered by the volume's contributors as relevant to this age-old question include task switching, masking, cross-modal interference, object substitution, the psychological refractory period, serial vs. parallel processing, low-level vision, and the attentional blink, with approaches to these topics coming from behavioural, neurophysiological, and neuropsychological perspectives. It is my sincere belief that those wishing to receive a comprehensive introduction to this topic, in addition to those seeking to extend their knowledge, will benefit from reading the book.

At the risk of redundancy, I will reiterate the questions posed in Chapter 1, which I wrote, to entice the reader to read the book for answers to them. *What is the time-course of attention? What are the consequences of this time-course? Are there demonstrable neural events that can account for it? What can the results of temporal attention experiments add to our knowledge about whether there are modality-specific attentional mechanisms or one supra modal mechanism? What can such results add to our knowledge of the interaction between early- and late-processing, task demands, and response-selection mechanisms? And finally, what influence do low-level perceptual mechanisms have on the time-course of attention?* The answers to these questions are addressed in detail in the 11 chapters that follow my introductory one. The authors were chosen for their ability to contribute insights to these issues, based on knowledge accumulated through their own research contributions in this domain. I believe we have been successful in our endeavour, but of course it is the reader who will ultimately judge.

My own interest in the temporal aspects of dual-task interference arises from research in which I and a number of the present authors have played a role, examining the consequences of attention to a specified target. We have determined that attention directed to such a target for later report has the consequence of preventing accurate report of a second target occurring within approximately a half second of the first. This finding should immediately prompt the question, 'But then, how is it we are able to continuously perceive our environment?' To which I (and others, for example those investigating the phenomenon known as 'change blindness') respond, 'But how do you know for sure that we do?' This inability to correctly report the second of two targets occurring in close temporal proximity, which is described in detail in the present volume, has been termed the 'attentional blink', whose label is meant to analogise the phenomenon to an eye blink, during which time perception is prevented. Thanks to Jane Raymond for such a clever label! Unlike an eye blink, however, the attentional blink has been shown to represent an inability to accurately *report* the second target, but not an inability to *perceive* it! In the 10 years since my and my co-authors' (Raymond and Arnell) first report of the attentional blink outcome and method used to achieve it,[1] I am personally gratified that the phenomenon has generated so much interest. While only one of the foci of the present volume, a considerable number of the contributors have described their own research on this topic, which extends not only our knowledge of its boundary conditions, but also its relevance to other temporal attentional phenomenon.

I have immensely enjoyed reading literature from, and being a part of, the past 10 years of work on this book's topic and look forward to what the next 10 years will bring. The advent of modern neuroimaging tools, coupled with the increased precision of data collected from patients with precisely localized lesions, affords an exciting future.

Bangor
January 2001   K. S.

---

[1] The first report of a phenomenon conceptually similar to the attentional blink was reported by Broadbent and Broadbent (1987). See Chapter 1 for the full reference.

Contents

Contributors

Alan Allport   Department of Experimental Psychology, University of Oxford, Oxford OX1 3UD, UK.

Karen M. Arnell   Department of Psychology, North Dakota State University, Fargo, North Dakota, 58105, USA.

Marvin M. Chun   Department of Psychology, Vanderbilt University, Nashville, TN 37203, USA.

Jacquelyn M. Crebolder   Defence and Civil Institute of Environmental Medicine, P.O. Box 2000, 1133 Sheppard Ave W., Toronto, Ontario M3M-389.

Roberto Dell'Acqua   Department of Human Sciences, University of Ferrara, Italy.

Vincent Di Lollo   Department of Psychology, University of British Columbia, Vancouver, BC, Canada, V6T 1Z4.

James T. Enns   Department of Psychology, University of British Columbia, Vancouver, BC, Canada, V6T 1Z4.

Shulan Hsieh   National Chung-Cheng University, Taiwan

Masud Husain   Division of Neuroscience and Psychological Medicine, Imperial College School of Medicine, Charing Cross Hospital, London W6 8RF.

Pierre Jolicœur   Department of Psychology, University of Waterloo, Waterloo, Ontario, Canada, N2L 3G1.

Jun-ichiro Kawahara   Department of Psychology, Hiroshima University, Japan.

Steven J. Luck   Department of Psychology, University of Iowa, Iowa City, IA 52242-1407, USA.

Cathleen M. Moore   Department of Psychology, Pennsylvania State University, University Park, PA 16802, USA.

Harold E. Pashler   Department of Psychology, University of California, San Diego, CA, USA.

Mary C. Potter   Department of Brain and Cognitive Sciences, Massachusetts Institute of Technology, Cambridge, MA, USA

Jane E. Raymond   School of Psychology, University of Wales, Bangor LL57 2AS, UK.

Eric Ruthruff   NASA-Ames Research Center, Moffett Field, CA 94035, USA.

Kimron Shapiro   Center for Cognitive Neuroscience, The University of Wales, Bangor LL57 2AS, UK.

Troy A. W. Visser   Department of Psychology, University of British Columbia, Vancouver, Vancouver, BC, Canada, V6T 1Z4.

Edward K. Vogel Department of Psychology, University of Iowa, Iowa City, IA 52242-1407, USA.

Robert Ward Center for Cognitive Neuroscience, University of Wales, Bangor LL57 2DG, UK.

Jeremy M. Wolfe Center for Ophthalmic Research, Brigham and Women's Hospital, Boston, MA 02115, USA.

1

Temporal methods for studying attention: how did we get here and where are we going?

Kimron Shapiro

Abstract

In this chapter I present an overview of current research investigating temporal processing and place such research in a historical perspective. The current interest in temporal processing has its roots in the study of the psychological refractory period (PRP), as well as in the general notion of dual-task interference, as studied primarily in the visuospatial domain. Links between the temporal and spatial domains of attention are explored. As the introductory chapter to the present book, one of its main purposes is to set the stage for the chapters which follow. Since a number of authors are focusing on the attentional blink (AB) phenomenon, I use this chapter to describe some of the early work in this area leading up to and including present views on the theoretical underpinning of the AB. I conclude by suggesting that the study of temporal processing requires behavioural, neuropsychological, and electrophysiological approaches and is useful for elucidating the nature of multi-modal dual-task interference, as well as the related areas of motion perception, masking, and task switching.

Introduction

Research reports examining the temporal distribution, or timecourse, of attention have been quite numerous in the last few years, as is clearly revealed when one examines the reference sections of the individual chapters in this book. In part, this may be attributable to the considerable popularity of the general topic, attention. To an equal extent, however, the number of such studies probably reflects the recent renewal of interest in concurrent dual-task performance—a topic with a significant history, the relevant parts of which will be reviewed in the present chapter. By 'timecourse of attention', I refer to the *temporal* availability of whatever property (or properties) of the brain is or are responsible for enhancing perception. At this early juncture, it is imperative to remain agnostic with regard to which properties, or to what type of perception, reference is made. I and various other contributors address this issue in detail in the present volume.

The temporal nature of attention is often contrasted to its spatial nature, which, while a useful distinction in some respects, is potentially misleading in others. One of the aims of this chapter is to promote more comparisons of these two dimensions of attention. I will come back to this issue later in the present chapter, and it is discussed in detail in Chapters 9 and 10 by Moore and Wolfe and by Ward. It may be useful to note that models from other areas of study, for example animal learning (Rescorla and Wagner 1972), have been advanced that similarly track the time-course of neural events leading, in this particular case, to the formation of associations between stimuli. Rescorla and Wagner's model describes the formation of excitatory and inhibitory neural connections, each leading to an association between stimulus events. Comparisons between models of this type that are advanced to account for the findings from research with animals and those that are advanced to account for similar findings with human participants may prove fruitful.

Before going on to outline the plan for the present chapter, I would first like to pose some of the questions that the present volume intends to answer, as well as to describe the methods used to answer those questions, in an attempt to justify 'yet another book' on a topic that has received more than its fair share of 'attention'.

What is the timecourse of attention? This question goes to the heart of a number of the fundamental issues discussed: for example, do we continuously process information or does our processing ability 'ebb and flow'? And if the latter, what are the temporal constraints under which the system operates? Can humans exert control over such availability?

What are the consequences of the timecourse of attention? The issues here have to do with the (implied) failure to sustain attention for the task(s) that demand it. Does 'unattended' information fail to be processed, or is it that it is processed, but resides at a level unavailable to consciousness, though still able to exert an effect on behaviour? Does such availability define the very nature of consciousness? Various of these issues are set out in this present introductory chapter.

Are there demonstrable neural events that can account for this timecourse? In other words, should we seek a 'signature' for attention? And if so, should we look in a particular anatomical or functional area of the brain, or across different areas? Two chapters in the present volume (chapter 12 by Husain; and Chapter 7 by Luck and Vogel) explore these issues.

What can the results of temporal attention experiments add to our knowledge about whether there are modality-specific attentional mechanisms or one supramodal attentional mechanism? This question has been asked in the past and has been a recent focus for investigations into cross-modal attentional issues using other paradigms, for example spatial cueing. The results of such investigations (cf. Spence and Driver 1996) have not been completely unambiguous. Similarly, the results of similar investigations using temporal attentional paradigms have also been mixed, though, as the present Chapter 8 by Arnell and 5 by Jolicœur *et al.* will attest, we are moving strongly toward a consensus.

What can temporal attention results add to our knowledge of the interaction between early- and late-processing, task demands, and response-selection mechanisms? This is an 'age-old' question in cognitive psychology, and if the reader of the present work thinks it is going to be resolved here, s/he will be disappointed. On a more positive note, Chapters 5 by Jolicœur *et al.*, 3 by Allport and Hsieh, 2 by Chun and Potter, 7 by Luck and Vogel, and 6 by Ruthruff and Pashler provide fresh insights into these age-old questions.

What influence do low-level perceptual mechanisms have on the timecourse of attention? The interface between perception and attention has always been a source of controversy, with proponents on both sides of the issue as to whether attention can influence low-level processing and vice versa. Two chapters in the present volume (Chapter 4 by Enns *et al.*; Chapter 11 by Raymond) deal with this issue, and in so doing underscore the importance of continuing to study the boundary of these two interacting disciplines.

Before going on to provide a brief review of the literature leading up to much of the research described in the present volume, I would like to note that there are three primary (dual-task) methods that will be described. First, the paradigm for studying what is referred to as the psychological refractory period or PRP has been in use for a number of years, and is described in detail elsewhere (cf. Pashler 1984). In the typical PRP paradigm, two tasks, each requiring a response, are separated by a stimulus onset asynchrony (SOA), varying from trial to trial. In the standard PRP task, a speeded response must be executed to both tasks and, as will be discussed as an important issue by various authors in the present volume, neither target typically is masked. A second method to be discussed here is that of task switching (cf. e.g. Chapter 3 by Allport and Hsieh in the present volume and Allport, Styles, and Hsieh 1994). In the generic task-switching paradigm, the task requires the participant to perform one task, and then rapidly to reconfigure whatever is necessary to perform a second, different task after a short temporal interval. Finally, the third method, and one that is discussed in considerable detail by nearly all the present contributors, is that of the rapid serial visual presentation (RSVP) procedure used in the attentional blink (AB) paradigm. Since this method has been used less than the others and forms a central method of many of the studies described here, its history will be reviewed in detail in the section that follows.

Historical view of temporal processing methods using RSVP

The investigation of temporal processing using the attentional blink paradigm typically involves the use of a procedure known as rapid serial visual processing, or RSVP for short. The RSVP procedure presents successive visual stimuli, one item at a time in a fixed location, at rates between 6 and 20 items per second. One (single-task) or two (dual-task) target(s) in the stream have to be detected or identified. For a discussion of the results of single-task RSVP studies and more detail than is provided here on dual-task studies, the reader is referred to Shapiro and Raymond (1994). Although the present discussion will focus on the results of dual-task studies, a few summary statements regarding single-task studies are important to set the stage for certain critical issues affecting dual-task experiments.

The results of single-task RSVP experiments indicate that (1) target identification requires the conjoining of the target-defining characteristic (for example, a white target letter in a stream of black non-target letters) with the to-be-reported feature (such as the letter 'R'); (2) feature conjunction in this task takes approximately 100 ms; and (3) the task requires attention. One might assume that once a target is identified, attentional mechanisms would be free to begin analysing subsequent stimuli. However dual-task RSVP research strongly suggests that this is not the case. Rather, it appears that after target identification is presumably complete, large deficits in the processing of subsequent stimuli are found for up to 700 ms later. We will return to this outcome and its implications after discussing the history of dual-task experiments.

The primary purpose of dual-task RSVP experiments is to track the timecourse of events succeeding selection of a single target. D. E. Broadbent and M. H. P. Broadbent (1987; Experiment 3) were the first to demonstrate a long-lasting interference effect following target identification, using a dual-task RSVP procedure. Participants were required to identify a target and a probe word (each defined by flanking hyphens) that were embedded at different serial positions within a sequence of unrelated lower-case words. The SOA between successive words was 120 ms, with an ISI of 20 ms. The number of items between the target and probe words varied from trial to trial. They found that when target and probe were temporally adjacent, subjects could produce a correct response to either the target or the probe, but not to both. As the temporal proximity of target and probe was decreased, the probability of correctly identifying the probe word on target-correct trials increased. This rose from a low value of 0.1 for target-probe intervals of less than 400 ms up to an asymptote of 0.7 for intervals of 720 ms or longer. Not only were subjects unable to identify the probe correctly when it was presented within 400 ms post-target; they frequently reported being unaware that it had been embedded in the stimulus stream. D. E. Broadbent and M. P. H. Broadbent (1987) proposed that the interference producing poor probe performance occurred at the target-identification stage. They suggested that once the key feature of the target was detected, a slow identification process was initiated that produced a long-lasting interference effect.

Using a variant on the dual-task RSVP procedure, Reeves and Sperling (1986) and Weichselgartner and Sperling (1987) also observed large deficits in the processing of post-target items. In both studies, highly practised subjects were presented with an RSVP stream of digits and asked to identify a highlighted or boxed digit (target) and to name the next three (post-target) digits. Reeves and Sperling presented the post-target digits at a location to the right of the target and pre-target items, while Weichselgartner and Sperling presented all items at the same location. Both studies found that subjects' reports generally consisted of the target, the first post-target (+1) item, and items presented about 300 to 400 ms after the target. The items presented in the interval between 100 and 300 ms post-target were rarely reported. They noted, as did Raymond *et al.* (1992), that the perceived temporal order of items recalled was non-veridical. Using this complex response requirement, it is not clear whether subjects were unable to process the items perceptually during this interval, or whether they were unable to store and/or retrieve these items from memory for later recall. In either event, a deficit in the ability to process successive post-target items to an output stage was reported. Although Reeves and Sperling attributed this post-target processing deficit in recall to a shift in the spatial allocation of attention, similar-sized effects were observed by Weichselgartner and Sperling, indicating that the post-target processing deficit occurs independently of a spatial shift in attention.

To account for the post-target processing deficit, Weichselgartner and Sperling (1987) proposed that two attentional processes were initiated by target detection. One is a fast, short-lived, automatic process producing effortless identification of the target and, occasionally, the +1 item. This process is terminated after about 100 ms, and is unaffected by task difficulty. The other process is a slow, effortful, controlled process that mediates identification of items appearing between 200 and 300 ms post-target. This process is acknowledged to be affected by factors normally seen to affect attention, such as practice, expectation, stimulus probability, and target signal-to-noise ratio. The post-target processing deficit reflects the interval after the automatic process has subsided and before the controlled process has gained sufficient power. Nakayama and Mackeben (1989) proposed a similar dual-process model of attention to account for cueing effects.

Dual-target deficits and the attentional blink

A more recent method for studying dual-target deficits is the attentional blink (AB) paradigm (cf. Raymond *et al.* 1992; Shapiro *et al.* 1994), which is modelled very closely on that of the RSVP method described earlier (e.g. Broadbent and Broadbent 1987). This procedure is described in considerable detail in the Raymond *et al.* and Shapiro *et al.* studies just cited; but a brief description of the procedure is provided here before I go on to summarize AB findings since the first study, reported in 1992. Such a summary seems warranted in this introductory chapter, as many of the chapters in the present volume discuss new attentional blink findings.

As stated, the AB paradigm is predicated on the RSVP method. In the typical visual AB experiment, a stream of between 16 to 24 letters are presented, one at a time, at fixation, at a rate of 11 items per second (15 ms 'on'; 75 ms 'off'). In the dual-target condition, two targets are required to be processed for report when the stream has ended. The first target (T1) is distinguished by being the only white letter in the non-target stream of otherwise black letters and the participant is requested to report its identity, which can be one of any of the letters in the alphabet, except the letter 'X'. The second target or probe (T2), which is always presented after the first, when it occurs (50 per cent of trials), is fully specified as the letter 'X', with the participant required to report its presence or absence, also at the end of the stream. The two targets are separated by a varying stimulus onset asynchrony (SOA) of between 90 to 720 ms. The use of the SOA range enables one to chart the availability of attention, as indexed by the percentage of correct detections of T2 (on trials when T1 was identified correctly). In the single-target, or baseline, condition, participants are instructed to ignore T1 and perform only the T2 task. Ideally, the single- and dual-target conditions are run as a 'within-subjects' design, with each participant's control performance subtracted from their performance in the experimental condition.

The 'signature' interaction of the AB can be seen in Fig. 1.1. Whereas the single-target condition reveals no effect of SOA, the dual-target condition shows a loss of ability to report T2 correctly somewhere between approximately 100 ms and 500 ms after T1. This finding has been replicated many times, with different stimulus types (digits, symbols) and locations (T1 and T2 displayed in different spatial locations) and different procedural (i.e., T1 and T2 masked, separated by a varying SOA, but without a non-target stimulus stream) and 'display' (i.e., timing) parameters. One point to which I will return later, but that it is important to note here, is that the interaction just described can take another characteristic form. The figure above accurately portrays a typical non-significant difference between the two conditions at Serial Position 1. This 'Lag-1 sparing' effect, as it has been called, which is characterized by the u-shaped function shown below, on occasion may be replaced by a *linear* function, where the first serial position of the dual-target condition does reveal a significant difference from its single-target counterpart. The conditions under which this outcome occurs are discussed in detail in Chapter 4 of the present book by Enns *et al.* and in a recent paper (Visser *et al.* 1999). Suffice it to say here that which of the two functions is obtained appears to depend on the number and

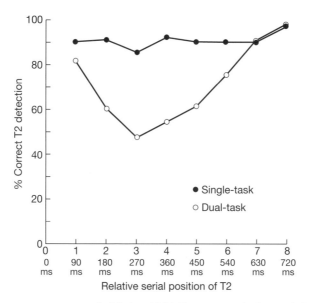

Fig. 1.1 The group mean percentage of trials in which T2 was correctly detected plotted as a function of the relative serial position of T2 in Experiment 2 of Raymond *et al.* (1992). Closed symbols represent data obtained in the (single-task) control condition, in which subjects were told to ignore T1. Open symbols represent data obtained in the (dual-task) experimental condition, in which subjects were told to identify T1 and T2.

type of attentional demands required to switch from performing the T1 to performing the T2 task. A further description of the AB task and a number of the recent findings from this task can be found in two recent chapters in which this author has been involved (Shapiro and Terry 1998; Shapiro and Luck 1999).

Overview of attentional blink findings

Past findings

Before embarking on a discussion of what I consider to be important findings from AB research since the initial report of the phenomenon in 1992, it is important to point out that many of the contributors to the present volume will be addressing this same issue. Thus I will try to avoid treading on their territory by attempting to discuss only those issues that they do not cover, with perhaps the odd exception regarding a point on which I may not fully agree with one of the contributors (i.e., I get to 'go' first).

In the first paper, published in 1992, on the topic of the attentional blink I and my co-investigators argued that the model to best explain the AB outcome was an 'inhibitory' model (Raymond *et al.* 1992). We suggested that mental operations required to perform the first target task were so demanding that an inhibitory process was initiated to facilitate target identification. This facilitation was accomplished by a cessation of the processing of further stimuli for the period of time that we refer to as the attentional blink. To test this model, I and my colleagues (Shapiro *et al.* 1994) performed a series of experiments where we varied the difficulty of the first target, keeping the second target task-constant. The logic was that this should manipulate the degree of inhibition proportionate to the difficulty of the first task. For example, in one experiment we required participants for the first target task to detect if the black letter 'S' (presented on 50 per cent of trials) was displayed; a difficult task in the context of a stream of black letters. At the other end of the difficulty spectrum, we required participants (first target task) to detect if a white letter (presented on 50 per cent of trials) was displayed. The white letter 'S' was the same from trial to trial, affording participants the easy task of having to detect a simple luminance difference. The degree of difficulty of the T1 task was assessed in these and other manipulations lying between these two extremes of difficulty by measuring d'. We then correlated the magnitude of the 'blink' (expressed as the area above the AB curve—the curve defined as the percentage of correct T2 judgements plotted as a function of SOA) with the difficulty measure d', and found only a very minor correlation. We interpreted this as evidence against an inhibitory model—at least an inhibitory model defined as one where a fixed amount of inhibition was generated by the T1 task (see also McLaughlan *et al.* (2001; Ward *et al.*, 1997). It is important to note that this correlation was based on relatively few data, and that a subsequent analysis (Seiffert and Di Lollo 1997) looking at considerably more experiments, and thus with a greater power for detecting a significant correlation, revealed a significant effect of T1 difficulty on the magnitude of the AB phenomenon.

In another experiment designed to test the inhibitory model, we asked participants to search for their own name, or another name, as the second target (T2) judgement in an AB task involving streams of words (Shapiro *et al.* 1997a). The logic was that if the T1 task fully inhibited processing of stimulus items occurring in the AB interval, then there should be no advantage in a task requiring the highly overlearned response demanded by searching for one's own name. The results of this experiment suggested, however, that there was a significant advantage in this condition, revealing a greatly attenuated AB effect. Not only does this outcome on a strict interpretation of it refute the inhibitory model; perhaps more importantly, it reveals that significant processing of the second target does occur.

In a subsequent experiment (Shapiro *et al.* 1997*b*) we went on to demonstrate further not only that T2 is processed, but that it is processed to a level where it can semantically prime an item occurring after the AB interval. The effect in this experiment was a weak one; and in a final attempt to make the same point my colleagues and I conducted another experiment (Luck *et al.* 1996) where we measured event-related potentials (ERPs) to a word occurring in the AB interval. This was accomplished by using the N400 mismatch potential, where a semantic context is established at the beginning of a trial, and then either an unexpected stimulus violating that expectation is presented, eliciting a large N400, or an expected stimulus congruent with the semantic context occurs, eliciting a smaller N400. The logic is such that since a word (occurring in the AB interval) must be identified before it can be compared to a context set prior to the trial, the presence of an N400 during the AB interval would be evidence of semantic processing. And indeed this was the conclusion we drew from the results of the experiment.

At the time, to explain the AB phenomenon, I argued that the evidence collected pointed to a late-selection mechanism where interference between the target and the probe was the source of the AB. Further, I argued that both the target and the probe are processed, but that these two items compete with each other during either entry into or retrieval from a short-term visual sensory buffer. Differential weighting of T1 over T2 (owing to its being the first task) yielded successful T1 retrieval at the expense of T2.

Before I present data directly supporting a visual interference account of the AB, two related points must first be made, which taken together rule out an alternative explanation. The alternative explanation is that the AB is due to response competition between the two stimuli; and, in ruling this out, I will argue that the AB is due to the consequences that follow from selecting an object for processing.

In a report I undertook with Robert Ward and John Duncan (Ward *et al.* 1996) we wanted to rule out the possibility that the AB is due to requiring the subject (s) to make two competing responses: one to T1 and one to T2. To distinguish between these two explanations, in the first experiment we manipulated the number of T1 attributes to be reported, whereas in the second we manipulated the number of T1 objects. Under certain circumstances interference in dual-task paradigms has been shown to be attributable to a bottleneck in response selection; in other words, when two tasks occur in rapid succession, humans seem unable to process the second response until the first is fully processed. Such an outcome has been referred to as the psychological refractory period, or PRP. To study whether such an explanation could account for the AB, we created an experimental manipulation where S was required to respond to either zero, one, or two attributes of the target, followed by a response to the probe. If multiple responses are the basis of the interference we have been observing, then interference should increase with the number of responses. On the other hand, if, as object-based theories predict, selecting an object brings with it all its attributes, then increasing the difficulty of the target task should only exert an effect when the number of objects, as opposed to attributes, is manipulated. The results of this experiment are easy to summarize: the number of attributes to be identified makes no difference to the magnitude of the AB effect.

In a related experiment, we manipulated the number of objects, rather than attributes, to test our hypothesis that object selection and the consequences that follow are responsible for the AB. The method was very similar to that just described, except that in this experiment the participant was required to make judgements about zero, one, or two objects rather than zero, one, or two attributes. The results revealed that two-object identification produces a deeper and longer AB effect than one-object identification.

The next important experiment was to test directly the model suggested above, that is, that the AB is affected by the similarity of competing visual objects. Additionally, we wished to address the question of whether this competition occurs at a perceptual (feature) or conceptual (meaning) level. To test this, Isaak, Shapiro, and Martin (1999) used the method previously described, where only the items critical to the AB effect are presented (T, T+1, P, P+1); but also a further T+2 was presented, to extend the range of the number of possible competitors. We randomly presented all factorial combinations of letter and false-font stimuli, yielding a minimum of two and a maximum of five letters per trial. The results of this experiment reveal that letters interfere more with letters than false fonts interfere with letters and thus, produce a greater AB effect. This finding is very similar to a results reported by Taylor and Hamm (1997) comparing letters and digits for interference. From these results we can conclude that AB magnitude is a function of the number of competitors—but the competitors must be from within the same category. Further support for similarity within the same category comes from an experiment by Ward *et al.* (1997), where we compared an identical T1 and T2 with a less similar T1/T2 combination, to find a greater AB in the identical condition. Note that this particular condition of identical targets is the stimulus condition referred to as repetition blindness.

Recent findings

Turning now to studies most recently completed, I would like to begin by discussing a line of experiments examining the temporal dynamics of attention in individuals with hemispatial visual neglect. Visual neglect is a common disorder following right-hemisphere lesions due to stroke, affecting over 70 per cent of patients with such lesions. These individuals are unaware of people or objects in their contralesional, or left, visual field. Prominent theories to explain visual neglect share the fundamental assumption of a spatial attentional deficit; for example, Posner and his colleagues (1987; 1984) suggested that neglect patients have a direction-specific impairment of disengaging attention from a stimulus on the right when they are required to shift attention to the left. My colleagues and I wanted to show that the impairment experienced by individuals with visual neglect might have a temporal attentional component to it: i.e., they might have an increased disability to disengage from a stimulus independent of the need to move attention from one location to another (Husain *et al.* 1997).

The AB paradigm as it has been described earlier was used with individuals suffering from visual neglect and with two control populations, one age-matched and the other lesion volume-matched. The two fundamental differences between the procedure described earlier and that used in the present study were that (1) the rate of presentation was reduced from 11 items to 5.5 items per second, and (2) ten post-T1 items, rather than eight, were presented. The lesion sites in the neglect group are shown in Fig. 1.2—four patients had lesions in the inferior frontal lobe and three in the inferior parietal, and one had suffered a haemorrhage of the basal ganglia.

The results of these two control groups revealed a normal profile in both single- and dual-task performance, as shown in Fig. 1.3. On the other hand, the performance of neglect patients did not return to baseline until nearly 1.5 s, demonstrating an AB lasting approximately three times as long as that of controls. Performance did not differ between patients with parietal, frontal, or basal ganglia lesions, and performance in the single task was not significantly different from that in either control group.

This study suggests that a significant factor in neglect patients' inability to attend to stimuli in the neglected hemifield may be a difficulty in disengaging attention from a target occurring in the unaffected hemifield. Such a conclusion is supported by the findings of di Pellegrino and

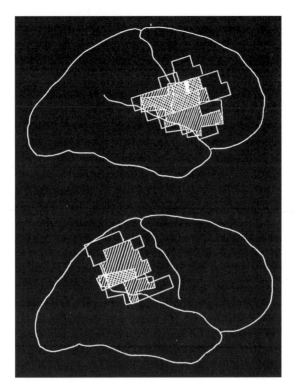

Fig. 1.2 Extent of cortical lesions in seven patients with left-sided visual neglect. Single hatching represents the region of overlap of 2 lesions; cross-hatching shows the overlap of 3 lesions; solid white is the zone of overlap of 4 lesions. The two lesion foci are located in the inferior frontal lobe and the inferior parietal lobe. One patient suffered a haemorrhage of the basal ganglia without cortical involvement (lesion not shown).

his colleagues (di Pellegrino *et al.* 1997). The relevant finding from di Pellegrino *et al.* is that approximately 600 ms had to elapse between the first target's presentation on the contralateral side and the second target's presentation on the ipsilateral side before participants could identify the first target with a significant degree of accuracy. The conclusion drawn was that, for this particular population, stimuli appearing in the ipsilateral field captured attention to such a degree that contralaterally appearing stimuli had to appear significantly beforehand in order not to be overshadowed.

In an attempt to understand further the neuroanatomical basis of the AB phenomenon, a recent investigation (by Shapiro *et al.* 2000) examined patients with lesions in three anatomical areas known to be involved in attentional control for their contribution to the AB effect. The lesions were localized to one dorsal stream area, the superior parietal lobe (SPL), and to two ventral stream areas, the inferior parietal lobe (IPL) and the temporal lobe (TL). The most important point is that the AB effect is considerably smaller in the SPL group than in either of the other two groups. On this basis, we suggest that there is a difference in the degree to which different neuroanatomical areas are involved in the dual-target task demands required by the attentional blink paradigm. Whereas participants with a lesion of the SPL area reveal an AB function similar to that of non-patients, recovering by approximately 500 ms, those with lesions

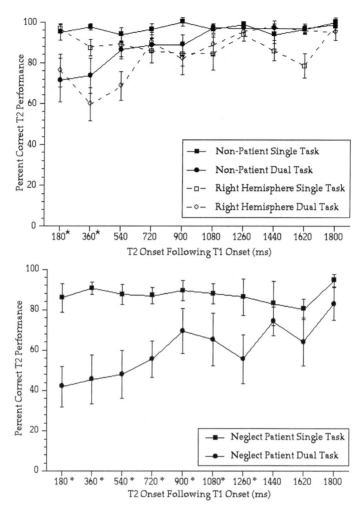

Fig. 1.3 *Upper.* Performance of normal individuals without stroke (filled symbols) and right-hemisphere stroke patients without neglect (open symbols). In both groups, T2 detection on the dual-target task (circle symbols) was significantly different from that on the single-target (square symbols) task for T1–T2 intervals up to 360 ms (indicated by starred ms designations on the abscissa; $p<0.05$). Error bars represent standard errors of the mean. T1 was correctly identified on 86 per cent and 81 per cent of dual-target trials respectively by normal volunteers and those with stroke. *Lower.* Performance of patients with visual neglect. Detection of T2 on the dual-target task (circle symbols) was significantly different from that on the single-target task (square symbols) for T1–T2 intervals less than 1440 ms (indicated by starred ms designations on the abscissa; $p <0.05$). Furthermore, the magnitude of visual unawareness during this attentional blink was much greater than in stroke patients without neglect or normal individuals. T1 was correctly identified on 72 per cent of dual-target trials.

of the IPL and LAT show an AB function that is protracted in duration, failing to recover until approximately 700 ms. This outcome is consistent with Milner and Goodale's (1993) view that the IPL may be a structure whose function is more in keeping with those commonly associated with the ventral stream; in particular, those associated with the function of object identification as required by the AB task. It is also consistent with Wojciulik and Kanwisher's (1999) revelation that the intraparietal sulcus mediates a variety of attentional functions, including those demanded by the RSVP paradigm tested by these investigators.

In another recent line of research, my colleagues and I have been exploring a finding from a research report in which I was involved (Shapiro *et al.* 1994). In this paper, we found that a T1 target task defined by a variable 'gap' in the non-target distractor stream was sufficient significantly to attenuate the AB effect. In this experiment, the T1 task was to report whether a 'short' or a 'long' gap occurred, and then to report the presence or absence of the letter 'X', as was the task for the original AB experiment described earlier in this chapter. A short gap was created by omitting one of the letters in the non-target stream, whereas a long gap was created by omitting two letters. This yielded a gap duration of either 80 ms or 160 ms, respectively. Although the T1 task proved quite difficult for subjects, the results are striking in their attenuation of the AB effect. Although my first interpretation of this was that the results could be explained in terms of a non-object (gap) failing to compete with an object ('X' judgement), there is another factor that distinguishes this T1 task from previous tasks. The distinguishing characteristic is that the gap task has become a duration judgement, rather than an object identification judgement *per se*.

To test the notion that the attenuation of the AB effect was due to the nature of the judgement having to be made on T1, we created an experimental situation where the T1 judgement became one where participants had to judge whether the longer letter in the stream was 'slightly' longer than the other non-target letters or 'more than slightly' longer (Sheppard *et al.*, submitted). T1 was thus differentiated from the other items in the stream solely by virtue of its duration (i.e., it too was black, as were the non-target letters). The durations by which the target letter was lengthened matched those used in the gap task. In effect, subjects now were making a duration judgement on an 'object', rather than on a 'non-object' (i.e., a gap). The results of this experiment as shown in Fig. 1.4 revealed a dramatically attenuated AB effect, consistent with the attenuation witnessed in the gap task. Moreover, in the next experiment, when we asked subjects to select the target by duration (as in the previous experiment just described), but to report the identity of the longer letter T1, a typical AB effect resulted. Although these results suggest that a duration judgement does not yield a divided attention deficit, such a conclusion may be premature. In a subsequent experiment we lengthened the gap in the stimulus stream, requiring participants to make a T1 judgement of whether a gap of three or five 'items' (i.e., 330 vs. 550 ms) occurred. The results of this manipulation yielded an AB of approximately equal magnitude to that typically seen in an 'object' identification task. There are two possible explanations for this outcome. First, the increased gap duration makes the gap itself appear more event-like, perhaps forcing a different kind of attentional allocation. Second, the increased gap duration may be serving to segregate the non-target stream into two discrete parts, again forcing the kind of attentional allocation that yields a typical divided-attention deficit. In support of this latter assertion, one's percept of the shorter gap duration judgement (1 vs. 2 letters omitted) appears to be that of a continuous stream, briefly interrupted, whereas the percept of the longer gap judgement (3 vs. 5 letters omitted) appears to be one of two separate events.

I would like now to describe another recent line of research, examining the attentional blink to targets other than visual. In a series of experiments, my colleagues and I examined the potential of an AB to be elicited by vibrotactile targets (Hillstrom *et al.*, submitted). These targets were generated by means of electric current to a standard hearing aid, mounted on a sponge, on which the subject's finger rested. The application of the electric current caused the membrane of the hearing aid to vibrate, by which means we were able to control three qualities of the

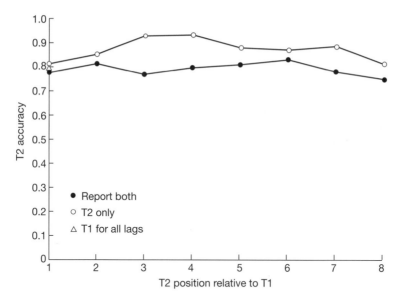

Fig. 1.4   Mean percentage of trials in which T2 was correctly detected plotted as a function of the serial position of T2 for the single-target condition (open circle symbols) and for the dual-target condition (filled circle symbols). Mean T1 performance is shown by the triangle symbol.

stimulus: duration, frequency, and intensity. In a set of six experiments, we examined a number of these attributes to define the T1 and T2 targets. The data from one of these experiments is presented in Fig. 1.5, where we required subjects to select T1 on the basis of duration (longer) and then report its frequency (high vs. low) and required subjects to select T2 on the basis of intensity (greater than the other stream elements) and again report frequency. The results of this manipulation do not reveal an AB, suggesting that perhaps the tactile modality is constructed in such a way as to facilitate integration of stimuli during the time period where the AB is normally seen. The other experiments in this series draw the same conclusion, though with some combinations of target tasks we found a main effect of Lag in both single- and dual-target conditions, indicating the possibility that subjects are experiencing some degree of dual-task deficit and simply cannot ignore the first target as instructed.

   One experiment in this series deserves special mention. In Experiment 7, we changed both target tasks from one of detecting frequency, intensity, or duration, to one of detecting whether the target occurred to either of two fingers,one on each hand. Thus subjects were required to judge whether T1 *and* T2 occurred to the 'left' or to the 'right' finger. As shown in Fig. 1.6, in this particular condition we saw the AB 'signature', an interaction between Condition (single- vs. dual-target) and Lag, indicating a more standard AB outcome. We speculate that forcing the vibrotactile judgements to be spatially mapped may have been responsible for this outcome.

   Turning now to a final line of recent research, my colleague and I have been experimenting with the concept of the object file and its relationship to the AB phenomenon. Various investigators have suggested that attention to an object opens a hypothetical 'object file' containing information about the object (Kahneman *et al.* 1992). The file is updated as the object moves or

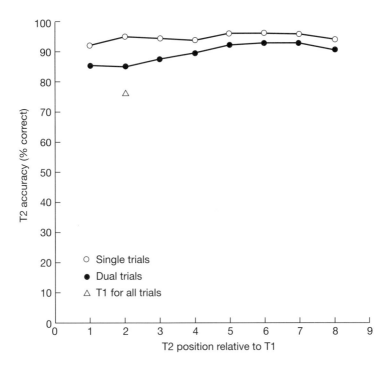

Fig. 1.5 Mean percentage of trials in which T2 was correctly detected plotted as a function of the serial position of T2 for the single-target condition (open symbols) and for the dual-target condition (filled symbols). Mean T1 performance is shown by the triangle symbol. T1 was longer in duration than non-targets and T2 was higher in intensity than non-targets, and participants reported the relative frequency (high or low) of the targets.

changes. If a subsequent object cannot be incorporated into the active 'file', a new file must be opened, yielding an attentional cost. An object-file account of the AB suggests that T1 opens a 'file' to which T2 normally cannot be added. Thus the prediction is that, if T2 could be incorporated into T1's object file, an AB might not occur. The method chosen to instantiate this was to link T1 to T2 by means of 'morphing' technology. Such a technology smoothly changes a stimulus from one state to another, such that the changes are only incrementally perceptible.

Francis Kellie and I 'morphed' a smoking pipe into a saucepan across the duration of the stimulus stream, as shown in Panel A of Fig. 1.7 (Kellie *et al.* 1999). T1 and T2 were defined by placing a patterned stimulus over two pre-selected stimuli in this morphing stream and asking subjects to make a judgement as to whether the patterns constituted large or small dots in the T1 task, and large or small squares in the T2 task. For a control comparison, as shown in Panel B of Fig. 1.7, we replaced the morphing stream with a series of randomly pre-selected objects, but again used the same targets defined for the experimental (morphing) condition. Apart from a small change in the stimulus parameters, this experiment basically used the same method as other AB experiments earlier described. The results of this experiment (shown in Fig. 1.8) are that, whereas the control condition revealed a typical AB outcome, the experimental condition showed a dramatic reduction in AB magnitude. This result suggests that when T1 and T2 are

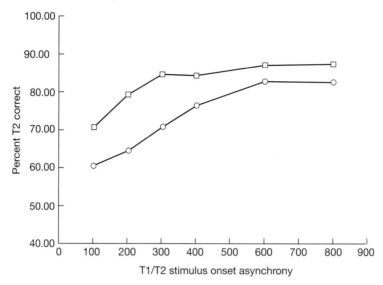

Fig. 1.6 Accuracy of reporting the presence of T2 at each T2 position (relative to the T1 position) in Experiment 7, both when only T2 was reported (square symbols) and when both T1 and T2 were reported (circle symbols).

incorporated into the same object file, a divided attention deficit does not occur. We intend to go on to examine other targets in this same context to gain a more thorough understanding of what constitutes an object file.

A look ahead

The next and final section of this introductory chapter will briefly summarize the contributions to this volume.

Task switching and AB
Chun and Potter (Chapter 2)
Chun and Potter suggest two distinct types of information processing. First, there is a visual stage, where there is a limit on consolidating information into a durable form of storage. This is a modality-specific limitation. Second, there is a non-modality-specific stage, where a change in selective set is required between two targets. In turn, this suggests a more central limitation. Chun and Potter go on to argue that Lag-1 sparing is a good marker for when task switching occurs, with reference made to the meta-analysis by Di Lollo and colleagues (Visser *et al.* 1999).

Allport and Hsieh (Chapter 3)
These authors argue that task switching appears to be different from the basic AB effect. They believe task-switching effects are due to the *number* of stimulus items between T1 and T2, and are independent of whether or not a T1 task has to be performed. In this way, these authors

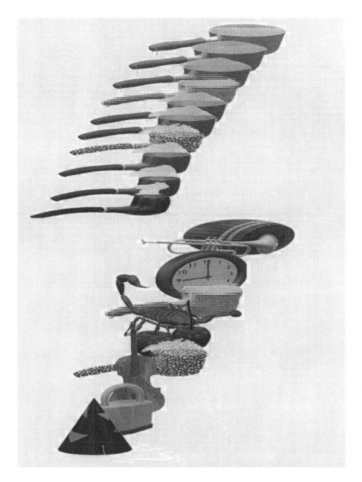

Fig. 1.7 *Upper.* Morphing stream showing T1 and T2 as the parts of the stream overlaid with dots (T1) and squares (T2). *Lower.* Control stream showing T1 and T2 as the parts of the stream overlaid with dots (T1) and squares (T2).

agree with Chun and Potter that task switching may be responsible for a component of many AB experiments, particularly those that employ non-visual targets. Their chapter goes on to suggest that the traditional view of interaction between top-down and bottom-up control may be too limiting and in need of being replaced by a notion of interactivity.

Enns et al. (Chapter 4)

Enns and his colleagues discuss the role of masking in the AB, citing the different requirements demanded by T1 vs. T2. Their chapter goes on to lay out the mechanism that they refer to as 'object substitution' as an account of the AB. In so doing, they discuss the importance of task switching in the AB, citing new data that show that an attention switch between T1 and T2 will cause an AB, even when T2 is not masked. Their argument casts doubt on the dwell-time hypothesis (Duncan *et al.* 1994), suggesting that the AB does not reflect processing requirements of T1. Alternatively, they suggest their re-entrant theory can explain the mechanism of 'object substitution', which they offer as an account of the AB.

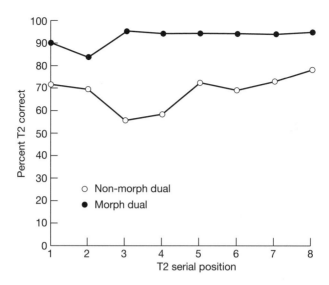

Fig. 1.8 Data from morphing experiment. Dual-target performance in the morphing condition (filled circle symbols) and in the non-morphing (control) condition (open circle symbols). Note that single-target performance is not shown.

Central interference and AB

Jolicœur et al. (Chapter 5)

Jolicœur *et al.* test various predictions from PRP models on the AB. What they find is that typical operations that postpone the bottleneck using the PRP paradigm also do so when the AB task is speeded. These include: (1) carry-over to Task 2 of pre-bottleneck and bottleneck Task 1 manipulations; (2) correlation between T1 and T2 at short SOAs; (3) absorption of the pre-bottleneck Task 2 effect; and (4) additivity of bottleneck and post-bottleneck Task 2 effect. In their view, these results rule out a capacity-sharing model.

Ruthruff and Pashler (Chapter 6)

Ruthruff and Pashler discuss basic differences between PRP and AB and conclude that, whereas both tasks represent some kind of central bottleneck, it is probably not the same one, though each may be tapping the same resource. They argue that PRP is thought to be an issue of response selection, whereas AB bottleneck is believed to arise from a perceptual limitation. To test this idea, they set up a series of experiments to examine the notion that the bottleneck of central resources is common to both. The conclusion is drawn that there appear to be central bottlenecks common to both, but since AB shows less of an effect, there may also be a perceptual bottleneck unique to AB, but not PRP.

Luck and Vogel (Chapter 7)

These investigators use ERP methods to compare AB and PRP in an attempt to establish the locus of the bottleneck. They arrive at the conclusion that the AB reflects input-level interference, whereas PRP reflects output-level interference. Furthermore, they suggest that these two forms of interference are not completely independent and reflect limitations in accessing working memory.

Arnell (Chapter 8)

Arnell employs behavioural methods to assess the cross-modal nature of AB. She finds an AB effect both within and between vision and audition. Her carefully-employed techniques avoid most sources of confound, such as task switching. Moreover, her chapter provides an excellent review of the literature on the relationship between task switching and the AB.

Serial vs. parallel models

Moore and Wolfe (Chapter 9)

Moore and Wolfe's chapter discusses serial vs. parallel models of visual attention, with arguments and support presented carefully for both sides of the debate. The authors adopt a compromise position, suggesting that search can be both serial and parallel at the same time. Moreover, they argue it is non-productive to attempt to resolve this dichotomy.

Ward (Chapter 10)

Ward's paper also discusses the serial vs. parallel debate, but takes a firm stance on the side of the 'parallel camp', arguing that there is no firm evidence for serial models. Ward's review of the literature is comprehensive, with new methods of assessing serial vs. parallel presented.

Perception and attention

Raymond (Chapter 11)

Raymond's chapter presents an analysis of the links between perception and attention, using data from recent research to support her argument that attention acts to provide coherent perceptual representations of the world. She presents data from an experiment where motion creates a single 'object-file', enabling reduction in the magnitude of the dual-task cost typically seen in attentional blink experiments.

Neuropsychology of attention

Husain (Chapter 12)

Husain's chapter discusses visual attention theories underlying neglect, extinction and simultanagnosia. He maintains that 'spatial' theories are not sufficient to explain these neuropsychological phenomena, and suggests the best way forward is to look at the *timecourse* of attention. This author then goes on to pose an interesting theory that attempts to unify our understanding of these three disorders.

The future of research into the time-course of attention

It is the present author's opinion that research into the time-course of attention will continue to play an important role in our understanding of the more general role played by attention in regulating the flow of information to higher cortical centres. One does not have to look very far into the crystal ball to determine that such a role will probably be played out on other than a purely behavioural stage. For example, neuropsychological approaches will most certainly play a larger part in the understanding of the temporal aspects of attention. Moreover, it is also likely that functional imaging studies will become more commonplace than they are at the time this book is being written. I am aware of only one functional imaging study looking at the attentional blink, which is a study by Marois *et al.* 2000. In the first condition of this study, they analysed data from healthy participants, who were scanned while performing a single-target (T1) RSVP task under two conditions. In one condition, they were required to identify one of three possible target letters, whereas in the other they performed the same task but the target was not masked by the letter normally succeeding it in the RSVP stream. Two distinct areas were more active during the former condition than during the latter: the intra-parietal sulcus (IPS) and the lateral frontal area. In a second condition, designed to look for haemodynamic changes associated with the consequence of attending to T1, participants were subjected to a standard AB paradigm, requiring them to identify T1 from the same three possible alternatives as before, and then to detect whether the letter 'X' occurred in the remaining RSVP stream. Participants' haemodynamic responses were compared for instances when T2 was detected and for those trials when T2 went undetected. The structure differentiating these two trial types was the precuneus, or the medial part of the posterior parietal cortex, suggesting that parietal cortex plays a role in the AB, as it does in more conventional spatial cueing and localization tasks.

Acknowledgement

Funding for most of the studies reported in this chapter were provided by the Wellcome Trust in a joint project grant to the author Dr Masud Husain. Their assistance is gratefully acknowledged.

References

Allport, D. A., Styles, E. A., and Hsieh, S. (1994). Shifting intentional set: exploring the dynamic control of tasks. In C. Umilta and M. Moscovitch (eds.), *Attention and performance* XV (Vol. XV). Cambridge, MA: MIT Press.

Broadbent, D. E., and Broadbent, M. H. (1987). From detection to identification: response to multiple targets in rapid serial visual presentation. *Perception and Psychophysics*, **42**(2), 105–13.

di Pellegrino, G., Basso, G., and Frassinetti, F. (1997). Spatial extinction on double asynchronous stimulation. *Neuropsychologia*, **35**(9), 1215–23.

Duncan, J., Ward, R., and Shapiro, K. L. (1994). Direct measurement of attentional dwell time in human vision. *Nature*, **369**(6478), 313–15.

Hillstrom, A. P., Shapiro, K., and Spence, C. (submitted). Attentional and perceptual limitions in processing sequentially presented vibrotactile stimuli. *Perception and Psychophysics*.

Husain, M., Shapiro, K., Martin, J., and Kennard, C. (1997). Abnormal temporal dynamics of visual attention in spatial neglect patients. *Nature*, **385**(6612), 154–6.

Isaak, M. I., Shapiro, K. L., and Martin, J. (1999). The attentional blink reflects retrieval competition among multiple RSVP items: tests of the interference model. *Journal of Experimental Psychology: Human Perception and Performance*, **25**, 1774–92.

Kahneman, D., Treisman, A., and Gibbs, B. J. (1992). The reviewing of object files: object-specific integration of information. *Cognitive Psychology*, **24**(2), 175–219.

Kellie, F. J., Shapiro, K., and Hillstrom, A. P. (1999). Targets in a common-object file reduce the attentional blink. Paper presented at the Psychonomic Society, Los Angeles.

Luck, S. J., Vogel, E. K., and Shapiro, K. L. (1996). Word meanings can be accessed but not reported during the attentional blink. *Nature*, **383**(6601), 616–18.

McLaughlan, E. N., Shore, D. I., and Klein, R. M. (2001). The attentional blink is immune to masking-induced data limits. *Quarterly Journal of Experimental Psychology*, **58a**, 169–96.

Marois, R., Chun, M. M., and Gore, J. C. (2000). Neural correlates of the attentional blink. *Neuron*, **28**, 299–308.

Milner, A. D., and Goodale, M. A. (1993). Visual pathways to perception and action. In T. P. Hicks, S. Molotchnikoff and T. Ono (eds), *Progress in Brain Research* (Vol. 95, pp. 317–37). Amsterdam: Elsevier.

Nakayama, K., and Mackeben, M. (1989). Sustained and transient components of focal visual attention. *Vision Research*, **29**(11), 1631–47.

Pashler, H. (1984). Processing stages in overlapping tasks: evidence for a central bottleneck. *Journal of Experimental Psychology: Human Perception and Performance*, **10**, 358–77.

Posner, M. I., Walker, J. A., Friedrich, F. J., and Rafal, R. D. (1984). Effects of parietal lobe injury on covert orienting of visual attention. *Journal of Neuroscience*, **4**, 1863–74.

Posner, M. I., Walker, J. A., Friedrich, F. A., and Rafal, R. D. (1987). How do the parietal lobes direct covert attention. *Neuropsychologia*, **25**, 135–45.

Raymond, J. E., Shapiro, K. L., and Arnell, K. M. (1992). Temporary suppression of visual processing in an RSVP task: an attentional blink? *Journal of Experimental Psychology Human Perception and Performance*, **18**(3), 849–60.

Reeves, A., and Sperling, G. (1986). Attention gating in short term visual memory. *Psychological Review*, **93**, 180–206.

Rescorla, R. A., and Wagner, A. R. (1972). A theory of Pavlovian conditioning: variations in the effectiveness of reinforcement and non-reinforcement. In A. H. Black and W. F. Prokasy (eds.), *Classical conditioning II*. New York: Appleton-Century-Crofts.

Seiffert, A. E., and Di Lollo, V. (1997). Low-level masking in the attentional blink. *Journal of Experimental Psychology: Human Perception and Performance*, **23**(4), 1061–73.

Shapiro, K. L., and Luck, S. J. (1999). The attentional blink: a front-end mechanism for fleeting memories. In V. Coltheart (ed.), *Fleeting memories: cognition of brief visual stimuli*, pp. 95–118. (MIT Press/Bradford Books Series in Cognitive Psychology). Cambridge, MA: MIT Press.

Shapiro, K. L., and Raymond, J. E. (1994). Temporal allocation of visual attention: inhibition or interference? In D. Dagenbach and T. H. Carr (eds), *Inhibitory processes in attention, memory, and language*, pp. 151–88. San Diego: Academic Press.

Shapiro, K. L., and Terry, K. (1998). The attentional blink: the eyes have it (but so does the brain). In R. Wright (ed.), *Visual Attention*, pp. 306–29. Oxford: Oxford University Press.

Shapiro, K. L., Raymond, J. E., and Arnell, K. M. (1994). Attention to visual pattern information produces the attentional blink in rapid serial visual presentation. *Journal of Experimental Psychology Human Perception and Performance*, **20**(2), 357–71.

Shapiro, K. L., Caldwell, J., and Sorensen, R. E. (1997a). Personal names and the attentional blink: a visual 'cocktail party' effect. *Journal of Experimental Psychology: Human Perception and Performance*, **23**(2), 504–14.

Shapiro, K., Driver, J., Ward, R., and Sorensen, R. E. (1997b). Priming from the attentional blink: a failure to extract visual tokens but not visual types. *Psychological Science*, **8**(2), 95–100.

Shapiro, K., Husain, M., Hillstrom, A. P., and Kennard, C. (submitted). Probing the parietal locus of the attentional blink.

Sheppard, D., Duncan, J., Shapiro, K., and Hillstrom, A. P. (submitted). New perceptual events trigger the attentional blink. *Psychological Science*.

Spence, C., and Driver, J. (1996). Audiovisual links in endogenous covert spatial attention. *Journal of Experimental Psychology Human Perception and Performance*, **22**(4), 1005–30.

Taylor, T. L., and Hamm, J. (1997). Category effects in temporal visual search. *Canadian Journal of Experimental Psychology*, **51**, 36–46.

Visser, T. A. W., Bischof, W. F., and Di Lollo, V. (1999). Attentional switching in spatial and nonspatial domains: evidence from the attentional blink. *Psychological Bulletin*, **125**(4), 458–69.

Ward, R., Duncan, J., and Shapiro, K. (1996). The slow time-course of visual attention. *Cognitive Psychology*, **30**(1), 79–109.

Ward, R., Duncan, J., and Shapiro, K. (1997). Effects of similarity, difficulty, and nontarget presentation on the time course of visual attention. *Perception and Psychophysics*, **59**(4), 593–600.

Weichselgartner, E., and Sperling, G. (1987). Dynamics of automatic and controlled visual attention. *Science*, **238**, 778–80.

Wojciulik, E., and Kanwisher, N. (1999). The generality of parietal involvement in visual attention. *Neuron*, **23**(4), 747–64.

2

The attentional blink and task switching within and across modalities

Marvin M. Chun and Mary C. Potter

Abstract

The attentional blink (AB) is a robust deficit obtained for a second visual target (T2) appearing within 200–600 ms of a correctly identified first target (T1). In most AB studies both targets appear among distractors in a rapid serial visual presentation (RSVP), and a key variable is the lag or SOA between the two targets. In the present chapter we review research bearing on the basis for the AB deficit and related deficits, including cross-modal versions of the AB procedure and tasks that require a task switch (a switch in target criterion) between T1 and T2. We conclude that the standard AB deficit is restricted to visual targets and can be distinguished from an additional deficit that results from a task switch between T1 and T2. The latter effect is found with cross-modal and auditory stimuli as well as visual stimuli, and is additive with the AB effect (when both targets are visual). We propose that the standard AB effect occurs at a different stage of processing than the more central task-switching deficit and shows features that distinguish it from the latter. The visual AB effect represents a limit in the speed with which visual targets—which are vulnerable to masking—can be consolidated into working memory or awareness: a second target may be lost while queuing for access to the consolidation process. We review studies that show a clear dissociation between AB and task-switching deficits, consistent with complementary findings of Allport and Hsieh (this volume, chapter 3) on criterion shifting in RSVP target search. The evidence suggests that there are multiple bottlenecks in processing that individually or together limit performance when two target stimuli must be processed within a brief space of time.

There are clear limitations to the ability to process perceptual information, but what is the nature of these limitations? This is a fundamental enterprise of inquiry in the field of cognitive psychology. A classic debate concerns the locus of attentional bottlenecks along the information-processing stream (Broadbent 1958; Deutsch and Deutsch 1963). According to these accounts, all sensory events that require a response must pass through some limited-capacity bottleneck. However, it is also possible that multiple attentional bottlenecks exist throughout the information-processing stream. As such, capacity limitations at one stage of the system may constrain performance independently of limitations at other stages of the system. Hence, an important goal for researchers is to dissect the human information-processing stream by examining which tasks interfere with which.

The endeavour to analyse and understand the architecture of perceptual and cognitive processing relies heavily on the use of dual-task paradigms (Pashler 1994). In dual-task paradigms, observers are presented with two tasks that must be performed concurrently or in rapid succession. For example, subjects could be asked to report two targets from a display (Duncan 1980). Typically, interference occurs between the two tasks, and such performance decrements are revealing of the attentional demands of the tasks performed. In sum, dual-task interference illuminates capacity limitations within the information-processing stream.

One of the most striking and compelling examples of dual-task interference is the attentional blink. In this paradigm, subjects are asked to identify and report two visual targets presented in rapid succession at various intervals (stimulus onset asynchronies, SOA) from each other. These targets appear amidst a rapid serial visual presentation (RSVP) sequence of visual distractors. The items are presented at rates of 8 to 12 items per second, pushing perceptual/cognitive processing mechanisms to the limit. While subjects are typically good at identifying and reporting the first target (T1) in such sequences, they exhibit a dramatic impairment in reporting the second target (T2) when it appears within half a second of the first (see Fig. 2.1). This deficit has been called the 'attentional blink' (AB: Raymond *et al.* 1992), and it has been observed in numerous labs using a variety of paradigms (Broadbent and Broadbent 1987; Chun and Potter 1995; Duncan *et al.* 1994; Grandison *et al.* 1997; Jolicœur 1998; Maki *et al.* 1997*b*; Seiffert and Di Lollo 1997; Weichselgartner and Sperling 1987). The AB paradigm is remarkable for several reasons. First, it is exceptionally robust, resulting in drops in target identification performance from 87 per cent to 30 per cent correct (e.g. Chun and Potter 1995). Second, the temporal lag (SOA) manipulations allow researchers to map a precise time course of interference. Performance on T2 is typically lowest at Lag 2 and gradually improves with increasing lag, asymptoting at Lags 6 or 7 (500–700 ms after T1 onset). Finally, the paradigm is such that essentially any type of task can be presented during the blink, making it a versatile tool for examining dual-task interference. For example, Joseph, Chun, and Nakayama (1997) used this paradigm to demonstrate that attention is needed to consciously report 'pre-attentive' orientation pop-out features. In this chapter, we will review many other variations of this AB paradigm.

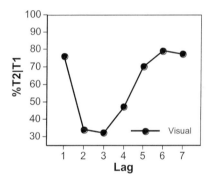

Fig. 2.1   The attentional blink effect: correct report of the second target (T2, given correct report of the first target, T1) as a function of lag; the stimulus onset asynchrony (SOA) was 100 ms. (From Chun and Potter 1995, Experiment 1. Reprinted with permission from the American Psychological Association)

The attentional blink demonstrates capacity limitations for consolidating visual information into working memory and awareness. A number of theories have been proposed to explain this phenomenon (for example: Chun 1997*b*; Chun and Potter 1995; Jolicœur 1998, 1999*a*; Maki *et al.* 1997*a*; Raymond *et al.* 1992; Raymond *et al.* 1995; Shapiro and Raymond 1994; Ward *et al.* 1996), and there is now substantial convergence among several of these (e.g. Shapiro *et al.* 1997*b*). We proposed that target processing in RSVP proceeds in two stages (Chun and Potter 1995), expanding suggestions by Broadbent and Broadbent (1987) and Duncan (1980). According to the two-stage model, every RSVP event (at least at rates of 10 items/s or slower) is rapidly identified in Stage 1, allowing for immediate detection. Analysis in Stage 1 may be fairly sophisticated and complete, such that visual types (stimulus identities) may be momentarily activated for all items in an RSVP sequence (Luck *et al.* 1996; Potter 1975, 1976, 1993,

1999; Shapiro *et al.* 1997a). However, we proposed that these Stage 1 representations may be subject to rapid forgetting or erasure as subsequent RSVP stimuli are presented (Giesbrecht and Di Lollo 1998). Successful target report requires further processing and consolidation of these initially activated target representations. This is achieved through Stage 2, a capacity-limited operation. While Stage 2 is occupied with a target, a second target cannot be consolidated (Chun and Potter 1995; Jolicœur 1998). Thus a T2 appearing at short lags during bottleneck processing for T1 is likely to be missed, producing the AB effect. The spared level of performance at Lag 1 suggests that both T1 and the immediately following item (whether a distractor or T2) are processed together (Chun and Potter 1995; Raymond *et al.* 1992). When the following item is T2, then both targets are likely to be successfully processed, producing Lag 1 sparing as seen in Fig. 2.1.

Other models of AB agree that items occurring during the blink are momentarily identified (Stage 1) and then are lost before entering a later stage of processing (Stage 2) that is limited in capacity (Shapiro *et al.* 1997b). These models differ, however, in how to characterize the limited-capacity process of Stage 2. Simply put, does the attentional blink reveal a single multi-purpose fundamental bottleneck that any type of information must pass through to be reported and used for action? Or is this attentional limitation restricted to visual targets? The answer to this question is important for understanding the locus of the attentional blink, and it is of broader significance for understanding the attentional capacity and architecture of information processing in general. Shapiro and Raymond (1994) proposed that the main source of interference in AB occurs as competition for retrieval between the two targets, as well as their trailing distractors, in visual short-term memory (VSTM). Hence, the interference model proposes that the AB must be a visual effect. Chun and Potter (1995) proposed that initial interference is inherently perceptual (between T1 and the +1 item), but that resolving this perceptual interference to identify the target slows down other processing stages such as consolidation into working memory. This theory also predicts that AB should be strictly visual. Finally, Jolicœur (1998) proposed that T1 reporting engages short-term consolidation processes that consume central resources. This theory predicts that AB does not have to be visual, as short-term memory consolidation processes may occupy central resources needed to consolidate any type of target, regardless of modality. In addition, this central mechanism has a late locus, sharing capacity with other important cognitive operations such as response selection (Jolicœur 1998, 1999a).

Thus, the main question confronting the field is whether dual-task performance in AB tasks can be described in terms of a central bottleneck or as an attentional limitation restricted to the visual modality. The most direct way to tackle this question is to investigate whether AB effects can be obtained for other modalities, and especially between targets from different modalities. If the AB has a central locus, then it should be observed for the auditory modality and between visual and auditory modalities. If it is specific to the visual system, it should not be observed for the auditory modality, and also not for cross-modal targets.

While the logic is straightforward, the empirical evidence is decidedly mixed, at least when viewed superficially. We will review some existing work that examines AB for targets presented within and across the visual and auditory modalities. Then we will discuss why we think discrepancies exist between studies published to this date. We will conclude with a model that outlines an architecture of human information processing that can explain these discrepancies, as well as pointing to future issues that deserve further empirical investigation.

Dual-task interference for visual, auditory, and cross-modal targets

Potter, Chun, Banks, and Muckenhoupt (1998) directly compared search performance for visual and auditory targets presented in rapid sequences. Subjects were asked to detect and report two letter targets embedded in a stream of digit distractors. Rapid auditory presentation (RAP) sequences were generated using compressed-speech versions of a set of letters and digits. Visual RSVP sequences were similar to those used in previous attentional blink studies (Chun and Potter 1995). Each modality was tested across blocks within subjects. The results are illustrated in Fig. 2.2. A robust AB effect was obtained for visual targets, replicating previous studies. In contrast, no auditory AB was observed. Performance was flat across lag, unlike the marked lag effect in visual search. This pattern was replicated across two experiments. In one experiment, target performance in each modality was equated for difficulty (Experiment 1, Potter *et al.*), by using slightly faster rates for visual stimuli (120 ms per item) than auditory stimuli (135 ms per item). In a second experiment, the two modalities were matched for rate of presentation (135 ms per item); this resulted in different baseline performance levels for the two modalities. In both experiments, visual AB was obtained and auditory AB was not. As Potter *et al.* put it, 'Auditory attention apparently does not blink' (p. 982). These results suggest that the capacity limitations in the visual modality are distinct from any present in the auditory modality.

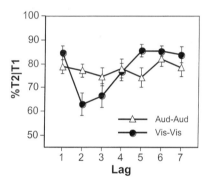

Fig. 2.2   Correct report of T2 (given correct report of T1) as a function of lag, for visual stimuli (SOA = 120 ms) and auditory stimuli (SOA = 135 ms). Vertical bars show standard error of the mean. (From Potter *et al.* 1998, Experiment 1. Reprinted with permission from the American Psychological Association)

What about cross-modal interference at a later stage of processing? In an additional experiment, Potter *et al.* (1998) examined whether a target in one modality interfered with the processing of a target in a different modality. On the basis of the within-modality studies described above, one could expect that cross-modal interference would not be obtained. This was indeed the case. As is shown in Fig. 2.3, visual T1 targets did not interfere with auditory T2 targets and vice versa. This further confirms the hypothesis that the AB phenomenon is restricted to the visual modality, revealing capacity limitations that are specific to the processing of visual

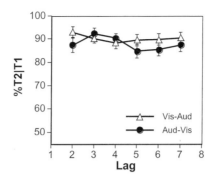

Fig. 2.3 Correct report of T2 (given correct report of T1) as a function of lag, for the two cross-modal conditions; the auditory stimuli had SOAs of 135 ms, the visual stimuli, 120 ms. Vertical bars show standard error of the mean. (From Potter *et al.* 1998, Experiment 3. Reprinted with permission from the American Psychological Association)

stimuli. These results argue against an architecture that proposes that AB is caused by a central bottleneck that is located at a late, amodal stage of information processing.

Although a clear story emerges from the Potter *et al.* (1998) study, a number of studies from other labs revealed conflicting results. For example, Duncan, Martens, and Ward (1997) observed an AB pattern for auditory targets, as well as the expected within-modality AB for visual targets. Interestingly, and consistently with Potter *et al.*'s study, they did not obtain cross-modal interference between the two modalities. Thus, at the very least, capacity limitations within each modality appear to be independent of each other. Still, an explanation is needed for why Duncan *et al.* observed auditory target interference while Potter *et al.* did not.

To further complicate matters, yet another distinct pattern of results was obtained in a series of experiments by Arnell and Jolicœur (1999). In contrast to both Potter *et al.* and Duncan *et al.*, Arnell and Jolicœur observed visual AB, auditory AB, *and* cross-modal AB between visual and auditory targets. In addition, Jolicœur and his colleagues (Jolicœur 1999*b*; Jolicœur & Dell' Acqua 1998) have demonstrated interference effects between visual and auditory tasks in a variety of other paradigms. This pattern of results provides strong evidence for a central locus of AB that is limited in capacity, producing interference for targets appearing within and across both visual and auditory modalities.

What could explain these serious discrepancies? We will consider a hypothesis, based on the attentional demands of task switching, that appears to explain all the existing published data. But first, let us review one alternative hypothesis. The alternative account concerns parametric factors such as the rate of stimulus presentation or overall task difficulty. Arnell and Jolicœur (1999) have pointed out that Potter *et al.*'s (1998) study employed slower presentation rates than those used in Arnell and Jolicœur's study. As the rate of presentation is critical for observing the attentional blink, this could possibly explain the failure to observe auditory and cross-modal AB using Potter *et al.*'s protocol. Arnell and Jolicœur (Experiment 5) manipulated presentation rate and demonstrated that the AB deficit was eliminated for auditory targets when slower rates (135 ms/item), comparable to that used in Potter *et al.*, were used. However, rate cannot be the main explanation. First, even if rate is constant across studies, it cannot be a psychophysically equalizing factor because performance in RSVP and RAP tasks is dependent on many other variables, such as the visibility or audibility of the stimuli. Second, in an unpublished experiment, Potter *et al.* were unable to observe auditory AB even when a faster rate of 120 ms per item was used (Fig. 2.4; Lag Effect, $F < 1$), a rate at which Arnell *et al.* obtained an auditory AB effect.

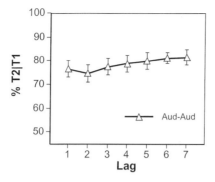

Fig. 2.4   Correct report of T2 (given correct report of T1) as a function of lag, for auditory stimuli (SOA = 120 ms). Vertical bars show standard error of the mean. (Unpublished experiment reported in Potter *et al.* 1998, p. 984.)

Finally, Potter *et al.* obtained strong dissociations between the visual and auditory modalities even when the stimuli were controlled for a critical parameter, stimulus identification difficulty. There is extensive evidence, reviewed in Seiffert and Di Lollo (1997), showing that stimulus identification difficulty is an important factor in AB. (Note that AB is only influenced by the difficulty of processing patterned stimuli (Shapiro *et al.* 1994)). Jolicœur (1998, 1999a,b) has additionally shown that the magnitude of AB correlates with the time it takes subjects to perform the T1 task. Thus, in our view, it would have been difficult to obtain the observed all-or-nothing dissociation in the time course between visual and auditory modalities if the processing capacities for each modality were not independent of each other.

We propose instead that all of the existing published data on dual-target performance for visual, auditory, and cross-modal targets can be understood as reflecting one or both of two distinct effects: (1) a vision-specific AB effect, and (2) a task-switch deficit. Task-switch deficits occur in the AB procedure when subjects must switch from one perceptual set for T1 to another for T2. Such task-switch costs are distinct from the processing limitations that cause AB for visual targets. What is task switching? People's interactions with the environment are guided by internally generated task sets (Allport *et al.* 1994), such as 'search for a letter while ignoring the digits'. Such task sets guide the observer's attention to incoming sensory information. Task switching occurs when the subject must discard the current task set and replace it with a new one. Such task switching incurs a severe cost in performance. As an example, consider a classic study by Jersild (1927; see also Spector and Biederman 1976). Subjects were given lists of two-digit numbers and asked to perform an addition (or a subtraction) task on all the numbers in the list. In the task-switch condition, they had alternately to add or subtract a constant number from successive items. Subjects were significantly slower in the task-switch condition, which produced large RT costs of the order of several hundred milliseconds per item.

Task switching is a high-level executive function (Allport *et al.* 1994; Meiran 1996; Rogers and Monsell 1995), with a relatively late locus in the information-processing stream. It is likely to be amodal, and may draw upon central resources used for other cognitive operations such as response selection and perhaps also short-term memory consolidation.

We argue that many findings that have been interpreted as auditory AB or cross-modal AB can be reinterpreted as reflecting task-switch deficits. In contrast, Potter *et al.*'s (1998) experiments presented a consistent task for T1 and T2. Subjects were asked to detect and report two letter targets (T1 and T2), ignoring all the extraneous digit distractors. No auditory or cross-modal deficits were obtained under these conditions. However, Arnell and Jolicœur (1999) employed a procedure that involved a salient switch in task set between T1 and T2. In their study, subjects were asked to monitor a stream of letters for the presence of a single digit (T1). Then the task switched to letter detection for the presence of the letter X. Not only was there a categorical switch in target set from digit to letters; subjects had to switch from ignoring the letter stream to attending to the stream of letters for the presence of the X probe. This produced deficits for secondary targets regardless of modality. Such conditions are optimal for observing task-switch costs (Allport *et al.* 1994; Meiran 1996; Rogers and Monsell 1995). For example, Allport *et al.* (1994) asked subjects to search for animal names among non-animal names, and then to switch their search to non-animals among animals. A significant deficit was observed for a target that occurred within 3 items of the task switch. Thus, in a target search task similar to that used in AB paradigms, a task switch can also produce deficits for secondary targets.

Thus all the published experiments to date that showed visual, auditory and cross-modal AB (Arnell and Jolicœur 1999; Jolicœur 1999*b*) may reflect capacity limitations for task switching, and such task-switch costs are distinct from the bottleneck responsible for AB of visual targets. This explains why no auditory or cross-modal AB was observed in Potter *et al.*'s study (1998; see Figs. 2.2 and 2.3), which did not involve a task switch between T1 and T2. As more compelling evidence, Potter *et al.* replicated Arnell and Jolicœur's results when a task switch was introduced between T1 and T2, using the same stimuli and apparatus that did not produce dual-task interference effects in the absence of a task switch. These results were obtained when the T1 task was to detect a digit among letters and the T2 task was to detect an X probe among letters. The results are shown in Fig. 2.5; now a visual AB, auditory AB, and cross-modal interference effects were observed. Contrast this with Fig. 2.6, which shows the results of a parallel experiment in which subjects did not switch tasks, but looked or listened for two letters among digits. Here, with no task switch, only the visual–visual condition showed a significant AB deficit. This provides strong support for the role of task switching as the decisive factor producing discrepancies between studies. Note that the task-switching deficit and the visual AB effect appear to be additive: the visual AB effect in Fig. 2.5 is larger than that in Fig. 2.6.

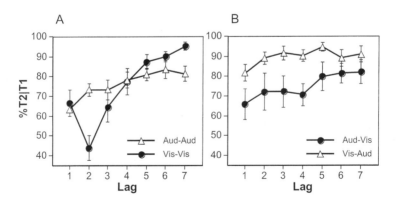

Fig. 2.5 Correct report of T2 (given correct report of T1) as a function of lag, with two simultaneous sequences of visual and auditory items (SOA = 120 ms) and instructions to identify T1 and detect 'X' (T2) in specified modalities, for (A) the groups with two visual or two auditory targets, and (B) the groups with one visual and one auditory target. Only trials on which an 'X' was presented are included. Vertical bars show standard error of the mean. (From Potter *et al.* 1998, Experiment 4. Reprinted with permission from the American Psychological Association)

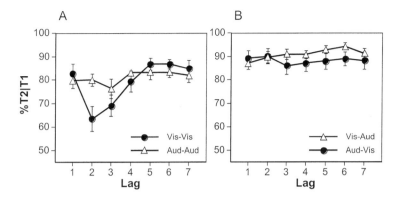

Fig. 2.6 Correct report of T2 (given correct report of T1) as a function of lag, with two simultaneous sequences of visual and auditory items (SOA = 120 ms), when the task was to report two letters among digit distractors. The results are shown separately for each of the four conditions, VV, AA, VA, and AV. Vertical bars show standard error of the mean. (From Potter *et al.* 1998, Experiment 5. Reprinted with permission from the American Psychological Association)

A theoretical account

On the basis of these findings, we propose an architecture of human information processing that posits separate capacity limitations for the visual modality as well as an amodal central bottleneck. This is illustrated in Fig. 2.7. First, the visual processing system is limited in capacity, and AB occurs as visual targets interfere with other visual targets appearing close to them in time (Broadbent and Broadbent 1987; Chun and Potter 1995; Duncan *et al.* 1994; Raymond *et al.* 1992). However, these processing limitations are restricted to the visual modality, so that in the absence of other task demands visual targets do not interfere with auditory targets and vice versa. AB interference between multiple visual targets is obtained only when two conditions are satisfied, and these two conditions reflect characteristics specific to the visual modality. First, there is high capacity for initial identification of the visual input if it is presented for about 100 ms an item, but these Stage 1 representations are ephemeral and subject to interference and substitution from subsequent visual events (Chun and Potter 1995; Giesbrecht and Di Lollo 1998; Potter 1975, 1993, 1999). Second, if attention is given to one of the items—for example, T1—then it can be encoded and consolidated in Stage 2, but at the expense of the immediately following items. They will be processed in Stage 1, but must wait for Stage 2 processing of the earlier item to be completed. During the wait, the items following T1 may be lost, resulting in the AB deficit. In sum, both of these conditions, rapid identification that is subjected to visual or conceptual masking (Chun and Potter 1995; Grandison *et al.* 1997; Moore *et al.* 1996; Raymond *et al.* 1992; Seiffert and Di Lollo 1997), and a limited-capacity, attention-demanding consolidation process, must be present to observe AB.

   The characteristics of the auditory module are less clear. First, it is plausible to assume that there are capacity limitations for processing simultaneous or rapid sequential auditory events (cf. e.g. Treisman and Davies 1973). That is, auditory targets do interfere with each other, as is suggested by the lower level of performance in the dual-target conditions versus the single-target condition baseline in the studies reviewed in this chapter (Arnell and Jolicœur 1999; Duncan *et al.* 1997; Potter *et al.* 1998). The question is whether these dual-target deficits for auditory stimuli are caused by modality-specific capacity limitations for auditory targets, and

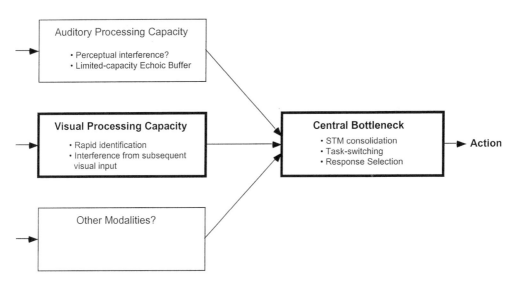

Fig. 2.7 Model of the structure of information processing to explain dual-target performance in visual, auditory, and cross-modal attentional blink, task switching, and short-term memory consolidation paradigms.

whether such limitations result in AB effects with a timecourse similar to that observed for visual targets. The evidence on auditory AB is mixed. Potter *et al.* did not observe AB lag effects for auditory targets; performance, although lower for a second target, was flat across lags. Arnell and Jolicœur (1999) demonstrated auditory AB; but this can be interpreted as reflecting task-switching deficits, and hence as engaging a different capacity limitation, to be described below. Finally, Duncan *et al.* demonstrated an auditory AB effect that is very similar to that observed for visual targets, although it was symmetrical for T1 and T2 at short intervals, unlike the visual AB. However, it is unclear how to characterize the nature of interference in Duncan *et al.*'s task. Although there were different sets of targets in the two channels that the subjects monitored, both were words, which involved no task switch. However, in the auditory case, the two channels differed in voice pitch. Hence, their task may have required a switch in perceptual set defined over pitch. We cannot fully determine whether switching one's auditory perceptual set from one pitch to another constitutes a 'task switch', although we are inclined to interpret it as a task switch on the basis of the fact that different features are used for selection. But further research is required on this issue, and new findings will help characterize the nature of capacity limitations for auditory stimuli.

One property of auditory processing that can be clearly distinguished from visual processing concerns masking properties and durability over time, critical factors for observing AB. In contrast to visual information, which is easily overwritten and replaced by subsequent visual events (Breitmeyer 1984; Enns and Di Lollo 1997; Sperling 1960), auditory stimuli appear to be more resistant to such masking interference from subsequent auditory events. The echoic buffer preserves auditory information over 1 to 2 s, presumably because many properties of sound depend on temporally extended information. If auditory stimuli are not as easily backward-masked as visual stimuli, then one would not expect to obtain an auditory AB, inasmuch as visual AB is dependent on the erasure of T2 representations by subsequent masking stimuli (Chun and Potter 1995; Giesbrecht and Di Lollo 1998).

Information from the visual and auditory modalities converges to a central bottleneck of information processing, required for most overt behaviours (see Fig. 2.7). Following Jolicœur (1998, 1999a) and Pashler (1994), we will call this the central processor. Capacity limitations in the central processor are distinct from those that may exist in the separate perceptual modalities. Interference at the central stage of processing leads to decrements in performance for any type of target, regardless of modality and regardless of task. The important question is what types of cognitive operations engage this central processor. Work by Jolicœur and his colleagues (1998, 1999a,b) has demonstrated that short-term consolidation engages this central processor, leading to interference for subsequent targets, including targets from different modalities. Thus, when subjects must encode information for later report, this causes interference with central processing of subsequent events (Jolicœur and Dell'Acqua 1998). Increasing the load of information that needs to be consolidated produces concomitant increases in interference for subsequent targets. According to Jolicœur, the short-term consolidation mechanism is engaged whenever information about target stimuli must be encoded for subsequent report. This is consistent with the operations of Stage 2 in Chun and Potter's (1995) model; but the two-stage model postulates that AB is only observed when processing is slowed down by the attentional requirements of identifying and individuating a visual target that is immediately followed by a mask or distractor. If, instead, the target is followed by a blank interval of 100 ms, there is little or no AB for a subsequent target.

The central processor is used for other important cognitive operations. For example, speeded response selection is also controlled by this central processor. Hence, increased AB is obtained when the response to T1 must be made immediately rather than delayed until after T2 is presented (Jolicœur 1998). The number of response alternatives also affects this processor's capacity. As the number of alternatives increases for speeded response tasks, increased interference is observed (Jolicœur 1999a). These results lead to the conclusion that the psychological refractory period (Pashler 1994), which is the result of a response-selection bottleneck, shares the same central processor as short-term consolidation. This produces dual-task interference whenever either one of these cognitive operations is engaged.

To this list of central operations, we add the attentional demands of task switching. Task switching involves remapping of stimulus–response associations, and hence may share capacity with response-selection mechanisms. Supporting a central locus, we obtained cross-modal dual-target deficits when a task switch was introduced (Potter et al. 1998).

We further propose that the AB should be distinguished from task-switching costs. There are several lines of evidence that point to a dissociation. Task-switch costs are standardly measured by an increase in RT when the next trial in a sequence requires a switch (Allport et al. 1994; Rogers and Monsell 1995; Meiran 1996). Such trials are separated by 1 s or more, well beyond the 500 ms range of AB interference. Using an RSVP version of a task-switch experiment in which the criterion that defines the target is changed unpredictably during a sequence, Allport and Hsieh (this volume, Chapter 3) have shown that a task-switch deficit on the next target item occurs immediately. The task-switch cost was present regardless of whether there was an earlier target, and it diminished as a function of the number of non-target distractors preceding the target, not as a function of the simple passage of time. Neither of these properties is true of AB. The AB is dependent on processing of T1 (Chun and Potter 1995; Jolicœur 1998; Joseph et al. 1997; Luck et al. 1996; Raymond et al. 1992), and is largely gone at an SOA of 500 ms or greater, whether or not the interval between T1 and T2 is filled with distractors (as long as each target is followed by a mask).

Conversely, the AB can be obtained in the absence of a task-switch. Chun and Potter (1995) replicated all Raymond *et al.*'s findings using a design that did not require a criterion shift between T1 and T2 (search for two letters amongst non-letter distractors). The AB appears to be restricted to the visual modality, as no AB was observed when the targets were auditory or cross-modal (one visual, one auditory). In contrast, task-switching deficits are found regardless of modality when a criterion shift is introduced between T1 and T2 (Potter *et al.* 1998).

Providing further direct support for a distinction between AB and task-switching costs, Chun and Jiang (1999) demonstrated a double dissociation between AB and task switching using identical stimuli. Subjects performed shape-discrimination tasks for targets consisting of over-lapping visual shapes (see Fig. 2.8). The two target shapes, which were coloured white, appeared within an RSVP sequence of black distractors on a grey background. As usual, the lag between T1 and T2 was varied to assess the timecourse of dual-target interference. Each target was composed of two shapes, and subjects were asked to perform one of two tasks for each target. For the circle task, subjects were asked to attend to and report the orientation (up, left, right, or down) of the gap in the circle. For the diamond task, subjects reported the direction of elongation, which was always different from the reportable direction of the circle.

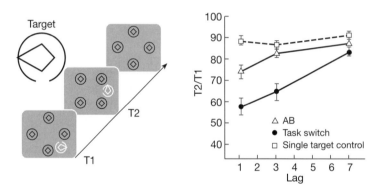

Fig. 2.8 Illustration of paradigm to test AB and task switching using identical physical stimuli (from Chun and Jiang 1999). A sample target stimulus is shown in the inset of the left panel; each target is composed of two shapes (circle and diamond). Two of these targets, coloured white, were presented in an RSVP sequence of black distractors (left panel). The right panel shows correct report of T2 (given correct report of T1) as a function of lag for the AB, Task Switch, and Single target control conditions.

There were two main conditions in the primary experiment. In the AB condition, the task was consistent between T1 and T2 (attend to the circle shape in T1 and the circle shape in T2). In the task-switch condition, the perceptual set was different for the two targets (attend to the circle shape in T1 and the diamond shape in T2). With respect to the robust AB effect obtained in the AB condition (poorer performance at shorter lags), the task-switch condition produced a further 20 per cent decrement in performance for T2 appearing within half a second of T1. Hence, task switching added a cost in performance, because subjects had to go from attending to a circle while ignoring the diamond in T1 to attending to the diamond while ignoring the circle in T2. Such perceptual set reversals classically produce task-switch costs (Meiran 1996;

Rogers and Monsell 1995). In two additional experiments, Chun and Jiang dissociated AB and task-switch costs. Task-switch costs were observed without AB when the distractors were simplified to minimize target–distractor interference, known to be a critical factor for observing AB (Chun 1997b; Chun and Potter 1995). This pattern was reversed when the target stimuli were manipulated so that the two overlapping shape stimuli did not compete for different responses. When these 'neutral' targets were presented amongst interfering distractors, AB was observed without an additional cost for task switching. Hence, AB and task switching are doubly dissociable, which is consistent with our proposal that the capacity-limited mechanisms involved may be placed at different stages of information processing.

To summarize, we propose that visual AB reflects sequential processing limitations that are restricted to visual stimuli, although more central, amodal limitations may add to those of visual AB. Our model is consistent with Jolicœur's proposal that a central bottleneck exists and that this operates over amodal representations of target events (Arnell and Jolicœur 1999; Jolicœur 1999b). We support the notion that the central processor is important for short-term consolidation, response selection, and long-term memory retrieval (Jolicœur and Dell'Acqua 1998; Jolicœur 1999a; Pashler 1994). We add task coordination (switching) to this list of cognitive operations mediated by the central processor. Our account differs only in how to define the loci of the bottleneck process responsible for visual AB.

First, we suggest that Stage 2 either precedes or forms a distinct subcomponent of the central processor that is separate from the other bottleneck processes responsible for response selection and task switching. Otherwise, it would be difficult to obtain double dissociations between AB and task switching (Chun and Jiang 1999). Also, such a distinction is consistent with the fact that AB can be observed without PRP and vice versa, a pattern that can be observed across published studies in the literature. Although the loci are distinct, it is important to note that the two processes form a single chain of information processing, which is why one can observe both effects within the same task (Jolicœur 1998, 1999a).

Second, Stage 2 differs from other central processes in its sensitivity to visual interference. The efficacy of Stage 2 and the subsequent interference effects it produces are tightly influenced by the degree of perceptual interference on T1 (Chun and Potter 1995; Grandison et al. 1997; Moore et al. 1996; Raymond et al. 1992; Seiffert and Di Lollo 1997). In contrast, manipulations of the target's perceptual integrity do not have direct impact on central processor operations such as response selection (Pashler 1984; Pashler and Johnston 1989). In addition, a visual T2 is uniquely vulnerable to any delay in entering the central processor (Chun and Potter 1995; Giesbrecht and Di Lollo 1998), in that it suffers a deficit from STM consolidation of T1 even without an additional delay due to task switching or response selection.

How can task switches be defined?

The notion of task switching is important for our framework and for distinguishing when dual-target interference (especially cross-modal interference) will occur. So it will be useful to operationalize what counts as a task switch.

All target search tasks are defined by a perceptual set for a particular target or class of targets. Task switches can be simply defined as a switch in perceptual set from one target to another. For instance, switching the category that defines a target (for example, from digit to letter) involves a task switch. However, an important condition is that switches in perceptual set should also involve a reversal in target-response and distractor-response mappings—for example, searching for a digit among letters (T1), and then searching for a particular letter (T2). The requirement to ignore the letters to detect T1 makes it more difficult to switch to that set in a subsequent search for T2, as is shown in Chun and Jiang's (1999) study described above.

We propose that there are performance 'markers' that reveal whether a particular task involves a task switch or not. One useful heuristic is to examine whether subjects can effectively ignore the T1 task when two tasks are presented in succession. For instance, in a task where T1 is defined by luminance and the second target is defined as X probe detection, Raymond *et al.* (1992) and numerous other labs have shown that the T1 task can be effectively ignored with minimal interference to the T2 task. We argue that this is because the two tasks are different enough to allow this. Another example can be found in a study by Joseph *et al.* (1997). The T1 task was white letter detection, and the T2 task was orientation pop-out detection. The two tasks were very different, and subjects had no difficulty ignoring the T1 task to perform the T2 task alone in the control condition. Contrast these examples with Chun and Potter's task, which involved no task switch. When searching for two letters amongst digits, subjects have great difficulty reporting just the second letter. Apparently this task requires subjects to register but then ignore the first letter, something they find almost impossible to do because of the high similarity between the two tasks. Another example involves a search for two coloured letters. Although one may expect subjects to have no problem ignoring the first coloured letter and reporting the second coloured letter, there is in fact a significant interference effect from the 'ignored' T1 (Chun 1997a; Jiang and Chun, in press). That is, subjects are unable to filter T1 if it meets the search description required to detect T2 (see also Folk *et al.* 2000).

Note that we are not claiming that previous demonstrations of dual-task deficits in AB studies can all be interpreted in terms of task switching. For example, in Raymond *et al.*'s (1992) original study, there is clearly a visual AB component, as is evidenced by the fact that the presence of AB was highly sensitive to perceptual manipulations (such as the insertion of a blank after T1). Also note that the degree of task switching involved in that task was minimal. Even if the first target task was to detect a 'white' letter, subjects could still focus attention on attending to letters. Thus, no reversal in categorical set was required in the Raymond *et al.* study, in contrast to the task employed by Arnell and Jolicœur, which did impose a categorical set reversal between T1 and T2 (i.e., search for digit T1 among letters, and then search for 'X' T2). Moreover, we propose that the visual AB effect is present even with a task switch: the two effects add, so that a visual 'blink' is increased when there is a task switch.

Another useful marker for the presence or absence of task switching was first suggested by Potter *et al.* (1998): lag-1-sparing is characteristic of AB, whereas the task-switching deficit is maximal at lag 1 (as observed in experiments reported by Potter *et al.* and also in Allport and Hsieh's studies in Chapter 3 of the present volume). Lag 1 sparing refers to the relatively high level of performance for T2 when it appears immediately after T1 (see Fig. 2.1). In a comprehensive literature review, Visser, Bischof, and Di Lollo (1999) demonstrated that the degree of Lag-1 sparing observed was dependent on the number of dimensional (category, modality, etc.) switches imposed between T1 and T2. For example, when no switches imposed (as in Chun and Potter's search for two letters among digits task), 44 out of 46 experiments demonstrated lag-1 sparing. However, when a task required more than one type of switch, (e.g., digit to letter and identification to detection, as in Arnell and Jolicœur (1999)), 29 out of 32 experiments *failed* to produce lag-1 sparing. Further details may be found in Visser *et al.* (1999), but the important conclusion is that lag-1 sparing is a sensitive index of the degree of task similarity between T1 and T2. The greater the similarity, the greater the lag-1 sparing observed. Thus, in a letter-identification task consistent for T1 and T2 (Chun and Potter 1995), lag-1 performance was similar to the asymptotic baseline performance. As an extreme example in the other direction, Joseph *et al.*'s task required a switch in location (centre to periphery), task (letter identification to orientation pop-out detection), and category (letter to Gabor patch). This resulted in a monotonic function of T2 performance across lag, with the poorest performance at Lag 1. In still other cases (e.g., Potter *et al.* 1998) in which there was a task switch but the targets were both visual, there was an intermediate degree of lag-1 sparing.

Summary

Interference between two targets can occur at one or several distinct stages within the human information-processing stream. Standard AB may be restricted to capacity limitations and stimulus-masking properties that are specific to the visual modality, independent of capacity restrictions in other modalities. For both visual and auditory targets, interference may be obtained within modality and across modalities if the task taxes attentional demands of short-term memory consolidation, response selection, or task switching. In particular, we propose that the cognitive operation of task switching explains why and when dual-target interference occurs between auditory and visual target pairs in AB paradigms.

References

Allport, A., and Hsieh, S. (this volume, Chapter 3). Task-switching: using RSVP methods to study an experimenter-cued shift of set. In K. Shapiro (ed.), *The limits of attention: temporal constraints on human information processing*. Oxford: Oxford University Press.

Allport, D. A., Styles, E. A., and Hsieh, S. (1994). Shifting intentional set: exploring the dynamic control of tasks. In C. Umilta and M. Moscovitch (eds), *Attention and performance XV*, pp. 421–52. Hillsdale, NJ: Erlbaum.

Arnell, K. M., and Jolicœur, P. (1999). The attentional blink across stimulus modalities: evidence for central processing limitations. *Journal of Experimental Psychology: Human Perception and Performance*, **25**, 630–48.

Breitmeyer, B. (1984). *Visual masking: an integrative approach*. Oxford: Oxford University Press.

Broadbent, D. E. (1958). *Perception and communication*. London: Pergamon Press.

Broadbent, D. E., and Broadbent, M. H. (1987). From detection to identification: response to multiple targets in rapid serial visual presentation. *Perception and Psychophysics*, **42**(2), 105–13.

Chun, M. M. (1997*a*). Binding errors are redistributed by the attentional blink. *Perception and Psychophysics*, **59**, 1191–9.

Chun, M. M. (1997*b*). Types and tokens in visual processing: a double dissociation between the attentional blink and repetition blindness. *Journal of Experimental Psychology: Human Perception and Performance*, **23**, 738–55.

Chun, M. M., and Jiang, Y. (1999). Task switching and the attentional blink. In V. Di Lollo (Chair), *Attentional Switching*. Symposium conducted at the annual meeting of the Psychonomic Society, Los Angeles, CA.

Chun, M. M., and Potter, M. C. (1995). A two-stage model for multiple target detection in rapid serial visual presentation. *Journal of Experimental Psychology: Human Perception and Performance*, **21**(1), 109–27.

Deutsch, J. A. and Deutsch, D. (1963). Attention: some theoretical considerations. *Psychological Review*, **70**, 80–90.

Duncan, J. (1980). The locus of interference in the perception of simultaneous stimuli. *Psychological Review*, **87**(3), 272–300.

Duncan, J., Ward, R., and Shapiro, K. L. (1994). Direct measurement of attentional dwell time in human vision. *Nature*, **369**(6478), 313–15.

Duncan, J., Martens, S. and Ward, R. (1997). Restricted attentional capacity within but not between sensory modalities. *Nature*, **387**, 808–10.

Enns, J. T., and Di Lollo, V. (1997). Object substitution: a new form of masking in unattended visual locations. *Psychological Science*, **8**(2), 135–9.

Folk, C. L., Leber, A., and Egeth, H. (2000). Made you blink! Contingent attentional capture in an RSVP task. Paper presented at the annual meeting of the Psychonomic Society, New Orleans, LA.

Giesbrecht, B., and Di Lollo, V. (1998). Beyond the attentional blink: visual masking by object substitution. *Journal of Experimental Psychology: Human Perception and Performance*, **24**, 1454–66.

Grandison, T. D., Ghirardelli, T. G. and Egeth, H. E. (1997). Beyond similarity: masking of the target is sufficient to cause the attentional blink. *Perception and Psychophysics*, **59**, 266–74.

Jersild, A. T. (1927). Mental set and shift. *Archives of Psychology, (whole no.)* **89**.

Jiang, Y., and Chun, M. M. (in press). The influence of temporal selection on spatial selection and distractor interference: an attentional blink study. *Journal of Experimental Psychology: Human Perception and Performance*.

Jolicœur, P. (1998). Modulation of the attentional blink by on-line response selection: evidence from speeded and unspeeded Task-sub-1 decisions. *Memory and Cognition*, **26**(5), 1014–32.

Jolicœur, P. (1999a). Concurrent response selection demands modulate the attentional blink. *Journal of Experimental Psychology: Human Perception and Performance*, **25**, 1097–1113.

Jolicœur, P. (1999b). Restricted attentional capacity between sensory modalities. *Psychonomic Bulletin and Review*, **6**, 87–92.

Jolicœur, P., and Dell' Acqua, R. (1998). The demonstration of short-term consolidation. *Cognitive Psychology*, **36**(2), 138–202.

Joseph, J. S., Chun, M. M., and Nakayama, K. (1997). Attentional requirements in a 'preattentive' feature search task. *Nature*, **387** (June 19), 805–8.

Luck, S. J., Vogel, E. K., and Shapiro, K. L. (1996). Word meanings can be accessed but not reported during the attentional blink. *Nature*, **383** (17 October), 616–18.

Maki, W. S., Couture, T., Frigen, K., and Lien, D. (1997a). Sources of the attentional blink during rapid serial visual presentation: perceptual interference and retrieval competition. *Journal of Experimental Psychology: Human Perception and Performance*, **23**, 1393–1411.

Maki, W. S., Frigen, K., and Paulson, K. (1997b). Associative priming by targets and distractors during rapid serial visual presentation: does word meaning survive the attentional blink? *Journal of Experimental Psychology: Human Perception and Performance*, **23**, 1014–34.

Meiran, N. (1996). Reconfiguration of processing mode prior to task performance. *Journal of Experimental Psychology: Learning, Memory, and Cognition*, **22**, 1423–42.

Moore, C. M., Egeth, H., Berglan, L. R., and Luck, S. J. (1996). Are attentional dwell times inconsistent with serial visual search? *Psychonomic Bulletin and Review*, **3**, 360–5.

Pashler, H. (1984). Processing stages in overlapping tasks: evidence for a central bottleneck. *Journal of Experimental Psychology: Human Perception and Performance*, **10**, 358–77.

Pashler, H. (1994). Dual-task interference in simple tasks: data and theory. *Psychological Bulletin*, **116**, 220–44.

Pashler, H., and Johnston, J. C. (1989). Chronometric evidence for central postponement in temporally overlapping tasks. *The Quarterly Journal of Experimental Psychology*, **41A**, 19–45.

Potter, M. C. (1975). Meaning in visual search. *Science*, **187**, 965–6.

Potter, M. C. (1976). Short-term conceptual memory for pictures. *Journal of Experimental Psychology: Human Learning and Memory*, **2**(5), 509–22.

Potter, M. C. (1993). Very short-term conceptual memory. *Memory and Cognition*, **21**(2), 156–61.

Potter, M. C. (1999). Understanding sentences and scenes: the role of Conceptual Short Term Memory. In V. Coltheart (ed.), *Fleeting memories*, pp. 13–46. Cambridge, MA: MIT Press.

Potter, M. C., Chun, M. M., Banks, B., and Muckenhoupt, M. (1998). Two attentional deficits in serial target search: The visual attentional blink and an amodal task-switch deficit. *Journal of Experimental Psychology. Learning, Memory, and Cognition*, **25**, 979–92.

Raymond, J. E., Shapiro, K. L., and Arnell, K. M. (1992). Temporary suppression of visual processing in an RSVP task: an attentional blink? *Journal of Experimental Psychology: Human Perception and Performance*, **18**(3), 849–60.

Raymond, J. E., Shapiro, K. L. and Arnell, K. M. (1995). Similarity determines the attentional blink. *Journal of Experimental Psychology: Human Perception and Performance*, **21**(3), 653–62.

Rogers, R. D., and Monsell, S. (1995). Costs of a predictable switch between simple cognitive tasks. *Journal of Experimental Psychology: General*, **124**, 207–31.

Seiffert, A. E., and Di Lollo, V. (1997). Low-level masking in the attentional blink. *Journal of Experimental Psychology: Human Perception and Performance*, **23**, 1061–73.

Shapiro, K. L., and Raymond, J. E. (1994). Temporal allocation of visual attention: inhibition or interference? In D. Dagenbach and T. H. Carr (eds.), *Inhibitory mechanisms in attention, memory and language*, pp. 151–88. Boston, MA: Academic Press.

Shapiro, K. L., Raymond, J. E., and Arnell, K. M. (1994). Attention to visual pattern information produces the attentional blink in rapid serial visual presentation. *Journal of Experimental Psychology: Human Perception and Performance*, **20**(2), 357–71.

Shapiro, K., Driver, J., Ward, R., and Sorensen, R. E. (1997*a*). Priming from the attentional blink: a failure to extract visual tokens but not visual types. *Psychological Science*, **8**(2), 95–100.

Shapiro, K. L., Arnell, K. M., and Raymond, J. E. (1997*b*). The attentional blink: a view on attention and glimpse on consciousness. *Trends in Cognitive Science*, **1**, 291–6.

Spector, A., and Biederman, I. (1976). Mental set and shift revisited. *American Journal of Psychology*, **89**, 669–79.

Sperling, G. (1960). The information available in brief visual presentations. *Psychological Monographs: General and Applied*, **74**(11), 1–29.

Treisman, A. M., and Davies, A. (1973). Divided attention to ear and eye. In S. Kornblum (ed.), *Attention and performance IV*, pp. 101–17. New York: Academic Press.

Visser, T. A. W., Bischoff, W. F., and Di Lollo, V. (1999). Attentional switching in spatial and non-spatial domains: evidence from the attentional blink. *Psychological Bulletin*, **125**, 458–69.

Ward, R., Duncan, J., and Shapiro, K. (1996). The slow time-course of visual attention. *Cognitive Psychology*, **30**, 79–109.

Weichselgartner, E., and Sperling, G. (1987). Dynamics of automatic and controlled visual attention. *Science*, **238**(4828), 778–80.

3

Task-switching: using RSVP methods to study an experimenter-cued shift of set

Alan Allport and Shulan Hsieh

Abstract

We describe an RSVP method that we have developed to study the timecourse of experimenter-cued shifts of set, in rapid, sequential visual search tasks. A number of critical design features distinguish this method from the better-known second-target paradigm, used to study the 'attentional blink' (AB) effect. These design features enable us to study the performance costs of a shift in target criterion, uncontaminated by any second-target 'AB' effects.

   This chapter reviews a series of experiments in which we have begun to explore this method. In four of these experiments, subjects monitored word stimuli for targets specified in terms of either physical (word- or letter-shape) or semantic criteria. Other experiments used numerals, and target discriminations based on simple number semantics. In all these experiments, on a subset of trials a cue is presented, during the course of the RSVP sequence, that instructs the subject to stop monitoring for the original target criterion (T1), and immediately to start monitoring for a different target category (T2). Monitoring performance shows an abrupt and dramatic drop in accuracy immediately following the cue to shift target criterion, followed by a gradual, monotonic recovery over the subsequent 5–7 task-relevant stimuli.

   Three experiments contrast the effects of shifting criterion within vs. between different domains of representation, and of reversal *vs.* non-reversal (orthogonal) shifts. Three further experiments contrast the effect of simple time-lapse since the shift cue, *vs.* the number of task-relevant stimulus items, on the recovery of selective target-report accuracy. The data show, unambiguously, that recovery of performance from the initial criterion-shift (C-S) cost is a function of the number of successive task-relevant stimuli, not of simple time-lapse since the shift cue. This result sharply differentiates these task-shift phenomena from the second-target 'AB' effect, described by other contributors to this volume, and from the effects of simple shifts of attention in visual space. Our results have a number of implications for models of the voluntary (and involuntary) control of set.

Introduction

A method for studying 'task switching' in rapid sequential search tasks

Over the past 20 years, much research has been directed to the shifting (or 'orienting') of visual selective attention from one spatial location to another (Posner 1995). Much less research has been directed to the shifting of selective attention in other, non-spatial domains. In this chapter we describe an experimental procedure, designed to study the timecourse of non-spatial shifts of selective 'task set'. The method uses sequential search for pre-specified 'target' stimuli, among a rapid sequence of non-target stimuli. In some conditions, the target criterion is abruptly altered during the course of the sequence, and the procedure enables us to track the effects of this change of task on subsequent performance (Allport *et al.* 1994; Hsieh 1993, 1995; Hsieh and Allport 1994).

In outline, the method is as follows. Stimuli appear in rapid serial visual presentation (RSVP), and selective report is required of one (or, at most, two) target items in any one sequence. An RSVP sequence (a 'list') of from 10 to 30 visual stimuli is presented one after another in the same location. (In the experiments reviewed here, the stimuli were words or numerals, and presentation rates were typically between three and eight items per second.) Each list presentation constitutes one 'trial'. Subjects are instructed to monitor the sequence of stimuli, and to report the identity of any target item in the list that matches a given, pre-specified criterion, for example—in the experiments reviewed here—any string of letters written in italics; any word denoting an animal; any numeral greater than five; etc. In control (no-shift) lists, subjects continue to monitor for the same target criterion throughout the list. On criterion shift trials, at some unpredictable point in the list, the subject is cued to switch the target criterion abruptly from one pre-specified target category, T1, to another, T2. Criterion shift and no-shift lists are randomly interleaved, equiprobably, in every 'task switching' trial block. To cue the required change of criterion we use either a change in the location of the stimuli, or a change in the background colour of the screen. A target stimulus can occur, in different lists, at various possible points in the sequence, either before or after the cue to a shift of criterion. In 50 per cent of trials only one target is presented; in the other 50 per cent there are two targets. Naturally, the different possible target locations must be sampled, repeatedly, over many RSVP lists, in order to build up a picture of the subject's performance accuracy before and after a shift of target criterion.

What we find, in all our experiments, is an abrupt, and generally very large, drop in selective target-report acuracy, for targets occurring at or immediately after the instruction to shift the selection criterion, and then a gradual, monotonic recovery of selective report accuracy towards the pre-shift level, over a number of subsequent, post-shift target positions in the list. We shall refer to this sharp drop in performance, and subsequent gradual recovery, as the RSVP *criterion-shift (C-S) cost.*

Critical design features: separating the costs of a criterion shift from the 'attentional blink'

Other contributors to this volume (Chun and Potter, Chapter 2; Enns, Visser, Kawahara and Di Lollo, Chapter 4) review studies of the so-called 'attentional blink' (AB) phenomenon—that is, the temporary reduction in selective visual report of a second (masked) target item, following within some 200 to 600 ms after a first visual target (Broadbent and Broadbent 1987; Raymond *et al.* 1992; Shapiro *et al.* 1997). These authors discuss the possible links and/or independence between the AB and 'task switching' costs, within what we can call the 'two-target' or 'second-target' AB paradigm. They argue, compellingly, that many 'second-target' AB studies have confounded an AB effect with an additional task switching (C-S) cost. Evidently, the 'second-target' paradigm can provide a clean measure of the AB effect, uncontaminated by any C-S cost, when there is no shift of target criterion (or location) between T1 and T2.

On the other hand, the simple 'two-target' RSVP paradigm does not allow us to study C-S costs uncontaminated by any AB effects.

We would like to emphasize three simple but critical features that differentiate our experimental paradigm from the procedures used in these other 'two-target' AB studies. These procedural details, in our method, enable the costs of task switching to be observed in the absence of any confounding, second-target AB effects:

1. In our method, the cue instructing the subject to shift target criterion from T1 to T2 is a totally separate event from the occurrence of a first target. Moreover, in our experiments, the shift cue—and thus the earliest possible appearance of a second target item—is always separated from a preceding T1 target item (when there is one) by upwards of 500 ms. The shift cue remains present, or is iteratively re-presented, throughout the remainder of the stimulus list; hence it is not masked by any subsequent stimuli.
2. On 50 per cent of trials no T1 target is presented. Thus the C-S cost can be assessed on trials with no preceding T1 target. The purpose of including a T1 target on half the trials is simply to ensure that subjects continue to monitor for T1 until the possible appearance of a shift cue. As it turns out, in our data the C-S cost is identical on trials with or without a T1 target present.
3. On 50 per cent of trials there is no cue requiring a criterion shift; thus subjects continue to monitor for T1. Hence the subject cannot use the occurrence of a first target as a cue to shift to the T2 target criterion. T1 targets presented at later positions in these no-shift lists control for any tendency to shift to the T2 criterion in the absence of a shift cue. The data confirm that subjects continue to monitor efficiently for T1 throughout the no-shift lists.

There is one further design feature of the criterion-shift paradigm that we believe to be essential for proper experimental control. This is that the non-target ('distractor') stimuli for both T1 and T2 should not all be drawn from the same, *un*related distractor set: there must be some overlap, or cross-over, between T1 distractors and T2 targets, and/or (perhaps less critically) between T2 distractors and T1 targets. Thus, prior to a shift-cue (i.e., during search for T1), the distractors should include items that satisfy the (not-yet-valid) T2 target criterion; and/or *after* the onset of the shift-cue some distractors should satisfy the (no longer valid) T1 criterion. We refer to such distractor items (which would count as targets in terms of the other, currently *invalid*, selection criterion) as 'pseudo-targets'. Without this constraint, it would be possible, in principle, for the subject to monitor concurrently for both T1 and T2 target criteria, without any need to shift selective set during the RSVP list. In fact, in unpublished, exploratory experiments, Hsieh and Allport found little or no C-S cost when T1 and T2 were unrelated target categories, embedded in a uniform distractor set.

Returning to the traditional 'two target' paradigm, Visser, Bischoff and Di Lollo (in press) have argued that, in the absence of a concomitant switch of target criterion, the 'second target' AB effect follows a characteristically V-shaped time function ('lag-1 sparing'), whereas the RSVP criterion-shift cost is maximum at the onset of the shift and shows a monotonic recovery thereafter. (See also Enns *et al.*, this volume, Chapter 4.) Our data clearly support the latter proposition: a monotonic recovery function is found in all our criterion-shift data, that is, in conditions in which a classic 'second target' AB effect can be strictly excluded.

However, our results (Experiments 4, 5 and 6, below) provide an even more striking empirical separation between these two classes of phenomenon (Allport *et al.* 1994; Hsieh 1993, 1995). A fundamental property of the 'second target' AB effect is that it follows an 'endogenous' timecourse, arguably a function of the duration of T1 processing, independent of whether there are other ongoing stimulus events (besides an initial T1 mask) during the interval between the first and second target stimulus (see for instance, Duncan *et al.* 1994; Raymond *et al.* 1992; Ward *et al.*, 1996). In numerous studies, the AB, as a function of real-time lapse, appears to be completed within around 600 ms of the triggering event. In sharp contrast, in our experimental paradigm the course of recovery from the C-S cost does not appear to be a function of simple time-lapse. To anticipate results (Experiments 4, 5 and 6, below), recovery of efficient T2 performance following a criterion shift appears to be strictly stimulus-driven. That is, unlike the endogenously determined timecourse of the 'two-target' AB effect, the rate of recovery from a criterion shift appears to depend on the rate of presentation of task-relevant stimuli.

Aims of this chapter

We have three principal aims in writing this chapter. The first is simply to review a number of exploratory experiments, using the RSVP criterion-shift paradigm outlined above. The experiments were designed to address a variety of questions about the effects of a shift of 'task-set', in RSVP selective-report tasks. For example, we were curious to know how the performance cost of a shift of set would be affected by the 'level' or domain of processing required in one or both tasks (for instance, contrasting visual vs. semantic selection criteria); and how it would be affected by the 'distance' or extent of the criterion shift between the T1 and T2 target criteria. For example, might a shift within the same level of processing be easier than a shift between levels? Or might a complete reversal of target criterion be more demanding than only a partial shift? Also, as mentioned above, we wished to know to what extent the process of shifting set, in this paradigm, reflected the timecourse of an 'endogenous' control operation and to what extent it was 'exogenously' controlled. As we shall see, several of these questions received rather surprising answers.

Our interest in these questions is coupled with a continuing interest in studies of 'task switching' using more conventional, reaction time (RT) methods (Allport *et al.* 1994; Allport and Wylie, 1999; 2000; Meuter and Allport 1999; Wylie and Allport, 2000). The second aim of this chapter is therefore to provide a brief review of some of these RT studies of 'task switching', as a context in which to consider the effects of a shift of selective set in RSVP tasks. It is often assumed that the speed of response in 'task switching' experiments can provide a rather direct window on the time costs of top-down control processes that directly implement a shift of set. The results that we review below may suggest that cognitive control does not conform in such a simple way to folk-psychological assumptions; or—at least—that subtractive RT methods may not offer such a direct window on the underlying control processes.

The third aim of this chapter is closely related to the second: that is, to consider what, if anything, we have been able to discover about the nature of 'executive control' in a shift of selection criterion in RSVP selective-report tasks, and to what extent these results concur with inferences that can be derived from RT studies.

Studies of task switching using RT methods

RT task-switching paradigms employ sequences of speeded responses to task-stimuli, generally contrasting the case in which the same task is required on immediately successive RT trials ('*non-switch*' trials) with the case in which a different, competing or 'divergent' task was required on the preceding trial(s) ('*switch*' trials). Two different kinds of comparison have generally been made, as a measure of behavioural 'switching costs': (a) the experimenter may compare fixed task performance, in separate blocks in each of two (or more) tasks, with performance of the same tasks in regular alternation (cf. e.g., Jersild 1927; Spector and Biederman 1976); and (b) comparison of both switch and non-switch RTs can be made within the same 'switching' block of trials. Both measures typically yield large RT 'switching costs'. Rogers and Monsell (1995) and Meiran (1996) have argued that comparison (b) is preferable to (a), as a measure of switching costs, since in (a) there is a possible confounding of the costs of task-switching and of the need also to 'hold two tasks in mind'.

The most popular, and perhaps the most intuitively attractive, interpretation of these RT differences ('switching costs') is that they reflect the time needed for the task-set to be 'reconfigured', as an interpolated, stage-like control operation, from the task-set imposed on the immediately preceding trial to that needed for the current trial. These time costs are commonly subdivided, theoretically, into at least two components: (1) an 'endogenous' or preparatory component of task-switching, which can be executed in advance of the imperative task-stimulus, and (2) an 'exogenous' component, which can only be completed (rather puzzlingly?) after the task stimulus has been presented (e.g. Meiran 1996; Rogers and Monsell 1995; Rubinstein *et al.*, in press). In contrast, as we argue below, a closer look at the data suggests that neither of these RT differences reflects the duration of a stage-like process of 'task-set reconfiguration', where this is taken to mean a process of enabling/disabling competing stimulus–response (S–R) pathways.

Anticipatory task preparation

The evidence in favour of a preparatory process is that, if the switch of task is precued (and/or if there is a fully predictable sequence of task switching), the RT 'switching cost' can be systematically reduced as the available preparation interval increases. Most of this benefit of task pre-cuing (in terms of reduced 'switching costs') is gained, typically, within the first 600–700 ms after the cue onset (e.g., Allport *et al.* 1994; De Jong *et al.* (submitted); Fagot 1994; Rogers Monsell 1995; Meiran *et al.*, in press). In comparing 'task switching' effects in the RT and RSVP paradigms, an obvious question is whether the inferred timecourse of task preparation, following a switch cue, could be the same in these two different paradigms. In the RT paradigm, it seems natural to assume that the benefits of an increasing time interval since the task-cue reflect the intrinsic timecourse of an 'endogenous' control operation, triggered by the cue (Meiran 1996; Meiran *et al.*, in press; Rogers and Monsell 1995). As we shall see, however, in the RSVP paradigm at least, this assumption appears to be seriously at odds with the data.

Many RT studies of task switching find that between-task interference ('crosstalk') is increased in switching conditions compared to 'non-switch' or pure task performance (see Allport and Wylie, 1999, for review). An example of such crosstalk can be seen in switching from colour-naming to word-reading, in response to incongruent 'Stroop' stimuli (e.g., the word PINK written in green ink; Stroop 1935). Normally (without task-switching), the presence of an incongruent ink-colour has *no* effect on word-reading, i.e., there is no 'reverse Stroop' interference (MacLeod 1991). After switching from Stroop colour-naming to word-reading, however, very large 'reverse Stroop' interference effects can be found in the word-reading task (Allport *et al.* 1994; Allport and Wylie, 1999; 2000). Strikingly, however, even when there is a large reduction in the RT 'switching cost' over a task-preparation interval, there is typically no corresponding reduction in between-task interference ('crosstalk' or incongruity) effects (e.g., Allport *et al.* 1994; De Jong *et al.*, submitted; Fagot 1994). Crosstalk effects can be just as large at the end of the task-preparation process, in these RT experiments, as they were at the start. If anticipatory task preparation involved the reconfiguration of competing S–R pathways, engaging the new, task-relevant S–R mappings and effectively disabling or disengaging the pathways used in the preceding competing task, we should expect these between-task crosstalk effects to be significantly reduced, or even abolished, during the course of a preparation interval. However, no such reduction is found. This result implies also that, whatever it does, the preparatory process has not affected the relative 'attentional salience' of task-relevant vs. task-irrelevant stimulus attributes.

One hypothesis that appears to be consistent with the crosstalk effects is that anticipatory task-preparation, in the RT paradigm, may determine *which* task is performed (e.g., the appropriate task 'goal' is activated; hence the system can only 'settle' to response states that are consistent with this goal), but that the required S–R mappings are not effectively 'reconfigured', or engaged/disengaged, until after the imperative stimulus is presented (e.g., Fagot 1994; Rubinstein *et al.*, in press).

'Residual' switching costs and item-specific interference

In the RT paradigm, in practically all cases so far studied, the reduction in RT 'switching cost' over the course of a preparation interval is only partial, leaving a substantial 'residual switching cost' that is not reduced by further preparation time (Allport *et al.* 1994; De Jong *et al.*, submitted; Fagot 1994; Meiran 1996; Meiran *et al.*, in press; Rogers and Monsell 1995). According to the type of hypothesis just outlined, the residual switching cost reflects (or includes) the time cost of a further control operation, triggered by the task stimulus, which finally ensures that the appropriate processing pathways are respectively enabled/disabled (e.g., Rubinstein *et al.*, in press). In contrast, Allport *et al.* (1994) suggested that the residual RT switching cost represents continuing, pro-active interference arising from the preceding task.

More recently, Allport and Wylie have provided evidence that residual switching costs can include also a longer-term, negative transfer of learning between the competing sets of stimulus-to-response (S–R) associations; moreover, these learned S–R associations can be in large measure item-specific. Thus, if the same, particular stimulus item is presented for speeded response in both task contexts, S–R associations acquired in one task can interfere with execution of the other, competing response in the other task context. This effect is particularly acute on the first trial of a run, that is, typically, on a 'switch trial' (but see the 'restart' effect, below). In contrast, when the same general type of stimulus was used, as before, but with different, individual exemplars presented for the two tasks, the RT 'switching costs' were greatly reduced (Allport and Wylie, 2000). Further, the magnitude of these RT 'switching costs' appears to depend on the ratio of past trials in which the same stimuli were presented, respectively, for one or other of these competing tasks (Wylie and Allport, 2000). It is difficult to see how an account of residual RT 'switching costs', in terms of the time needed for a hypothetical control operation to shift 'task set' generally (i.e., in terms of the 'reconfiguration' of processing pathways, which should presumably affect *all* task-relevant stimuli equally) could account for these item-specific effects. Rather, as Allport and Wylie (1999; 2000; Wylie and Allport, 2000) have argued, residual RT 'switching costs' appear to depend in large part on retrieval-based interference ('long-term negative priming') from previously acquired response associations to the same stimuli that are triggered at the start of a new run of trials.

The 'restart' effect

Allport and Wylie (1999) and Wylie and Allport (2000) noted that, even with no explicit switch of task from one trial-block to the next, the first trial of each run (or 'block') of speeded RT trials was consistently slower (and had fewer errors) than all other trials in the run. They labelled this the 'restart' effect. Other authors have also noted this effect (De Jong *et al.*, submitted; Gopher *et al.*, 2000). What Allport and Wylie also found, however, was that the 'restart' trials showed even larger (e.g., ×2) RT costs in task B if the same stimulus had previously been responded to in task A. In other words, these 'restart' trials behaved very much like 'switch' trials, even though there was no switch of task, and the prior, competing task was last performed some considerable time before.

A look ahead

Clearly, there are many issues arising from these RT studies of task-switching that may alert us to possible analogues, and/or contrasts, in the RSVP criterion-shift paradigm. Perhaps most obvious among these is the apparent timecourse of anticipatory task-preparation, as found in the RT paradigm: how is this related (if at all) to the effect of increasing cue-to-target interval in RSVP criterion shift? Could these different measures reflect the same underlying process, with the same or a similar timecourse? Second, is there a 'residual' switching cost in the RSVP paradigm, even after recovery from the shift-cost has reached asymptote, as generally found in the RT switching paradigm? Third, does the gradual recovery from the RSVP shift-cost (*unlike* the RT task preparation benefit, in nearly all studies) reflect a gradual reduction in the 'attentional salience' of previously task-relevant stimuli?

Experiments

Before we embark on a review of these experiments, it may be helpful to sketch an outline model of what we imagine some of the processes in these RSVP selective-report tasks to be.

In the experiments we describe here, the stimuli are familiar, meaningful forms (words, numerals). We assume that, when a subject attentively monitors an RSVP sequence of such stimuli, each successive stimulus triggers a temporally overlapping processing cascade, that includes the activation of low-level visual-feature codes, orthographic (letter and word-level) codes, and also semantic feature codes. At rapid presentation rates, the activation of low-level codes, in response to a given stimulus item, is liable to overlap temporally with higher-level codes activated by an *earlier* stimulus item (or items) in the sequence. We imagine that this continuous (and continuously changing) temporal overlap makes the binding or integration of feature codes pertaining to the same stimulus 'event', at different levels, highly error-prone or even impossible at rapid presentation rates. Binding of feature codes into some form of inter-linked 'event file' (Hommel 1998) we also assume is needed for the event to be encoded as a recoverable item in short-term memory. Otherwise, so long as this relentless, stimulus-driven cascade continues, feature-codes activated by earlier stimuli in the sequence are liable to be overwritten (or 'masked') by the feature-codes generated by later stimuli. As a result, unless some further selection process is initiated, to 'grab' a subset of these ever-changing cognitive codes, each item (except perhaps for the last items in the list) will be overwritten by those coming later, and so, rapidly 'forgotten'.

The selective 'grab' operation, we suppose, confers an enhanced representational status, and enhanced persistence, on the selected, currently active codes, thereby enabling the binding or integration of (at least some of) the selected features. In conventional psychological terms, the selected feature-set is 'encoded into short-term memory'. If the selected feature-set includes a word-level code (or perhaps other, orthographic and/or semantic, feature-codes from which a word ('name') can be recovered), and short-term memory is not overloaded, then the corresponding stimulus item may be 'available for report' after the end of the RSVP list.

For example, suppose that (as in some of our experiments) the task is to report selectively any word in the RSVP list that is the name of a living thing. Correct target report will depend on detection of the criterial semantic feature(s), triggering selection ('grabbing') of other feature codes pertaining to the target word. We imagine that target detection, in this task, depends on setting up an attentional orienting trigger, or 'filter', that monitors the activation of certain semantic features, coding for 'living'. (That is, these semantic features are assigned high 'attentional salience'.) 'Orienting' of attention, in turn, cues the system to 'grab' the currently activated semantic feature-codes that triggered the attentional orienting, together with any other, currently active (semantic, orthographic, or 'name') codes that can be appropriately integrated with the trigger-features (i.e., that are consistent with them). The latter process is necessarily selective, and time-consuming, and its duration arguably constitutes the 'attentional blink'.

These suggestions are by no means original. They are indebted to a variety of 'two-stage' models of RSVP selective report, including Reeves and Sperling (1986), Weichselgartner and Sperling (1987), Duncan *et al.* (1994), Chun and Potter (1995; this volume, Chapter 2), Shapiro *et al.* (1997), and others. Our account differs from these, if at all, in its emphasis on the role of temporal overlap of feature-codes (at different levels) activated by different sequential stimuli; and in its possibly even greater reluctance to posit 'attention' as some kind of unitary mechanism, or causal agency (cf. Allport 1989, 1993).

Now, if the experimenter-specified target criterion changes in mid-sequence, from one set of criterial features to another, the principal change required is presumably in the attentional orienting 'filter' itself; other components of the selective report process can remain more or less unchanged. Thus we hoped that this experimental procedure would permit us to study the process of re-setting a selective filter, from one set of criterial target features to another, when cued by an external stimulus. In particular, we hoped to be able to explore the *timecourse* of such a process, under different conditions.

Experiment 1: shifting set between physical stimulus attributes (letter-size or shape) and between semantic attributes (object-size, living–non-living) and physical attributes

Experiment 1 will be described in some detail, as it illustrates a number of general features of the paradigm that we have used. We describe these general features first. Further details can be found in Hsieh and Allport (1994).

Subjects were presented with rapid sequences ('lists') of from 10 to 30 written words (or digits, in Experiments 5 and 6), one after another in the same location on a computer monitor. (In Experiment 1, each stimulus word was displayed for 115 ms, with a 15 ms blank interval before the onset of the next word—hence a stimulus onset asynchrony (SOA) of 130 ms and an overall presentation rate of about 7.7 stimuli per sec.) The subject's task was to monitor each list for the occurrence of up to two possible target words (digits), reporting these orally at the end of the list.

In all these experiments, cue words (in capital letters), specifying either one or two possible types of target category, were presented prior to the start of each list, for as long as the subject wished. One cue word, flanked at either side by three asterisks, appeared at the centre of the screen. This word specified the target category for which the subject was to begin searching (Target Category 1, or T1). In the shift blocks, a second cue word was shown, one line-space above, specifying a second possible target category (T2). The subjects were instructed that, as long as the RSVP sequence continued to appear in the central location (between the asterisks), they were to continue to monitor for T1; however, if at some point the sequence jumped to the location one line-space above, they should immediately stop monitoring for T1 and start monitoring for T2. Target categories for T1 and T2 remained the same over a block of 24 lists. A shift of target criterion from T1 to T2, cued by a shift of the stimulus location, occurred on 50 per cent of lists. Thus shift and non-shift trials were mixed unpredictably within a block. Without this constraint, or if a block consisted only of (or of a majority of) shift trials, subjects could start to shift their selective set immediately after detecting the first target.

The subject could rest between trials, before pressing a key on the keyboard to initiate the next trial, whereupon the two cue words for that trial were displayed. The subject could take as long as he or she wanted to study the cue words, before pressing the key again to start the rapid stimulus sequence. The word 'Ready' appeared for 1000 msec immediately after the subject's key press; then the RSVP sequence began. The target and non-target stimuli were selected randomly for each trial, without replacement.

Experiment 1 (and Experiments 4–6) also included control, non-shift blocks, in which the target criterion remained constant throughout the list. Prior to the start of each list, in these non-shift blocks, the same cue word was displayed both in the centre location (between asterisks) and one line-space below. In 50 per cent of these lists the stimulus location was shifted, at some point in the sequence, one line-space *down* the screen. Subjects were instructed to ignore this shift of stimulus location and to continue to monitor for T1. This condition ('Location-shift control') was of course included to control for the effect of a spatial shift of stimulus location during the list.

Stimulus-lists were constructed as follows. Each list was notionally divided into two immediately successive sublists. On half the trials, the first sublist contained one target stimulus, located randomly between the third and the tenth serial position from the start of the list; on the other 50 per cent of trials there was no target in sublist 1. A target stimulus in sublist 1 was always followed by at least three non-target stimuli before the onset of sublist 2. (Thus, in Experiment 1, there was at least 520 ms between the onset of the first and second targets). This was intended to minimize the possibility of an 'attentional blink', triggered by T1, affecting the detection of a second target. In trials with no target in sublist 1, there were randomly from 6 to 13 non-target stimuli in sublist 1 before the onset of sublist 2. Sublist 2 always contained one target stimulus, located equiprobably as the first, third or seventh item. At least four non-target stimuli followed the occurrence of the target item in sublist 2. On half the trials, sublist 2 was displayed one line-space above the location of sublist 1, cuing a shift of set to the T2 target criterion. The target and non-target stimuli were selected randomly for each list, from a pool of 92 object names, without stimulus repetition within a list.

In Experiment 1 we studied two main types of shift in selective set: (1) from one physical stimulus-attribute to another ('physical → physical'), and (2) from a semantically specified target category to a visual stimulus-attribute ('semantic → physical'). Words (in lower case) were presented in either large or small font, and in either italic or roman type. When monitoring in terms of physical stimulus-attributes, a target stimulus could be any word in italics, or (alternatively) any word in small type.

A shift of selective set from one physical stimulus-attribute to another ('physical → physical', P → P) was investigated in Experiment 1 as follows: Words in the RSVP lists appeared randomly, one after another, in either large or small font, and (independently of size) in either italic or roman type. In different conditions, the target item could be any word printed in italics, or any word printed in small type. Physical → physical attention shifts were tested in both directions: 'italics → small-letters' and 'small-letters → italics'. (Notice that these different target criteria were *orthogonal*. That is, a word in small type can be italic or roman; a word in italics can be in large or small type.) Control, non-shift blocks were 'small-letters → small-letters' and 'italics → italics'.

All the words were the names of familiar, physical objects; the word-sets were divided in terms of two (orthogonal) semantic categories: (1) living vs. non-living things; and (2) small vs. large physical objects. (A 'small' object was defined arbitrarily as any object whose largest dimension was smaller than the diameter of a 'soccer' football; a 'large' object was any object larger than a football. All the objects fell clearly into one or other size category, e.g., sardine, bracelet, vs. kangaroo, guitar.) When monitoring in terms of semantic attributes, a target stimulus could be any living thing or (alternatively) any 'small' object.

To investigate a shift of selective set from a semantic to a physical stimulus-attribute ('semantic → physical', S → P), T1 target items could be any word denoting a living thing or any word denoting a 'small object' (as defined above). Again, the target categories were orthogonal: a living thing can be 'large' or 'small'; a 'small' object can be living or not-living. There were four possible types of semantic → physical criterion shifts: 'living-thing' → small-letters', 'living-thing → italics', 'small-object → small-letters', and 'small-object → italics'. Thus, to facilitate the comparison between these different types of shift, T2 was always a physical stimulus attribute ('italics' or 'small-letters'). Figure 3.1 illustrates some example lists.

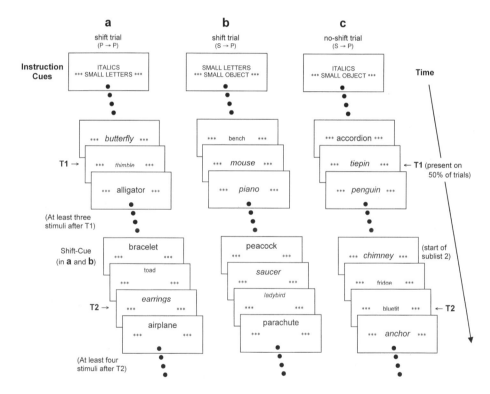

Fig. 3.1 The figure shows sections of three example RSVP lists, **a, b, c** from Experiment 3.1. In all three examples a T2 target is illustrated at target position 3. (All three examples also show a T1 target. In the experiment, however, a T1 target was present only in 50 per cent of lists.) Example **a** illustrates a criterion-shift trial for a 'physical → physical' (P→P) shift. The subject begins by monitoring for any word in small letters (T1); after the appearance of the shift cue (a change in stimulus location) the subject should ignore letter-size, and monitor for any word in italics (T2). Example **b** and **c** illustrate a shift trial and no-shift trial, respectively, from two conditions (S→P) in which T1 had a semantic target criterion, T2 a physical target criterion. Note that in example **b**, T2 (*'ladybird'*) is a 'double hit' target, satisfying both T1 and T2 criteria. See page 51, Experiment 3, for further discussion of 'double hit' targets.

Twelve subjects, aged 19 to 25 years, were tested individually. Each subject completed four blocks (of 24 trials each), under each of the eight different types of criterion-shift and non-shift conditions.

The results are shown in Fig. 3.2 (panel a) for the non-shift (location shift control) blocks, and (in panel b) for the criterion shift blocks. In both these types of block, lists in which no change of stimulus location occurred gave very stable levels of accuracy, for all target positions in the list, at around 80 per cent correct. (Comparison between *non-shift* trials in the criterion shift and no-shift blocks, respectively, suggests a small overhead (around 3 per cent) for maintaining conditional readiness for the T2 task. This contrast was marginally reliable.)

In the non-task-shift (control) blocks, a shift of stimulus location (but no shift of selection criterion) resulted in a small but significant drop in accuracy, of the order of 10 per cent, for targets occurring as the first stimulus item in the shifted location (target position 1), but had no effect on targets probed at later positions in the list. In the task-shift blocks (in which a shift of stimulus location cued a shift in the target selection criterion), by contrast, both physical → physical (P → P) and semantic → physical (S → P) shifts resulted in a very large drop in T2 identification accuracy following the shift cue, from around 80 per cent down to about 20 per cent, with a gradual, monotonic recovery in accuracy for targets probed later in the sequence, recovery being apparently complete by T2 position 7. The drop in T2 identification was highly significant both at post-shift positions 1 and 3; however, there was no significant difference between the magnitude of this RSVP shift cost for S → P vs. P → P shifts.

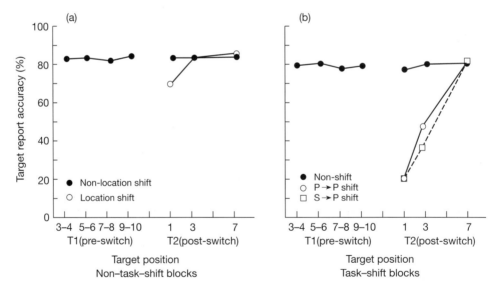

Fig. 3.2 Results of Experiment 1. Figure 3.2a shows the data from the Control condition, in which subjects were instructed to ignore any shift of stimulus location, and continue to monitor for T1. Figure 3.2b shows the data from the criterion-shift blocks: on no-shift trials and for a physical → physical (P→P) and semantic→physical (S→P) criterion shift.

In other words, in this experiment, a shift of selection criterion over a very large 'cognitive distance', namely from a high-level semantic attribute (e.g. living/non-living) to a simple visual attribute (e.g., letter size) resulted in no greater (and no smaller) C-S cost than did a shift from one relatively simple visual attribute to another. The fact that recovery from $P \rightarrow P$ and $S \rightarrow P$ shifts showed approximately the same timecourse may suggest that this procedure taps some relatively basic constraint on (non-spatial) shifts of attention, independent of the level of processing at which target selection is required.

The experimental paradigm is thus shown to yield clear-cut and highly consistent results. Selective report is grossly impaired immediately after a shift of set. However, T2 identification has returned to pre-shift (or no shift) levels of accuracy at least by post-shift position 7, that is by a target onset 780 ms after the shift cue. On the basis of these data, therefore, the apparent timecourse of recovery from the RSVP shift-cost appears broadly comparable to the timecourse of task-preparation benefits in the RT paradigm. (Clearly it is not possible to establish, from the data in this experiment, whether recovery might have been completed by an earlier T2 position, for instance, post-shift position 5.) On the other hand, there was no sign of a 'residual' shift-cost in the RSVP task.

In contrast to the very large drop in performance following a shift of non-spatial selection criterion, a shift of stimulus location alone resulted in only a small, and short-lived, dip in performance. Hence the C-S cost cannot be a simple artefact of the (exogenous) shift in spatial attention. In the next two experiments, therefore, we dispensed with this location shift control; however, the corresponding control condition was included again in Experiments 4, 5 and 6.

Experiment 2: reversal shifts of semantic target category

Experiment 1, above, investigated shifts within and between physical and semantic stimulus attributes, as target criteria. In Experiment 2, by contrast, we used semantic selection criteria only. Moreover, while Experiment 1 studied orthogonal shifts in target criterion, Experiment 2 studied *reversal* shifts. Thus, all stimulus items that were potential targets for T1 became non-targets following a shift to T2, and all the non-targets for T1 become potential targets for T2. Intuitively, it seems plausible that this complete reversal of target and non-target categories might impose an even greater demand on attentional control, on the shift trials, than the orthogonal shifts required in Experiment 1. As in Experiment 1, the stimuli were the names of physical objects, animate and inanimate. All were presented in lower case.

The semantic criteria for selection of target items were, respectively, 'small' and 'large' physical objects. (As before, a 'small' object was defined arbitrarily as any object whose largest dimension was smaller than the diameter of a 'soccer' football; a 'large' object was any object larger than a football. All the stimuli were names of objects that fell clearly into one or other of these size categories (e.g., thimble, mouse, vs. piano, donkey), 92 items in each category.) As before, cue words ('LARGE', 'SMALL') specifying T1 and T2 categories were displayed at the start of each trial, and subjects initiated the start of each stimulus list when they felt ready.

In this experiment we excluded the control, non-criterion-shift blocks, used in Experiment 1 to calibrate the effects of a location-shift alone. In all other major features the procedure and the construction of stimulus lists were similar to Experiment 1.

Experiment 2 differed from Experiment 1 in the following further respects. There were just two types of target-selection condition: (1) in which T1 was any large object, T2 any small object ('large' → 'small'), and (2) in which T1 was any small object, T2 any large object ('small' → 'large'). The target selection condition was held constant for a block of 16 trials. Blocks of type (1) and (2) alternated, for a total of 12 blocks (192 trials), giving 12 trials per target position per condition, for each subject. A shift of target criterion—cued as before, by a shift of stimulus location one line-space up the screen—occurred on 50 per cent of trials. Thus semantic shift and non-shift trials were mixed unpredictably in every block. Post-shift target positions 1, 3, 7, and 11 were probed, equiprobably. On 50 per cent of trials (both shift and no-shift) there were two targets present in the list (one in Sublist 1; one in Sublist 2); on the remaining 50 per cent only one target was present (in Sublist 2 only). Each stimulus word was displayed for 150 ms, with a 15 ms blank interval before the onset of the next word (SOA = 165 ms), giving a presentation rate of approximately six words per second. When two targets were presented in a list, the first target was always followed by at least three non-target words before the onset of Sublist 2. Thus there was a minimum SOA of 660 ms between the first and the second target stimulus. At this interval, detection of the second target is unlikely to be affected by any 'attentional blink' (AB) triggered by the first target, since reported AB effects are rarely detectable beyond around 600 ms after the triggering event.

Eighteen students at the University of Oxford served as paid volunteers. They were tested individually, and completed the required experimental conditions in a single session.

The results are shown in Fig. 3.3. Filled symbols show target identification when the semantic target criterion was any 'small' object; open symbols, when the specified target was any 'large' object. Pre-shift (and non-shift) target report accuracy was relatively stable, at close to

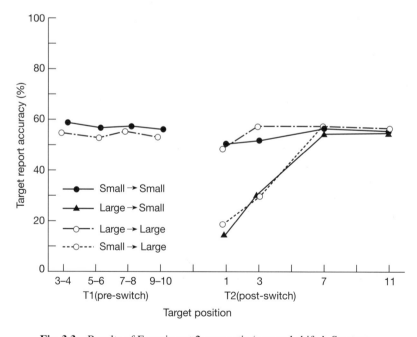

Fig. 3.3   Results of Experiment 2: semantic 'reversal shifts'. See text.

60 per cent, for both tasks. (This was lower than the performance level found in Experiment 1. In that experiment, post-shift performance relates to physical target selection only, i.e., letter-size or -shape. For semantic targets, in Experiment 1, 'small' vs. 'large' was also harder than 'living' vs. 'non-living'.) As can be seen in Fig. 3.3, however, on criterion-shift trials correct target reports dropped to about 15–20 per cent at post-shift target position 1; they were still only around 30 per cent at target position 3 (SOA of 330 ms after the shift cue); performance had returned to pre-shift level by target position 7 (at an SOA of 990 ms after the shift cue). Analyses of variance showed highly significant effects of shift vs. non-shift ($p<0.0001$), for both 'small' and 'large' semantic targets, and of T2 position ($p<0.0001$, again, in both cases), and a significant interaction ($p<0.0001$). Analysis of the non-shift trials alone showed no significant effect of target position in the list ($F = 1.07$). A further analysis showed that the presence vs. absence of a first target had no significant effect on T2 report accuracy, regardless of whether the first target was correctly reported or not.

These results thus confirm and extend the picture obtained in Experiment 1, to a full 'reversal' shift of target criterion. In a 'reversal' shift, all post-shift (*non*-target) stimuli would have been valid targets, according to the previously valid selection criterion for T1. Even in this 'reversal shift' case, however, performance has fully recovered to pre-shift (or non-shift) levels, at least by post-shift target positions 7 and 11. There appears to be no 'residual shift-cost', by post-shift position 7 (SOA of 990 ms after the shift cue). This suggests that the T1 selection bias (selective 'set'), which should trigger an attentional orienting response to any T1 target stimulus, has been effectively and fully disengaged—and the orienting response to a T2 target stimulus has been fully engaged—within the course of no more than six intervening, non-target stimuli. To the degree that the T1 selection bias was still operating after the shift cue, then what is now (according to the experimenter-defined, post-shift selection task) a *non*-target stimulus should still elicit an attentional orienting response. Indeed, this suggests a rather simple, candidate hypothesis to account for the slow recovery of the RSVP shift-cost, in effect by assimilating the RSVP shift-cost to the 'attentional blink' phenomenon, as follows.

As noted above, the size of the RSVP shift-cost, and its gradual recovery, was unaffected by the presence or absence of a prior T1 target stimulus. This conclusively rules out the possibility that the shift-cost represents an 'attentional blink' triggered by detection of the T1 target. However, a quite different hypothesis is that the timecourse of recovery from the RSVP shift-cost might correspond to an attentional blink, caused by an orienting response to the first (or any one) of the *non*-target stimuli occurring at or after the experimenter-cued shift of selection task, i.e., before the T1 selection set could be disengaged. Call this a 'pseudo-target' detection response. Note that the C–S cost at post-shift *position 1* could not be explained in this way, since the T2 target stimulus is not preceded by any stimulus that could elicit a 'pseudo-target' detection response.

The following three experiments provide some tests of this hypothesis.

Experiment 3: orthogonal vs. reversal shift

The main aim of Experiment 3 was to provide a direct comparison between a 'reversal' shift, as in Experiment 2, and a shift between two orthogonal target categories, as in Experiment 1. A further objective was to extend the 'reversal' shift to a different semantic category (animal vs. non-animal). Experiment 3 has been reported in detail by Hsieh and Allport (1994), as well as in Allport *et al.* (1994), and will be summarized briefly here.

All selection criteria were semantic. There were three target categories: (1) any member of the animal kingdom ('animal'); (2) any physical object not a member of the animal kingdom ('not-animal'); (3) any physical object smaller, in its largest dimension, than a football ('small'). Each target category contained 92 words, organized orthogonally with respect to size (large/small) vs. animacy (animal/not-animal).

There were just three types of T1 → T2 criterion shift: (1) a 'reversal' shift ('not-animal → 'animal'); (2) an orthogonal shift ('small' → 'animal'); and (3) another reversal shift ('animal' → 'not-animal'). A shift of target criterion (cued by a shift of stimulus location) occurred on 50 per cent of trials; on the remaining 50 per cent there was no location shift, and subjects continued to monitor for T1 throughout the list (non-shift trials). T1 and T2 were held constant for blocks of 16 successive trials. Cue words (e.g., 'SMALL'; 'ANIMAL') specifying T1 and T2 categories were displayed at the start of each trial, and subjects initiated the onset of each stimulus list by a key-press, when they felt ready. The 288 stimulus lists were presented in one test session. The apparatus, procedure, and the construction of stimulus lists were in all other respects the same as in the preceding experiments. Eighteen students at the University of Oxford served as paid volunteers.

The results are illustrated in Fig. 3.4. For full comparability between conditions, the critical data are for the 'animal' target category, in both shift and non-shift trials. Thus, the third type of T1 → T2 condition ('animal → not-animal') provided data for pre-shift and non-shift trials only. A preliminary analysis was carried out on target detection of T2 in each condition, with and without a preceding T1. As in Experiment 2, the presence or absence of a first target had no effects on T2 detection, in any of the conditions, regardless of whether the T1 target was correctly reported or not.

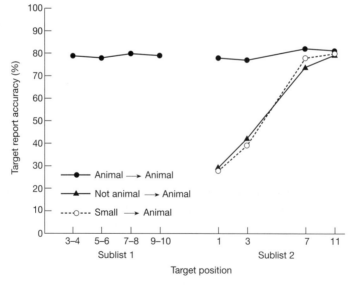

Fig. 3.4   Results of Experiment 3: comparison of a semantic 'reversal' shift (Not-animal → Animal) and an orthogonal shift (Small object → Animal).

Non-shift trials resulted in a stable level of selective report accuracy, at about 80 per cent correct. By contrast, both types of semantic criterion-shift (orthogonal shift: 'small → animal', and reversal shift: 'not-animal → animal') resulted in a large drop in T2 detection accuracy immediately following the shift cue, with a gradual recovery for the targets appearing at later positions in the second half of the RSVP sequence. A two-way ANOVA (type of shift × target position) on T2 performance showed a significant effect of target position ($p<.0001$), but *no* significant main effect of type of criterion-shift ($F=0.04$), and no significant interaction between position and type of shift ($F<1.0$). That is, in this experiment, both the magnitude of the RSVP shift-cost, and its subsequent recovery, were essentially the same in both a reversal and an orthogonal shift of selection criterion. This result is surprising for several reasons, not least because it appears to imply that the T1 selection bias can be disengaged more or less immediately, on the arrival of the shift cue. The argument is as follows. In a shift of target category between two orthogonally related categories (e.g., 'small → animal'), on average 50 per cent of the T2 target items ('animals') will also count as valid target items according to the preceding, T1 criterion ('small objects'). (Let us call a stimulus that satisfies *both* the T1 and the T2 target criteria (viz., in the present example, any *small* animal) a 'double hit' target item, while a stimulus that fits *only* the T2 criterion (in the present example, any *large* animal) can be referred to as a 'single hit' target item.) Thus, in an orthogonal shift from T1 to T2, if the T1 selection bias remained fully or even partially engaged after the shift cue, we might expect to see an advantage for the detection of 'double hit' over 'single hit' T2 targets, perhaps most obviously in post-shift target position 1. In a reversal shift, on the other hand, 'double hit' targets cannot occur. Consequently, unless the T1 selection bias can be disengaged immediately, we might expect to see an advantage (a smaller shift-cost) for orthogonal over reversal shifts. On the contrary, we found no difference. In a further analysis of the orthogonal shift data ('small → animal'), we directly compared T2 performance on 'single hit' and 'double hit' targets. There was no difference ($F=0.01$). If correct, this result apparently leads to the surprising inference that the T1 selection bias can be disengaged practically instantaneously. This would also appear to rule out the hypothesis, outlined at the end of Experiment 2, whereby the recovery phase of the RSVP shift-cost was attributed to an attentional blink, triggered by detection of a 'pseudo-target' while the T1 selection bias was not yet disengaged.

However, both conclusions should be treated with caution. In an unpublished study in our laboratory (Bramley 1995), we found a reliably better T2 report of 'double hit' targets in an orthogonal shift. Clearly, the topic needs further research.

Experiment 4. Recovery from the RSVP shift-cost: time vs. stimulus-items

The foregoing arguments rest on the assumption that the C-S cost, and its gradual recovery, depend on—and hence directly or indirectly reflect—the time needed for an 'endogenous', top-down control operation, which actively disengages the T1 selection set and engages the T2 selection set, cued or triggered by the onset of the shift-cue. This assumption seems so intuitively plausible, and so attractively familiar, that it is difficult to escape from its magnetic grasp. Applying this assumption to the results of Experiment 3, however, led us to the rather disconcerting conclusion that disengagement from the T1 selection set can be practically instantaneous. We may begin to question, therefore, whether the reasoning leading to this conclusion was faulty, or whether the basic assumptions, about the timecourse of an 'endogenous', top-down control operation, are the appropriate context in which to interpret the RSVP shift-cost.

Fundamental to the idea of any such control operation, generated by an autonomous, 'supervisory attentional system' (Norman and Shallice 1986), is the idea that the postulated control operation takes some determinate, real-time duration to execute or complete. (No doubt, different T1 → T2 shifts could have different, characteristic durations for the relevant control operation.) Moreover, the time taken by any such control operation should be independent of the amount of processing activity, the presentation rate of task-relevant stimuli, etc., in the lower-level processing systems. The same applies to any type of autonomous, top-down, attentional control system, which is (by hypothesis) functionally separable from the lower-level processing-systems that it controls. Indeed, this is part of what it means for the control system(s) to be 'autonomous', and 'functionally separable' from the lower-level systems, so that control system(s) and lower-level processing systems each have their own 'processing resources'. In contrast, if control system and lower-level ('controlled') systems shared the same, limited processing resources, we might predict that the greater the processing load on the lower-level systems (for instance, the faster they are paced by task-relevant stimuli), the longer a given control operation should take to complete. These contrasting predictions were tested in Experiments 4, 5 and 6.

Another characteristic that has been put forward as a fundamental distinguishing feature between top-down 'control processes' and lower-level 'controlled' processes (Shiffrin and Schneider 1977) is that lower-level ('controlled') processing is directly stimulus- or event-driven (cf. the 'triggering conditions' for schemas in 'contention scheduling', Norman and Shallice 1986; Cooper and Shallice, in press), whereas top-down control processes are 'endogenous'. Again, Experiments 4, 5 and 6 were designed to test whether the RSVP shift-cost, and its subsequent recovery, reflected the timecourse of an 'endogenous' control operation, independent of the pacing of task-relevant stimulus events.

The rationale of all three experiments was therefore to contrast a 'slow' and a 'fast' RSVP presentation rate, and to observe the effects of this manipulation on the rate of recovery of RSVP selective-report accuracy, following a shift of target criterion.

In Experiment 4, subjects performed a 'reversal shift' of semantic target-criterion. We used the same two target categories ('large' and 'small' objects) and the same word-sets as in Experiment 2. In the criterion-shift blocks ('large' → 'small'), on 50 per cent of trials there was an unpredictable shift of stimulus location, one line-space up the screen, cuing a shift of target criterion in the course of the list. T2 targets were presented, on different trials, either as the first, second, third or fifth word of sublist 2 (i.e., following the shift of stimulus location). On the remaining 50 per cent of trials in a block, subjects continued to monitor for 'large' objects throughout the list. We also included control, no-semantic-shift blocks (as in Experiment 1), to provide a location-shift control. On 50 per cent of these trials, at some point in the list the word sequence abruptly shifted one line-space down the screen. Subjects were instructed to ignore this shift of location and to continue searching for T1 as before.

There were two stimulus presentation rates, 'fast' and 'slow', such that, at the slow rate, the stimulus onset asynchrony (SOA) between successive words in the RSVP list was exactly twice the SOA of that at the fast rate. To prevent the accuracy of performance at the slow rate from

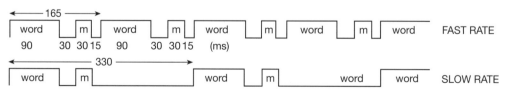

m: mask 'XCXCXCXCXC'

Fig. 3.5 Timing of stimulus sequences in Experiment 4: 'fast' and 'slow' presentation rates.

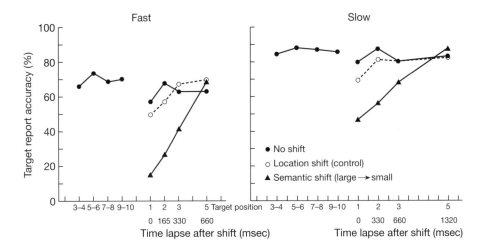

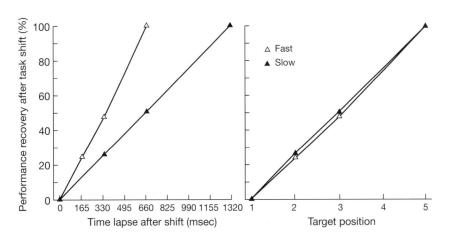

Fig. 3.6 Results of Experiment 4. The upper two panels show target report accuracy at 'fast' and 'slow' presentation rates, respectively, plotted as a function of real-time lapse. Data are shown for no-shift trials, and for the semantic criterion-shift and the location-shift controls. The lower two panels show the criterion-shift data normalized in terms of percentage recovery of performance toward the no-shift level. (See text.)

approaching 100 per cent, we included (at both presentation rates) a masking stimulus immediately following each word. The mask consisted of a row of upper case letters ('XCXCXCX-CXC'), which covered the full extent of all word-stimuli. The temporal sequence of stimuli is diagrammed in Fig. 3.5. A total of 18 subjects, recruited from the Oxford University subject panel, served in this experiment. Each subject performed eight trial blocks, 16 trials per block, making a total of 128 trials.

The results are illustrated in Fig. 3.6. As the top two panels of Fig. 3.6 show, a shift of semantic target criterion ('large' → 'small') resulted in a temporary but highly significant drop in monitoring accuracy for T2, as compared to the non-shift trials (fast rate: $F(1,17)=70.25$, $p<0.0001$; slow rate: $F(1,17)=77.42$, $p<0.0001$). The location-shift control trials also showed a reliable dip in T2 accuracy immediately following the shift ($p<0.01$ at both the fast and slow rates), but very much smaller in both amplitude and duration compared with the cost of a semantic criterion shift (Fig. 3.6). The difference between the semantic shift decrement and the location shift control was highly significant ($p<0.0001$, at both the fast and slow rates).

The main purpose of this experiment was to compare the timecourse of the RSVP shift-cost, as between the fast and slow presentation rates. The backward masking procedure was evidently successful in keeping selective report accuracy below ceiling, at the slow presentation rate. Nevertheless, as expected, at the fast presentation rate baseline performance on non-shift trials was less accurate, and post-shift T2 performance fell to even lower levels than at the slow rate. To enable a direct comparison between the timecourse of the shift-cost, at fast and slow presentation rates, the data obtained at the two different rates were normalized, as follows: For each subject, the T2 shift-cost was calculated relative to that subject's mean T2 accuracy on non-shift trials. The shift-cost at target position one was treated as 100 per cent, and the shift-cost at later target positions was then calculated relative to this initial 100 per cent shift cost. The two lower panels in Fig. 3.6 show the recovery of performance for T2, following the criterion shift, from target position 1 onwards: in the left panel as a function of real time lapse, in the right panel as a function of sequential target position. The results are extremely clear. Plotted as a function of time-lapse, fast and slow presentation show an entirely different timecourse. In contrast, when the same data are plotted as a function of the number of stimulus items following the shift cue, data for fast and slow presentation are practically identical.

These results appear to demonstrate that execution of an effective shift of selective set, in the RSVP selective-report task, does not proceed independently of the stimuli to which attention is to be directed—and which, by their location, continue to cue the change of task. Rather, the shift of set appears to be directly 'stimulus-driven'. Note that the effect of varying stimulus presentation-rate was exactly the opposite of that predicted by capacity-sharing models, which predict a slower attention shift at the faster stimulus rate.

In view of the possibly counter-intuitive nature of these results, we designed two further experiments to explore the same issue, and to establish the robustness of our findings. Several major changes were introduced: in the type of stimuli in the RSVP sequences, the target selection criteria, the use of orthogonal (vs. reversal) shifts, the cues used to signal a shift of target criterion, and the manipulations used to separate the effects of time-lapse vs. number of stimulus events. Further discussion is deferred until these experiments have been described.

In both Experiments 5 and 6, the task stimuli were single digits, rather than words, and the two (orthogonal) target criteria were (1) any digit which is larger than 5, and (2) any digit which is an even number. Experiment 5 used a small shift of location in the stimulus stream, as in previous experiments, as the cue to a shift of selection criterion; in Experiment 6 we used a change of background colour as the shift cue.

Experiment 5, RSVP shift-costs in digit monitoring: effects of time-lapse vs. task stimuli

In Experiment 4, the manipulation of 'fast' and 'slow' presentation rates confounded the rate of presentation of task-relevant stimuli (i.e., words denoting 'large' and 'small' physical objects) and the rate of presentation of *any* visual stimulus-events *in the changed location*, thus signalling the change of task. An important question, therefore, is whether the stimulus-driven character of the shift of set was driven by the successive, iterative cuing at the changed stimulus location, or by the number of task-relevant stimulus items. Accordingly, in Experiment 5 the

SOA between successive, potential target items (digits) was varied, in a 'slow' and a 'fast' rate, whereas the rate of presentation of visual stimuli, as such, was held constant. This arrangement was achieved as follows.

At both presentation rates there was a masking pattern ('X'), immediately following each digit; the combined duration of digit and mask was 150 ms. At the 'fast' rate, potential target stimuli (digits) followed one another at an SOA of 150 ms (i.e., about 6.7 digits per second); at the 'slow' rate, digits followed one another at an SOA of 300 ms (i.e., about 3.3 digits per second). In addition, for the 'slow' digit presentation rate, what would have been every second digit (in the fast rate) was replaced by a non-digit, filler stimulus (the letter 'C'). Both 'Cs' and digits were followed by the 'X' mask.

The task stimuli in Experiment 5 were the single digits 1–9, excluding 5. The two different target categories were (1) any digit that is an even number ('even'), and (2) any digit that is larger than 5 ('large'). The same stimuli and target categories were used also in Experiment 6. Hence these two experiments used orthogonal target categories for T1 and T2 ('large' → 'even'), rather than a 'reversal' shift. Cue words ('LARGE', 'EVEN') were presented prior to each list for as long as the subject wished, to indicate T1 and T2, as in the previous experiments. A shift in the location of the RSVP stimuli, one line-space up the screen, occurred unpredictably on 50 per cent of these lists, and signalled a shift of target criterion from T1 to T2. Experiment 5 also included blocks of trials with no criterion shift ('even' → 'even'). On these trials a shift of stimulus location, one line-space down the screen, occurred also on 50 per cent of these trials; subjects were instructed to ignore this shift of location and to continue searching for T1 as before (Location shift control). In both criterion shift and no-shift trials, the first sublist contained one target word (T1) on half the trials, located randomly as the third, fifth, seventh, or ninth item from the start of the list, followed by three or more non-target words before the onset of sublist 2. For those trials without a first target, randomly from six to twelve non-target words occurred before the onset of sublist 2. Sublist 2 always contained one target word, located equiprobably either as the first, second, third or fifth item. At least four non-target words followed the occurrence of T2.

Eighteen subjects, aged between 19 and 30 years, recruited from the National Chung-Cheng University, Taiwan, participated in this experiment, and were tested individually. All subjects completed eight blocks, four criterion shift, four no-shift, 16 trials per block, resulting in eight trials per target position per condition.

The results showed essentially the same pattern as Experiment 4, for fast and slow presentation rates. As in Experiments 1 to 4, the shift of target criterion resulted in a large drop in T2 report accuracy, for example, at the fast presentation rate, a drop from about 85 per cent on non-shift trials down to about 25 per cent correct at post-shift position 1 ($p<0.0001$). This was followed by a gradual recovery of performance over later target positions. Unlike Experiments 1 and 4, however, the location-shift control trials showed no significant drop in accuracy.

In order to compare the rate of recovery from the RSVP shift-cost at the fast and slow digit presentation rates, we applied the same normalization procedure as in Experiment 4. For each subject, the shift-cost at post-shift target position 1, relative to that subject's mean T2 accuracy on no-shift trials, was treated as 100 per cent. Thus, at post-shift target position 1, *recovery* from the shift-cost is, by definition, zero. The data, normalized in this way, are shown in Table 3.1, both as a function of serial position of the target and as a function of time since the shift cue. Clearly, at the slow rate of digit presentation, targets at later serial positions occur after longer and longer times have elapsed, relative to the fast rate. However, as can be seen in Table 3.1, these differences in elapsed time (i.e., between the fast and slow digit presentation rates, in the same column) had a negligible effect on T2 performance. In contrast, comparisons between target positions 2 and 3, and between 3 and 4, which can each be made for *identical* values of elapsed time since the onset of the shift cue, show very marked differences. An analysis of variance as a function of *serial position* of T2 showed a highly significant effect of serial position ($p<0.001$), but no main effect of presentation rate ($F=0.25$) and no significant interaction between serial position and rate ($F=0.00$).

Table 3.1 Post-shift recovery of T2 report per cent shown (a) as a function of serial position, (b) as a function of time elapsed since the shift cue (Expt. 5)

Elapsed time	Serial position of target 2				
	1	2	3	4	5
0 msec.	0				
150 msec.		26.6			
300 msec.		30.0	61.5		
450 msec.					
600 msec.			72.2		97.0
750 msec.					
900 msec.					
1050 msec.					
1200 msec.					100.0

Recall that 'slow' and 'fast' presentation rates refer only to the rate of presentation of successive *digit* stimuli, i.e., of potential target items. If we include stimuli of any kind, including the repeated 'filler' stimuli, then stimulus presentation rates were identical in both conditions. Experiment 5 was designed to test the hypothesis that the stimulus-driven character of recovery from the RSVP shift-cost (as observed in Experiment 4) depends on cumulative, repeated cuing of the intended criterion shift, by stimuli in the changed location—that is, by *any* physical stimuli (such as the repeating 'fillers'), regardless of whether or not they are potential target stimuli. The results of Experiment 5 offer no support for this hypothesis. They also offer no support to the hypothesis that the RSVP shift cost, and its recovery, reflect the timecourse of some real-time-dependent, 'endogenous' control operation, that depends merely on detection of the shift-cue for its initiation, and thereafter follows a timecourse of its own, independent of further external stimulation. On the contrary, the data suggest a predominantly or even entirely *stimulus-driven* process. Apparently, completion of an effective shift of selective set, in the RSVP monitoring context, is dependent on exogenous triggering by the type of (non-repetitive) task-relevant stimuli to which the T1/T2 selection criteria could be applied.

(Whether other, non-repetitive stimuli (e.g., pseudo-digits, other forms) might suffice for this triggering function is a question for further research.)

Experiment 6. Varying SOA within a single RSVP list: colour cuing of criterion shifts in digit monitoring

The preceding two experiments demonstrate, perhaps rather surprisingly, the stimulus-driven pacing of an effective shift of non-spatial selection bias, in RSVP word-monitoring and digit-monitoring tasks. A question arises, however, whether the stimulus-driven pacing occurs 'on-line', during the course of the criterion shift, or whether it might be entrained by the stimulus-presentation rate in the earlier section of the RSVP list—or even by the presentation rate in the entire block. Experiments 4 and 5 manipulated the rate of presentation of task-relevant and irrelevant stimuli as between different trial blocks; within a list, the rate of task-relevant stimulus presentation remained constant. Thus the rate of processing in the pre-shift section of an RSVP list might, in principle, pace the rate of other ongoing operations in the post-shift section of the same list. To test this idea, in Experiment 6, digit-presentation rate was varied within a list. Thus, long and short SOAs were randomly interleaved between successive digit stimuli. Each digit was presented for 100 ms followed by an 'X' mask (combined duration of digit followed by mask = 150 ms). An interval followed of either 0 or 150 ms before the next digit. As in Experiment 5, the 150 ms inter-digit interval was filled by a letter 'C', also followed by an 'X' mask.

A further aim of Experiment 6 was to explore a different procedure for cuing the shift of target criterion: viz. by means of a change of background colour. Might the pattern of results that we have found so far be dependent on the particular cuing method used, namely a shift of stimulus location? In Experiment 6 all stimuli were presented in the same location. Each stimulus appeared on a rectangular, coloured background. As before, the two selection tasks were (1) any digit which is an even number ('even'), and (2) any digit which is larger than 5 ('large'). The same two conditions were run as in Experiment 5, that is, criterion shift blocks ('large' → 'even'), and no-shift control blocks ('even' → 'even'). In the criterion shift blocks, digits appeared initially on a yellow background, and a shift of selection task (occurring unpredictably on 50 per cent of trials) was cued by a change to a red background. In the no-shift control blocks, on 50 per cent of trials the background colour changed to green. Subjects were instructed to ignore the latter, and to continue monitoring for T1.

Half the trials contained a target item in the first part of the RSVP list, located randomly at the third, fifth, seventh, ninth or eleventh digit position from the start of the list. All trials contained a target item in the second part of the list ('sublist 2'), located at either the first, second, third, fourth, or fifth serial digit position in sublist 2. In other respects the procedure was identical to that in Experiment 5. Twelve subjects were recruited from the National Chung-Cheng University.

The results appear very similar to those in the previous experiment, this time with colour cuing rather than spatial cuing of the attention shift. On no-shift trials, mean target report accuracy remained stable at around 90 per cent correct. On criterion shift trials, accuracy dropped abruptly on post-shift target position 1, and thereafter progressively recovered, approximately linearly, approaching the no-shift accuracy level by target position 5. Analysis of variance showed highly significant effects on T2 accuracy of shift vs. no-shift ($p<0.0001$) and of target position ($p<0.0001$). In the no-shift control condition, a change of the colour background did not result in a significant drop in accuracy.

The manipulation of SOA between successive digits means that a post-shift target at serial position 2 could occur either at 150 or 300 ms after the onset of the shift cue. Similarly, a target digit at serial position 3 could occur either 300, 450 or 600 ms after cue onset; and so on. This manipulation permits us, once again, to separate the effects of time elapsed since the shift cue from the effect of number of intervening, task-relevant stimulus items (digits). Thus, for each T2 target position in the RSVP sequence (1 through 5), the data for T2 report accuracy were divided according to the time elapsed since the shift cue, at which that target item had occurred. Table 3.2 shows the results. If elapsed time were the principal causal variable, then accuracy should be approximately constant across any horizontal row in the table. If the number of inter-

Table 3.2 Mean accuracy (%) of T2 report, shown (a) as a function of serial position, (b) as a function of time elapsed since the shift cue (Expt. 6)

Elapsed time	Serial position of target 2				
	1	2	3	4	5
0 msec.	67.4				
150 msec.		73.3			
300 msec.		73.2	81.1		
450 msec.			78.5	89.0	
600 msec.			82.0	86.3	93.8
750 msec.				86.2	91.2
900 msec.				85.7	94.0
1050 msec.					89.7
1200 msec.					85.7
Mean	67.4	73.2	80.5	86.8	90.9

vening stimulus items (digits) since the shift cue is the determining factor, then performance should remain approximately constant down any column. The table clearly shows that the number of stimulus items is the critical factor. There were no significant effects of elapsed time, when target position was held constant.

Thus, in three different experiments (Experiments 4, 5 and 6) we have found that the process of recovery of selective report accuracy, following a shift of target criterion, appears to be entirely stimulus-driven. Time lapse *per se* had no detectable effect. In other words, the 'time-course' of the RSVP shift cost is not properly speaking a function of real time at all. It follows that these data could not possibly reflect the timecourse of some 'endogenous' control operation, triggered by detection of the shift cue but from then on following its own, internally or endogenously determined, course. This conclusion has been tested using different stimuli, different tasks, different methods of cuing the intended shift of selection task, and different ways of manipulating time-lapse versus sequential target position.

General discussion

These experiments, we believe, demonstrate the usefulness of the criterion-shift paradigm for studying shifts of selective set, in the course of rapid sequential visual search tasks. However, we suggest that the criterion-shift paradigm, using RSVP selective report tasks (i.e., without time pressure for the subject's report), needs also to be complemented by on-line speeded response tasks (cf. Jolicœur *et al.* 2000). To our knowledge, such experiments remain to be done.

In this discussion, we focus on three topics, outlined at the start of this chapter: (1) the contrast between the C-S cost and the more widely studied 'attentional blink'; (2) the relationship between these C-S costs, as observed in RSVP selective report tasks, and the effects of task-switching in reaction time (RT) studies. Finally, (3) we shall discuss, briefly, some possible implications of these results (from both RSVP and RT approaches to 'task switching') for our understanding of 'executive control'.

RSVP criterion-shift (C-S) costs and the 'Attentional Blink' (AB)

As was noted in the Introduction, the standard 'two-target' or 'second-target' paradigm is appropriate for investigating the AB effect, with or (preferably) without a shift of target criterion between T1 and T2. On the other hand, it is difficult to observe the effects of a criterion shift, in the standard, RSVP 'two-target' paradigm, except in conjunction with (i.e., confounded with) an AB effect. (Cross-modal 'two-target' paradigms may be another matter.) As we have argued, the experimental procedure used in these six experiments (above) permits us to study the effects of an RSVP criterion-shift, uncontaminated by the additional costs of a second-target AB. These two classes of phenomenon appear to be very clearly separable, on a number of grounds.

First, as was found in all six experiments above, the C-S cost was identical, whether or not there was a previous target present in the list. This excludes, unambiguously, the simplest possible identification of the two types of effect, namely that the C-S cost in our experiments is merely an 'attentional blink' triggered by the first target, T1.

An alternative possible account might propose that the C-S cost is an AB effect, triggered by the shift cue itself. However, there is strong evidence that a long-lasting visual AB can be triggered only by a rapidly masked target stimulus (e.g. Raymond *et al.* 1992; Chun and Potter 1995, this volume, chapter 2, Enns *et al.*, this volume, chapter 4, Seiffert and Di Lollo 1997). In our experiments, the shift cue was, in effect, continuously re-presented throughout the whole of sublist 2; there was no possible masking of the shift cue. Thus the C-S cost, in our data, could not result from the same causal process as the previously studied, 'second-target' AB effect, triggered by the shift-cue. (Moreover, there was no 'lag-1 sparing' in the C-S cost. In our data, the C-S cost is always greatest at post-shift target position 1. Enns *et al.* (this volume, Chapter 4, Visser *et al.*, in press) show that 'lag-1 sparing' is a consistent characteristic of second-target (AB) costs, when there is no change of RSVP task.)

A more subtle hypothesis concerns the possible role of 'pseudo-targets' in triggering an 'attentional blink'. (See discussion of Experiment 2.) The idea is as follows. In all six experiments, above, the distractor stimuli presented in 'Sublist 2', after a shift cue, included items that would have counted as a target stimulus in terms of the previously valid, T1, target criterion. These are 'pseudo-targets'. If the selection bias in favour of T1 is still operating when the shift cue (location- or colour-change) is presented, it seems initially plausible that a 'pseudo-target', especially at the start of sublist 2, might be liable to trigger an 'attentional blink', which would then impair detection of valid T2 targets over, say, the following 200–600 ms. There are two principal arguments against this hypothesis. The first is that this hypothesis offers no explanation of the very large C-S costs at post-shift target position 1, since, by definition, these target stimuli cannot be preceded by any 'pseudo-target'. Hence, at best, the hypothesis could be relevant only to the later 'recovery' phase of the C-S cost. The second argument is based on the consistent finding from Experiments 4, 5 and 6 that the 'recovery' phase of the C-S cost, unlike the second-target AB effect, is not a simple function of time duration, but is exogenously driven (or paced) by successive, task-relevant stimuli. This appears to constitute a fundamental distinguishing characteristic between C-S costs and second-target ('AB') effects.

Thus, however plausible this hypothesis may seem at first sight, the recovery phase of the C-S cost, over post-shift target positions 2–5, could not be the product of a simple 'AB', triggered by the detection of a 'pseudo-target' at post-shift position 1. (It is possible that the presentation rates used in our experiments were simply too slow to elicit a regular 'AB' effect (cf. Arnell and Jolicœur, in press)).

Chun and Potter (this volume, chapter 2) have proposed another basic distinction between pure second-target ('AB') effects, uncontaminated by additional 'task-switching' costs, and more general criterion-shift costs. According to their analysis, pure AB effects are confined to intra-modal (visual–visual) sequences of T1 and T2; 'task-switching' costs, on the other hand, apply equally to cross-modal sequences. In the light of their proposal, it would be interesting to study criterion-shift costs, using the basic shift-cue paradigm developed here, but with cross-modal stimulus sequences in sublist 1 and sublist 2. We hope that other experimenters will take up this suggestion.

Criterion-shift (C-S) costs and RT task-switching effects

We noted earlier, when reviewing some of the effects found in RT 'task-switching' experiments (pp. 40–41), two consistent findings about anticipatory task preparation: (1) in most conditions there is still a large 'residual switching cost', even after an extended task-preparation interval following a shift cue; (2) the reduction in RT 'switching cost', as a function of increasing preparation time, is not accompanied by any corresponding reduction in between-task 'crosstalk' effects. The latter finding, we suggested, argues that the preparation process, in advance of an RT 'switch' trial, has not altered the relative 'attentional salience' of the previously task-relevant vs. task-irrelevant stimulus attributes. If so, however, such a preparation process seems unlikely to be directly related to the effective criterion-shift observed in our RSVP studies. In the criterion-shift paradigm, the successful monitoring for the post-shift target, T2, at post-shift target positions 5–7 onwards, seems very hard to understand without a shift of attentional salience from T1 to T2 target attributes. Moreover, unlike the majority of RT 'task-switching' studies, we found no evidence of any 'residual' C-S cost at these later T2 target positions. In both these respects, therefore, the pattern of results seems markedly different in discrete RT tasks and in RSVP selective report. (It is interesting, however, that in at least one study, when the switch cue for an upcoming RT trial was itself an exemplar (or included the set of exemplars) of the upcoming task-relevant stimuli, the RT 'switching cost' was reduced to near zero over the subsequent preparation interval (De Jong *et al.*, submitted, Experiment 3). This suggests a possible parallel, in the role of *exogenous* task-relevant cuing, in both the RT and the RSVP 'task-switching' paradigms.)

In the RT task-switching paradigm, it is generally recognized that a major component of the RT switching cost is 'exogenous', that is, it depends on triggering by an external task-stimulus (e.g., Meiran 1996; Meiran *et al.*, in press; Rogers and Monsell 1995). However, there are at least two contrasting accounts of the causal origin of these 'exogenous' RT costs. One type of account attributes them—in whole or in part—to the additional time needed for a further control operation, which completes the 'task-set reconfiguration' needed for the current (switch) trial (e.g. Rogers and Monsell 1995; Rubinstein *et al.*, in press). An alternative account attributes them predominantly to both long-term and short-term 'positive and negative priming' from the previous task (or tasks), which required a different set of responses to the same type of 'bivalent' task stimuli (e.g. Allport and Wylie, 1999, 2000). A critical experimental manipulation, in this context, is to present different individual stimulus-items for response, respectively, in tasks A and B, even though all stimuli are still drawn from the same general class of 'bivalent' stimuli. (Note that, to date, practically all RT 'task-switching' studies have used the same individual stimuli in both tasks.) We have now repeated this manipulation, using a number of different 'bivalent' stimulus-types (e.g., Allport and Wylie, 2000; Waszak *et al.* 1999). In these studies, the use of non-overlapping stimulus-items reduced the RT switching cost by a factor of two or more. It seems clear, therefore, that at least this *item-specific* component of the residual RT cost could not be attributed to any global resetting of S-R connection weights, or 'pertinence' values, etc., since such changes should affect the speed of response to *all* stimuli. Whether the greatly reduced RT 'switching cost' (which is found even with *different* individual stimulus items in tasks A and B) should be attributed to an active 'resetting' process, or to passive 'task-set inertia', remains to be established.

Evidently, item-specific interference (long-term 'positive and negative priming') can be an important contributor to RT 'switching costs'. However, we have not yet explored any possible, equivalent effects in the RSVP criterion-shift paradigm. In the studies described here, the *same* individual stimulus items appeared, either as 'targets' or 'distractors' (i.e., in 'varied mapping' assignments: Shiffrin and Schneider 1977), during search for both T1 and T2. It will be important to establish whether similar; item-specific stimulus manipulations ('consistent mappings') affect the C-S cost, in the same way as they affect RT switching costs. If so, this would suggest at least some underlying commonality in the mechanisms responsible for both C-S costs (in the RSVP paradigm) and RT 'switching costs' (in discrete reaction time studies). We hope that other experimenters will pursue this question.

Implications for models of 'executive control'

Discussions of 'executive control' in current cognitive neuroscience typically share a number of basic assumptions (e.g. Baddeley 1986, 1996; Cooper and Shallice, in press; Duncan 1995; Fuster 1995; Logan and Gordon, submitted; Miller, 1999; Norman and Shallice 1986; Passingham 1993; Posner and Raichle 1994; Shallice and Burgess 1996; Shiffrin and Schneider 1977; Wise *et al.* 1996; and many others). In particular, they assume a hierarchical relationship between a top-down 'control system' (or systems), whose operation is essentially 'endogenous' (that is, not directly stimulus-driven or event-driven), and a lower-level set of processing systems, which are, by contrast, 'exogenously' driven (that is, they are relatively passively activated by ongoing stimuli or events). This basic assumption leads to a particular interpretation of the idea of *task set* (and of other, so-called, 'attentional states'). According to this interpretation, the top-down control systems are able to 'set' the lower-level systems, in advance, into certain states of processing readiness, corresponding to the effective facilitation or enablement of some processing pathways, and the suppression or disablement of others. To some people, these basic assumptions have suggested the metaphor of a railway system: the control system(s) 'switch' the interconnecting points on the railway lines; the trains of stimulus-processing then automatically follow the pre-set lines (e.g. Rogers and Monsell 1995).

In this chapter we have presented a number of behavioural observations that do not appear to fit entirely comfortably with these assumptions. To begin with, in the RT studies of 'task switching', we have noted at least three sets of findings that appear problematic for this kind of view:

1. Anticipatory task preparation ('setting') does not generally have any effect on the amount of interference, or 'crosstalk', from the other task. It is not easy to see how this finding can be reconciled with the assumption that task 'setting', in advance, corresponds to a change in the 'pertinence', or attentional weighting, of task-relevant versus irrelevant attributes, or to the selective enabling/disabling of task-processing pathways ('switching the railway lines').

2. The existence of a large 'residual' component of the RT 'switching costs' also suggests that, contrary to these intuitions, some important part of the process of 'task switching' is itself stimulus-driven (i.e., is not 'endogenous').

3. The finding of a large item-specific component in RT 'switching costs' challenges the concept of 'task-set' as a *global* determinant of something called 'task readiness'. (To put this point another way, it is tempting to think of RT 'switching costs' as a direct measure of 'task readiness': naturally, on 'switch' trials the control system has not yet been able to get the lower-level processing systems 'set' for the new task, so 'getting set' will cost extra time. However, the finding that some individual stimulus items appear to be much better 'pre-pared-for' than others seriously undermines this simple interpretation of the RT switching costs.)

Now, to add to these awkward findings in the RT 'task-switching' literature, our studies of the RSVP criterion-shift seem to provide still further difficulties for these commonplace assumptions about (so-called) 'executive control'. Contrary to intuition, the results we have described suggest that the progression of an effective 'shift of set', in the RSVP selective report task, is not the product of some *endogenous*, top-down control operation but—on the contrary—is entirely *stimulus-driven*. (At least, this is so within the conditions of our experiments.) Together, these results lead us to question the assumptions, widespread in the literature, about a fundamental distinction between *endogenous* (or 'autonomous') supervisory control system(s), on the one hand, and subordinate processing-systems that are passively stimulus-driven, on the other.

'Control', we suggest, is a much more reciprocal business: subsystem (or network) A may modulate processes in subsystem B; but B also modulates A; and both are liable to be driven by ('exogenous') events coming from C. This is the simple idea of 'heterarchical' control, put forward long ago by Warren McCulloch (1965), and illustrated in a number of more recent approaches to understanding autonomous, behaving systems (e.g. Brooks 1994; Clark 1997; Dean 1998). Cognitive neuroscience, we believe, would be better served by metaphors of the latter kind, in attempting to understand the emergent nature of 'control', than by metaphors of a fixed hierarchy of top-down, 'endogenous' controllers, and passive 'slave systems' below.

References

Allport, A. (1989). Visual attention. In *Foundations of cognitive science* (ed. M. I. Posner). Cambridge, MA: MIT Press.

Allport, A. (1993). Attention and control: have we been asking the wrong questions? A critical review of 25 years. In *Attention and performance* XIV (ed. D. E. Meyer and S. Kornblum). Cambridge, MA: MIT Press.

Allport, A. and Wylie, G. (1999). Task-switching: positive and negative priming of task set. In *Attention, space and action: studies in cognitive neuroscience* (ed. G. W. Humphreys, J. Duncan and A. M. Treisman). Oxford: Oxford University Press.

Allport, A. and Wylie, G. (2000). Task-switching, stimulus-response bindings, and negative priming. In *Attention and Performance XVIII: Control of cognitive processes* (ed. S. Monsell and J. Driver). Cambridge, MA: MIT Press.

Allport, A., Styles, E. A. and Hsieh, S. (1994). Shifting intentional set: exploring the dynamic control of tasks. In *Attention and Performance XV* (ed. C. Umilta and M. Moscovitch). Cambridge, MA: MIT Press.

Arnell, K. and Jolicœur, P. (1999). The attentional blink across stimulus modalities: evidence for central processing limitations. *Journal of Experimental Psychology: Human Perception and Performance*.

Baddeley, A. D. (1986). *Working memory*. Oxford, Oxford University Press.

Baddeley, A. D. (1996). Exploring the central executive. *Quarterly Journal of Experimental Psychology*, **49A**, 5–28.

Bramley, T. (1995). Switching task set: investigating orthogonal shifts of semantic category. Unpublished MS, University of Oxford.

Broadbent, D. E. and Broadbent, M. H. (1987). From detection to identification: response to multiple targets in rapid serial visual presentation. *Perception and Psychophysics*, **42**, 105–13.

Brooks, R. (1994). Coherent behaviour from many adaptive processes. In *From animals to animats*, 3 (ed. D. Cliff *et al.*) Cambridge, MA: MIT Press.

Chun, M. M. and Potter, M. C. (1995). A two-stage model for multiple target detection in rapid serial visual presentation. *Journal of Experimental Psychology: Human Perception and Performance*, **21**, 109–27.

Chun, M. M. and Potter, M. C. (this volume, Chapter 2). The attentional blink and task switching within and across modalities.

Clark, A. (1997). *Being there: putting brain, body and world together again*. Cambridge, MA: MIT Press.

Cooper, R. and Shallice T. (in press). Contention scheduling and the control of routine activities. *Cognitive Neuropsychology*.

Dean, J. (1998). Animats and what they can tell us. *Trends in Cognitive Sciences*, **2**, 60–7.

De Jong, R., Emans, B., Eenshuistra, R. and Wagenmakers, E. J. (submitted). Structural and strategical determinants of in intentional task control.

Duncan, J. (1995). Attention, intelligence, and the frontal lobes. In *The cognitive neurosciences* (ed. M. S. Gazzaniga). Cambridge, MA: MIT Press.

Duncan, J., Ward, R. and Shapiro, K. L. (1994). Direct measurement of attentional dwell time. *Nature*, **369** (6478), 313–15.

Enns, J. T., Visser, T. A. W., Kawahara, J-I. and Di Lollo, V. (this volume, Chapter 4). Visual masking and task switching in the attentional blink.

Fagot, C. (1994). Chronometric investigations of task switching. Ph.D thesis, University of California, San Diego.

Fuster, J. M. (1995). *Memory in the cerebral cortex*. Cambridge, MA: MIT Press.

Gopher, D., Armony, L. and Greenshpan, Y. (2000). Switching tasks and attention policies. *Journal of Eperimental Psychology: General*, **129**, 308–39.

Hommel, B. (1998). Event files: evidence for automatic integration of stimulus–response episodes. *Visual Cognition*, **5**, 183–216.

Hsieh, S. (1993). Shifting task 'set: Exploring non-spatial aspects of intentional control. D. Phil. Thesis, University of Oxford.

Hsieh, S. (1995). Stimulus-driven or autonomous shift of attention? *Perceptual and Motor Skills*, **80**, 1187–99.

Hsieh, S. and Allport, A. (1994). Shifting attention in a rapid visual search paradigm. *Perceptual and Motor Skills*, **79**, 315–35.

Jersild, A. T. (1927). Mental set and shift. *Archives of Psychology*, **89**. whole no.

Jolicœur, P., Dell'Acqua, R. and Crebolder, J. (2000). Multitasking performance deficits: forging links between the attentional blink and the psychological refractory period. In *Attention and performance* XVIII: *Control of cognitive processes* (ed. S. Monsell and J. Driver). Cambridge, MA: MIT Press.

Logan, G. D. and Gordon, R. D. (submitted). Executive control of visual attention in dual-task situations.

MacLeod, C. M. (1991). Half a century of research on the Stroop effect: an integrative review. *Psychological Bulletin*, **109**, 163–203.

McCulloch, W. S. (1965). *Embodiments of mind*. Cambridge, MA: MIT Press.

Meiran, N. (1996). Reconfiguration of processing mode prior to task performance. *Journal of Experimental Psychology: Learning, Memory and Cognition*, **22**, 1423–42.

Meiran, N., Chorev, Z. and Sapir, A. (1999). Component processes in task switching. *Cognitive Psychology*.

Meuter, R. F. I. and Allport, A. (1999). Bilingual language switching and naming: asymmetrical costs of language selection. *Journal of Memory and Language*, **40**, 25–40.

Miller, E. K. (1999). Prefrontal cortex and the neural basis of executive functions. In *Attention, space and action: studies in cognitive neuroscience* (ed. G. W. Humphreys, J. Duncan and A. M. Treisman). Oxford University Press.

Norman, D. A. and Shallice, T. (1986). Attention to action: willed and automatic control of behaviour. In *Consciousness and self regulation: advances in research and theory* (ed. R. Davidson, G. Schwartz and D. Shapiro). New York: Plenum.

Passingham, R. (1993). *The frontal lobes and voluntary action*. Oxford University Press.

Posner, M. I. (1995). Attention in cognitive neuroscience. In *The cognitive neurosciences*, (ed. M. S. Gazzaniga). Cambridge, MA: MIT Press.

Posner, M. I. and Raichle, M. E. (1994). *Images of mind*. New York: W. H. Freeman.

Raymond, J. E., Shapiro, K. L. and Arnell, K. M. (1992). Temporary suppression of visual procesing in an RSVP task: an attentional blink? *Journal of Experimental Psychology: Human Perception and Performance*, **18**, 849–60.

Reeves, A. and Sperling, G. (1986). Attention gating in short term visual memory. *Psychological Review*, **93**, 180–206.

Rogers, R. D. and Monsell, S. (1995). The cost of a predictable switch between simple cognitive tasks. *Journal of Experimental Psychology: General*, **124**, 207–31.

Rubinstein, J. Meyer, D. E. and Evans, J. E. (in press). Executive control of cognitive processes in task switching. *Journal of Experimental Psychology: Human Perception and Performance*.

Seiffert, A. E. and Di Lollo, V. (1997). Low-level masking in the attentional blink. *Journal of Experimental Psychology: Human Perception and Performance*, **23**, 1061–73.

Shallice T. and Burgess, P. (1996). The domain of supervisory processes and temporal organization of behaviour. *Philosophical Transactions of the Royal Society of London*, **B351**, 1405–12.

Shapiro, K. L., Arnell, K. M. and Raymond, J. E. (1997). The attentional blink. *Trends in Cognitive Sciences*, **1**, 291–6.

Shiffrin, R. M. and Schneider, W. (1977). Controlled and automatic human information processing: II. Perceptual learning, automatic attending, and a general theory. *Psychological Review*, **84**, 127–90.

Spector, A. and Biederman, I. (1976). Mental set and shift revisited. *American Journal of Psychology*, **89**, 669–79.

Stroop, J. R. (1935). Studies of interference in serial verbal reactions. *Journal of Experimental Psychology*, **18**, 643–62.

Visser, T. A. W., Bischoff, W. F. and Di Lollo, V. (in press). Attentional switching in spatial and non-spatial domains: evidence from the attentional blink. *Psychological Bulletin*.

Ward, R., Duncan, J. and Shapiro, K. (1996). The slow time-course of visual attention. *Cognitive Psychology*, **30**, 79–109.

Waszak, F., Hommel, B. and Allport, A. (1999). Item-specific priming in task-switching. Poster presented at meeting of Max-Planck Institute, Munich, July, 1999.

Weichselgartner, E. and Sperling, G. (1987). Dynamics of automatic and controlled visual attention. *Science*, **238**, 778–80.

Wise, S. P., Murray, E. A. and Gerfen, C. R. (1996). The frontal-basal ganglia system in primates. *Critical Reviews in Neurobiology*, **10**, 317–56.

Wylie, G. and Allport, A. (2000). Task switching and the measurement of 'switch costs'. *Psychological Research*, **63**, 212–33.

4

Visual masking and task switching in the attentional blink

James T. Enns, Troy A. W. Visser, Jun-ichiro Kawahara, and Vincent Di Lollo

Abstract

When two targets are presented in rapid succession, identification of the first target is nearly perfect, but identification of the second target is impaired when it follows the first by less than about 500 ms. In the present chapter, we examine the role of two factors — visual masking and attentional switching — in producing this second-target deficit, commonly known as the *attentional* blink (AB). We begin by reporting that the AB is affected in different ways by how the first and second targets are masked. The AB occurs whether the first target is masked by interruption or by integration. In contrast, only interruption masking is effective for the second target. In the next section, we examine the role of attentional switching as revealed by the phenomenon of 'lag-1 sparing', which refers to a reduction in the magnitude of the AB when the second target is presented directly after the first. This 'sparing' points to the importance of attentional switching between sequential tasks in producing the AB. In the final section, we present new evidence showing that when the two targets have different attentional requirements, the AB occurs reliably even without a mask after the second target. Based on this evidence, we conclude that the AB may represent the temporal cost of reconfiguring the visual system in readiness for different sequential tasks.

Introduction

This chapter summarizes research, conducted in our laboratory over the past 5 years, on the phenomenon known as the attentional blink (AB). In one popular procedure used to induce the effect, observers attempt to identify two target letters inserted into a temporal stream of digits (Chun and Potter 1995; Raymond *et al.* 1992; Shapiro *et al.* 1994). Because all the items are displayed in the same location, one every 100 ms or so, this procedure is often referred to as *rapid serial visual presentation* (RSVP). The AB refers to a striking imbalance in the identification accuracy of the two targets. Although the first target can be identified almost perfectly, identification of the second target is substantially impaired, especially when it is presented with a temporal lag of 200–500 ms after the first target.

The same pattern of results is obtained using a simplified version of the task (Duncan *et al.* 1994; Ward *et al.* 1997), which we call the *two-target* procedure. In this case, the only items presented are the two targets, displayed in different screen locations at various temporal lags from each other, and each is followed by a pattern mask. As with the RSVP method, identification is nearly perfect for the first target, but is dramatically reduced for the second target when it lags the first by 200–500 ms. The perceptual and cognitive mechanisms responsible for this second-target deficit are the focus of this chapter.

In our research we have come to the conclusion that there are two quite separable aspects to the AB. One aspect concerns the contributions of visual masking. Because the methods for studying the AB typically involve the rapid presentation of visual patterns in the same spatial location, issues of masking cannot be ignored. Indeed, as we will show in summarizing our research, whether or not an AB is observed can depend critically on the kind of masking that is employed.

These studies have also revealed an important relationship between masking and attention. Namely, visual pattern perception is shown to be vulnerable to backward masking only when attention cannot be focused rapidly on a target stimulus. This means that in a typical AB experiment, the second target is inherently more vulnerable to backward masking than the first. By the very design of the task, the first target is presented when the observer is optimally prepared to process it (i.e., it is the first target to engage response selection and execution processes). In contrast, the second target is presented shortly after the first, so that if the observer is not fully prepared to process the second target, its representation will be replaced in low-level visual registers by a representaton of an item that follows the second target. Accuracy in reporting the second target therefore suffers as a direct consequence of what we call an *object substitution* process for unattended items.

The second aspect of the AB concerns attention, by which we mean the control and deployment of limited capacity mechanisms in the brain involved in the identification of patterns. As we will show, in the absence of any visual masking, whether the AB is observed or not depends critically on the degree to which attentional mechanisms must be reconfigured from the task designated for the first target to the task required for the second target. At one extreme, we will demonstrate that no AB occurs when the tasks for each target have similar attentional demands. At the other extreme, we will show that a large AB ensues, even in the absence of visual masking, if the attentional requirements differ sufficiently between the first and second targets. Because of these differences, we will argue that the AB is best understood neither as a measure of the time needed to perform the first task (i.e., the *attentional dwelltime* of Duncan *et al.* 1994), nor as an index of those second-target tasks that can be performed concurrently with the first-target task without attentional conflict (i.e., the *concurrent processing* measure of Joseph, Chun and Nakayama 1997). Instead, we believe the AB to be a temporal index of the cognitive cost of reconfiguring the visual processing system for the performance of one task versus another.

We present our story in three sections. In the first section we summarize our already-published research on the multiple roles of visual masking in the AB. In the second section we introduce the basis for our ideas on attentional reconfiguration through an examination of a curious and little-understood feature of AB studies: whether or not second targets presented in to temporal position immediately following the first target are subject to the blink. The reduction of AB for the second target in that position is refered to as *lag-1 sparing*, and it can be linked reliably to the differences in processing demands between the first and second target (Visser *et al.*, 1999). In the third section, we extend these ideas to the entire range of AB effects across temporal lag. This is the most speculative aspect of our story, since it will depend ultimately on the outcome of a series of experiments that are not yet complete. However, we argue this case by describing selected conditions from ongoing experiments that point to the AB as a measure of the cost of task switching, or attentional reconfiguration.

The role of masking in the AB

'Visual masking' refers to a reduction in the visibility of a stimulus, called the target, which occurs as a result of the presentation of a second stimulus, called the mask. Masks can be divided into two broad categories; those consisting of patterns that overlap the targets in space, often referred to as pattern masks, and those that do not (Breitmeyer 1984). In the present work, we emphasize the role of pattern masking in the AB.

Pattern masks present the visual system with at least two different kinds of spatio-temporal conflict (Breitmeyer 1984; Ganz 1975; Kahneman 1968; Scheerer 1973; Turvey 1973). One conflict occurs when the target and mask are perceived as part of a unitary pattern. In this case, masking occurs by a process of *integration*, and is akin to the addition of noise (the mask) to the signal (the target). Turvey (1973) referred to this as 'peripheral masking' to emphasize that it takes place at early levels of visual processing. A second kind of conflict arises when processing of a first pattern (the target) is interrupted by a second pattern (the mask) which appears in the same spatial location before the target has been fully processed. This conflict involves a competition for the higher-level mechanisms required for object recognition. Therefore, it is often referred to as masking by *interruption* or 'central masking' (Turvey 1973). Unlike integration masking, interruption masking occurs only when the mask follows the target in time, with target accuracy lowest at target–mask intervals that are greater than zero, and gradual improvement as interval increases (e.g. Bachman and Allik 1976; Turvey 1973).

Masking of the first target: any mask will do

Masking of the first target emerged as an important factor in early demonstrations of the AB (Raymond *et al.* 1992). These investigators found that the AB did not occur when the item directly following the first target (the '+1 item') was replaced by a blank screen. On the hypothesis that the +1 item acted as a mask, its omission allowed processing of the first target to continue without interference for the ensuing mask-free period.

In an initial experiment aimed at demystifying the role of first-target masking, Seiffert and Di Lollo (1997) examined the effects of integration masking by presenting the first target and the +1 item simultaneously and spatially superimposed. This combined stimulus was followed by a blank screen during the interval in which the +1 item would normally have appeared. A large AB was obtained as compared to a condition in which the first target was not masked (i.e. the +1 item was omitted), suggesting that integration masking of the first target was one way in which an AB could be obtained.

Having established a role for integration masking, Seiffert and Di Lollo (1997) went on to evaluate another form of masking. They did this by presenting the +1 item directly after the first target, but in a location that was spatially adjacent to where the target had appeared. This arrangement eliminated the possibility of masking by integration, but was suitable for metacontrast masking, in which processing of the target is interrupted by a trailing non-overlapping mask. Again, the magnitude of the AB was significantly greater than that produced when the first target was not masked, a result also found by Grandison, Ghirardelli, and Egeth (1997). Collectively, these results led Seiffert and Di Lollo to conclude that, in order to obtain an AB, the first target must be masked. The precise form of masking, however, is unimportant. An AB occurs whether the first target is masked by integration, interruption, or even metacontrast.

Masking of the first target is clearly important in the production of the AB. However, it is not immediately obvious why it should affect identification of the second target. Indeed, this is quite surprising, given that masking of the first target impairs identification of the second target more than identification of the first. The explanation proposed by both Seiffert and Di Lollo (1997) and Grandison *et al.* (1997) follows from the two-stage theory of the AB proposed by Chun and Potter (1995). This theory comprises an initial stage in which items are detected rapidly as potential targets, and a second capacity-limited stage in which these items are processed in greater detail for subsequent report. Access to Stage 2 is gained by items that have been identified as potential targets in Stage 1. Until processing of the first target is completed, however, no subsequent items can gain access to Stage 2. Therefore, if the second target is presented before Stage 2 is free, access to Stage 2 will be delayed. The AB occurs because the initial representation of second target decays or is overwritten by subsequent items during the delay. On this account, masking increases the length of time required to process the first target. Although this does not ultimately impair first-target accuracy, it does delay the admission of the second target into Stage 2. As the delay increases, the likelihood that the second target will decay or be overwritten in Stage 1 also increases (Grandison *et al.* 1997; Seiffert and Di Lollo 1997).

This account of the influence of first-target masking is consistent with the general relationship between difficulty of processing the first target and the AB posited earlier by Chun and Potter (1995). From this perspective, masking is simply one of many potential ways to make processing of the first target more difficult, and to increase the period of delay for the second target. One implication of this account is that the size of the AB should vary inversely with first-target accuracy. To evaluate this relationship, Seiffert and Di Lollo (1997) compared the percentage of correct responses on the first target with the magnitude of the AB using their own data as well as that from 26 other separate experiments. The results of this analysis are presented as a scatter diagram in Fig. 4.1. As predicted, there was a significant negative correlation between first target accuracy and the magnitude of the AB, $r = 0.73$, $p < .001$.

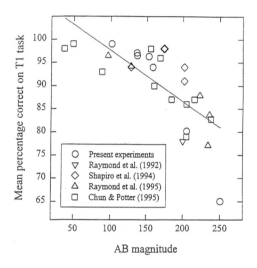

Fig. 4.1 Scatter diagram relating attentional blink (AB) magnitude to percentage of corrent responses on the first target. AB magnitude increases significantly with a reduction in first target (T1) accuracy from Seiffert and Di Lollo 1997).

The proposed relationship between difficulty of processing the first target and the magnitude of the AB has not gone unchallenged. Ward *et al.* (1997) noted that all the studies cited by Seiffert and Di Lollo (1997) manipulated difficulty through masking. This left open the possibility that other difficulty manipulations might not produce the same relationship. To test this, Ward *et al.* (1997) varied the difficulty of a size judgement for the first target. They found that although accuracy was reduced when the judgement task was made more difficult, the magnitude of the AB was unaffected. On the basis of this result, Ward *et al.* (1997) concluded that not all manipulations of first target difficulty lead to changes in the AB, and that earlier findings might be specific to masking.

Masking of the second target: interruption is essential

It has long been recognized that the second target must be masked in order for the AB to occur. Ostensibly, the purpose of masking has been to increase the difficulty of processing the second target, thereby bringing accuracy within a measurable range (e.g. Moore *et al.* 1996). If this were the principal function served by masking, then either integration or interruption masking should be sufficient, and an AB should be found using either procedure.

To investigate the two forms of masking, Giesbrecht and Di Lollo (1998) compared the magnitude of the AB when the second target was masked by interruption or by integration. Both targets were letters, presented in an RSVP stream of digit distractors. In the interruption-masking condition, the second target was followed by at least one digit that acted as a mask. In the integration-masking condition, the second target was presented simultaneously with a digit, with no additional items following in the RSVP stream. This brought accuracy of the second target within a measurable range, while eliminating interruption masking by trailing items.

When the second target was masked by interruption, a significant AB was obtained, with identification of the second target steadily improving beyond an inter-target interval of about 200 ms. This finding was consistent with results obtained in other studies (e.g. Chun and Potter 1995; Raymond *et al.* 1992; Shapiro *et al.* 1994). In contrast, when the second target was masked by integration, identification was impaired equally across all lags, but the lag-dependent deficit that is the signature of the AB was missing.

Similar results were obtained by Brehaut *et al.* (1999) using the two-target procedure. As is shown in Fig. 4.2, when the second target and the mask were presented simultaneously, a lag-dependent deficit was notably absent. However, as the interval between target and mask increased, an AB became increasingly apparent. On the basis of these findings, it is clear that interruption masking of the second target is more than a methodological convenience—it is necessary to obtain an AB.

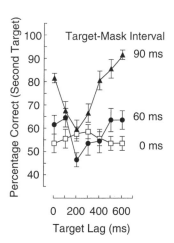

Fig. 4.2 Mean accuracy of second target identification under three target-mask intervals (0 ms, 60 ms, 90 ms). Lag-dependent effects increased as the interval between target and mask increased. Chance accuracy level in this task was 5 per cent (redrawn from Brehaut *et al.*, 1999).

We have argued that the two-stage theory (Chun and Potter 1995) provides a comprehensive account for these results (Brehaut *et al.* 1999; Giesbrecht and Di Lollo 1998). As was noted earlier, this theory specifies two sources of deterioration for the second target while it is delayed in Stage 1: passive decay and overwriting by temporally trailing items. The results of Giesbrecht and Di Lollo (1998) suggest that the main source of deterioration in Stage 1 is over-writing. Passive decay would be evidenced by impairment in accuracy at the shortest lag, fol-lowed by a gradual improvement over lags. If decay were an important determinant of accuracy, this trend should have been observed in the integration masking condition, in which there were no trailing items to erase the second target. However, the results in this condition show little or no evidence of such a trend. This suggests that passive decay does not play a major role in the deterioration of the second target.

In contrast, overwriting provides a consistent account of the results obtained with both inte-gration and interruption masking. With integration masking, accuracy was impaired because the pattern of the second target was impoverished, making it harder to extract the target from the noise. However, the magnitude of the impairment did not vary as a function of lag. This result can be explained within the two-stage theory by noting that the second target and the mask formed a unitary stimulus that remained available in Stage 1 because there was no trailing item to overwrite it. Thus, when the delay experienced by the second target ended, its representation entered Stage 2 and was processed to the extent allowable by its reduced stimulus quality. On the other hand, with interruption masking, second target accuracy did vary with lag. Accuracy was most impaired at lags of 200–300 ms, and improved progressively thereafter. According to the two-stage theory, accuracy was impaired at the shorter lags because the trailing mask erased the second target during the period of delay in Stage 1. At longer lags, the probability that the second target could enter Stage 2 before being erased by the mask increased, and accuracy improved accordingly.

Implicit within this account is an object-substitution theory of interruption masking. While it is delayed in Stage 1, the representation of the second target is vulnerable to overwriting by the trailing mask. When that happens, the representation of the mask replaces that of the second target and eventually gains access to Stage 2. As a consequence of this replacement, the mask is substituted for the second target as the object for eventual conscious registration. This object-substitution theory argues that the interruption mask does more than simply halt processing of the second target. Instead, because the representation of the target has been overwritten, Stage 2 mechanisms are left with only the mask to process.

A view of interruption masking akin to object substitution is well supported in the masking literature. There is ample evidence to show that when two targets are presented sequentially at an optimal interval, it is the second that is perceived to the detriment of the first (Bachmann and Allik 1976). This effect has been found to be more pronounced in unattended visual locations, suggesting that stimuli displayed outside the focus of attention are more likely to be delayed in Stage 1, thus remaining vulnerable to substitution over a longer period (Enns and Di Lollo 1997).

This tendency has been observed not only when attention is distributed over space, but also when it is distributed over time, as in the AB. In studies by Chun (1997) and Martin, Issak, and Shapiro (1995), an RSVP stream of letters was presented, with the two target letters being dis-tinguished from the rest of the stream by either colour (Chun 1997) or size (Martin *et al.* 1995). In both studies, the most prominent type of error consisted of misidentifying the second target as the item directly following it. These results are consistent with the hypothesis that while delayed in Stage 2, the second target is overwritten by the trailing item.

Summary of masking in the AB

Considered collectively, the results demonstrate that masking plays two separate roles in the AB. Masking of the first target, whether by integration, interruption, or metacontrast, introduces a delay in the processing of the second target. During this delay, the second target is vulnerable to being replaced by a trailing mask, with a probability that decreases as the temporal lag between the targets is increased. These findings suggest that the effect of first-target masking is to make its processing more difficult, and that several kinds of masking accomplish this equally well.

The second role played by visual masking is specific to the second target. Here, it is essential to use interruption masking in order to obtain the AB. Masking by integration, although sufficient to reduce second target accuracy, does not produce a lag-dependent effect. In addition to being of methodological interest, this finding points strongly to a visual process of considerable theoretical importance. This process, referred to as object substitution, is the overwriting or replacement of visual representations by trailing items, when those representations are unattended. Given how much of our visual world goes unattended, object substitution may play an extensive role in perception. Indeed, we have suggested (Brehaut *et al.* 1999) that substitution may underlie demonstrations of 'change blindness', referring to the finding that major changes in the visual world can go undetected if attention is misdirected away from the area of change (Rensink *et al.* 1997; Simons 1996).

Lag-1 sparing and the attentional blink

The AB points to the limits in the ability of the visual system to process information. When two targets are presented in rapid succession, processing resources required in common by both targets are less available for the second target until processing of the first has been completed. On this account, identification of the second target should be maximally impaired when it is presented directly after the first target, in the ordinal display position known as *Lag 1*. The deficit should then diminish as lag increases, reflecting the increased availability of resources previously deployed to the first target. However, this pattern of results is found in only about one-half of published AB experiments. In the other half, accuracy is largely unimpaired at lag 1, drops substantially at lags 2 and 3, and then gradually recovers. It is this pattern of improved accuracy at lag 1, followed by a pronounced deficit at longer lags, that is referred to as *lag-1 sparing* (Potter *et al.* 1998).

Lag-1 sparing has been attributed to the sluggish closing of an attentional gate (Chun and Potter 1995; Shapiro and Raymond 1994). The gate is said to open upon the presentation of the first target and then to close slowly, thus allowing a trailing stimulus to gain access to processing resources along with the first target. If the trailing item is the second target, then it is processed along with the first target and an AB is avoided. Although this explanation can account for the occurrence of lag-1 sparing, it cannot explain why lag-1 sparing does not occur, under identical rates of presentation, in about one-half of studies. These failures to obtain lag-1 sparing suggest that factors other than temporal contiguity must determine whether sequential stimuli give rise to the same or different attentional episodes. By uncovering these factors, the phenomenon of lag-1 sparing may become relevant to broader issues relating to distribution of attention and rapid changes in attentional set in both spatial and non-spatial domains.

In studying the dynamics of switching attentional sets, the main issue is how the cognitive system is reconfigured to cope with rapidly changing demands. The impact of such reconfiguration can be observed readily in studies of the AB, because observers must perform a rapid attentional switch from the first target to the second across a brief temporal gap. In particular, accuracy in the report of the second target at lag 1 indicates whether the intervening interval of approximately 100 ms is sufficient for an attentional switch to be made. Lag-1 sparing suggests that a switch was successful, with little detriment to second target accuracy. Absence of lag-1 sparing, on the other hand, suggests that a switch could not be made, perhaps because it required too great a reconfiguration of the system. On this reasoning, identifying the factors that influence lag-1 sparing should reveal factors that are relevant to switching of attentional sets.

Visser, Bischof, and Di Lollo (1999) conducted a systematic examination of lag-1 sparing in the AB literature. Finding over 100 separate experiments in which an AB had been reported, they devised a taxonomic scheme that specified four major classes of switches in attentional set between targets. These included switches in categorical identity (e.g. digits vs. letters), in task (e.g. detection vs. identification), in modality (e.g. auditory vs. visual), and in spatial location (e.g. central vs. peripherally-located targets). Visser et al. (1999) then tabulated the frequency of lag-1 sparing as a function of the number and types of switches in attentional set.

The results of this survey are presented in Fig. 4.3. The major trends can be characterized as follows. Lag-1 sparing occurred reliably when there were no switches between targets, or when the switch was unidimensional, involving either task or category alone. Lag-1 sparing was not found with switches in location or with concurrent switches involving two or more categories (e.g. switches of both task and category). As was noted by Visser et al. (1999), the presence of multiple switches seemed to act synergistically to prevent lag-1 sparing. For example, task and category switches alone yielded lag-1 sparing in 76 per cent of cases, but when these switches were implemented concurrently, lag-1 sparing occurred in only 18 per cent of cases. These results point to an important relationship between lag-1 sparing and attentional switches that are implemented between targets. At one extreme, lag-1 sparing occurs reliably when there is no switch in attentional set. At the other extreme, lag-1 sparing never occurs when there are multiple switches in attentional set.

This pattern of results can be explained by a revised version of the two-stage theory (Chun and Potter 1995). As noted earlier, this theory suggests that lag-1 sparing occurs when the first

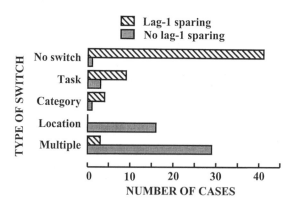

Fig. 4.3   Frequency of lag-1 sparing and no lag-1 sparing in the AB literature, plotted as a function of the type of task switch required between the first and second targets. The likelihood of lag-1 sparing decreases directly as a function of task switching (drawn on the basis of tabular data reported in Visser et al., 1999).

and second targets both fit through the same attentional gate. This gate opens upon the onset of the first target, and closes 150–200 ms later. If the second target enters the gate, both targets will become part of the same attentional episode, and both will gain access to high-level mechanisms required for stimulus identification and response planning. As is indicated by the large number of studies that have failed to find lag-1 sparing, however, temporal contiguity between targets cannot be the sole criterion for the second target to enter the same attentional gate. Rather the findings reviewed by Visser *et al.* suggest that an additional criterion is necessary in the form of a filter that controls access to the attentional gate.

The characteristics of this filter are indicated by the relationship between lag-1 sparing and the type of attentional switch implemented between the two targets. To enter the same attentional gate as the first target, the second target must arrive no more than 150–200 ms later, *and* it must match the characteristics of the input filter. If both of these conditions are met, the two targets will become part of the same attentional episode and will gain access to high-level processing mechanisms. In this event, lag-1 sparing will occur. If the second target does not pass the input filter, perhaps because it differs from the first target on some dimensions, an appropriate new filter needs to be set up. By the time this new filter is ready, the temporal gate will have closed, and the second target will remain in Stage 1, where it is vulnerable to masking. In this event, lag-1 sparing does not occur because the representation of the second target will have been replaced by the that of the mask by the time Stage 2 is again free.

To function optimally, the input filter must operate at a relatively low level in the visual system, where sensory signals flowing towards higher cortical centres can be monitored. At the same time, the filters must be responsive to rapid changes in attentional set and in response planning, which are functions normally associated with high-level structures in the prefrontal cortex (Goldman-Rakic 1987, 1988). These requirements suggest that the gating mechanism may be instantiated in more than one brain area, depending on the nature of the attentional set. For example, attending selectively to stimuli in motion would probably include gating circuitry in cortical areas V1 and V5. But, there are strong indications that input filters can be set up even more peripherally than in primary visual cortex. Among the likely candidates for subcortical gating mechanisms is the perigeniculate nucleus, which is a network interposed between the lateral geniculate nucleus and area V1. Sitting astride the main input pathways to the visual cortex, the perigeniculate nucleus is ideally suited for monitoring incoming sensory signals.

What is suggested by these considerations is an intelligent filtering system that is not restricted to physical features alone. Rather, the reconfiguring of a filter is probably part of a more comprehensive and goal-oriented process aimed at selecting those stimulus attributes that are useful for performing the task at hand. Such task-set reconfigurations have often been associated with higher cortical functioning (Monsell 1996). However, in view of the reciprocal neural connectivity between prefrontal cortex and other brain regions such as the perigeniculate nucleus, intelligent filtering, even at the earliest input level, seems to be a realistic possibility (Motter 1993; Roelfsema *et al.* 1998; Somers *et al.* 1999).

The idea of intelligent filtering invites a reversal in the usual conceptualization of events in visual information processing. Traditionally, such processing has been thought to occur in two discrete stages (cf. e.g. Neisser 1967). The first is an input stage of nearly unlimited capacity in which stimuli are encoded in parallel. This stage is considered to be largely stimulus-driven, and hence under exogenous control. The second is a resource-limited stage that processes stimuli serially under the direction of a unitary controlling mechanism. In contrast to the first stage, processing at this second stage is considered to be conceptually-driven, and hence under endogenous control.

By contrast with this viewpoint, the intelligent filtering proposed by Visser *et al.* suggests that the loci of exogenous and endogenous control be reversed. The input filters are reconfigured under the control of signals from brain areas as high as the prefrontal cortex, and their configuration determines whether any given stimulus can gain access to higher processing levels, or whether it is excluded. Given the role played by prefrontal cortex, it follows that the functioning of this input stage must be governed largely by endogenous, conceptually-driven signals.

Conversely, it would seem that the second stage is governed indirectly by exogenous stimulus-driven events. Consider that if an item passes the input filters, it gains access to this second stage, which consists of specialized feature- or task-specific processing modules. What processing module is activated by this item will depend on the nature of the stimulus. For example, stimuli in motion will activate motion-processing modules, but may not activate colour-processing modules. It follows from these considerations that the functioning of processing modules in this second, high-level stage must be governed in good part by exogenous, stimulus-bound factors.

The functional organization of the brain implied by this viewpoint is that of a number of independent, interconnected processors, each of which performs a specific function. These independent processors operate in parallel on incoming stimuli that have passed the input filters. This is consistent with view expressed by Allport, Antonis, and Reynolds (1972), Allport, Styles, and Hsieh (1994), Allport and Hsieh, this volume, Chapter 3, and Monsell (1996), and is congruent with the modular brain organization revealed in current neuroanatomical and neurophysiological studies (e.g. Felleman and Van Essen 1991; Posner and Raichle 1994).

Task switching and the attentional blink

From the foregoing discussion, two factors have emerged as important in the AB: backward masking of the second target, and attentional switching between targets. What is less clear from the extant literature is how these two factors combine in producing an AB. One way to examine their relationship is illustrated in Fig. 4.4, where the two factors are combined in a 2×2 table.

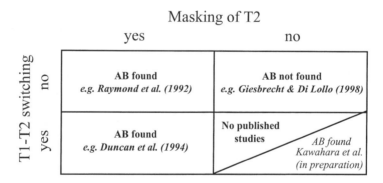

Fig. 4.4 A 2×2 table illustrating the results of AB experiments in which the influences of second-target masking and task switching were combined. The results for the experiment identified in the lower right cell are shown in Fig. 4.5

Readily noticeable in Fig. 4.4 is the absence of studies in which an attentional switch was implemented, but the second target was not masked. This lacuna is understandable because, in most studies of the AB, the second target has been masked for a practical reason: unless it was masked, accuracy was at ceiling, and an AB was no longer in evidence. An exception to this rule was seen when the second target, instead of being backward-masked, was degraded by visual noise (Brehaut *et al.*, 1999; Giesbrecht and Di Lollo 1998). Under these conditions, a lag-dependent deficit in second-target accuracy was not obtained, leading to the conclusion that backward masking of the second target was necessary for obtaining an AB.

This conclusion is consistent with the evidence from all other AB studies, but its generality is constrained in that it was based on a study in which there was no attentional switch between targets. Specifically, in the study by Giesbrecht and Di Lollo (1998), both targets were alphabetical characters presented in the same spatial location. For this reason, the conclusion reached in that study cannot be generalized to cases in which an attentional switch is required. What remains to be determined, then, is whether an AB occurs when an attentional switch is implemented between the two targets, but the second target is not followed by a masking stimulus.

A study designed for that purpose was carried out by Kawahara, Di Lollo, and Enns (1999). Its major objective was to determine whether an attentional switch between targets is sufficient to produce an AB, even if the second target is not masked. Observers viewed an RSVP stream of digit distractors displayed in the centre of the screen, and were required to identify both targets. The first target was always a letter, and the second was either another letter (no switch) or a diagonal line whose direction of tilt was to be identified (attribute switch). In addition, the second target was displayed either in the same location as the first (no switch), or unpredictably in one of 12 eccentric locations (location switch). To bring the level of second-target identification within a measurable range, the second target was embedded in random-dot noise, but was never followed by a mask. The design was a 2×2 between-subject factorial, in which the presence or absence of a switch in location was crossed with the presence or absence of a switch in attribute.

The results are shown in Fig. 4.5, averaged over 16 observers in each of the four groups. In agreement with the findings of Giesbrecht and Di Lollo (1998), no statistically significant AB was obtained when there was no switch between targets (Fig. 4.52). In contrast, a significant AB was obtained in all conditions involving attentional switches, despite the fact that the second target was never followed by a mask. In addition, the AB in Fig. 4.5d was reliably larger than that in Fig. 4.5b, suggesting that a concurrent switch in two dimensions (attribute and location) produced a larger AB than a switch in location alone. A compelling inference from these results is that an attentional switch between targets is sufficient for producing an AB, even when the second target is not followed by a mask. This information can be entered in the empty cell in Fig. 4.4, thus completing the 2×2 table. It is now apparent from Fig. 4.4 that an AB is obtained with any combination of second-target masking and attentional switching, but is not obtained when there is neither masking nor switching. We next consider the relative roles of attentional switching and masking in producing the AB.

On the role of attentional switching

It is clear from Fig. 4.4 that an attentional switch between targets leads to an AB even when the second target is not masked. Why is perception of the second target impaired by such a switch? Our answer to this question hinges on the assumption that an attentional switch entails a reconfiguration of the visual system. The idea of a system reconfiguration was outlined earlier

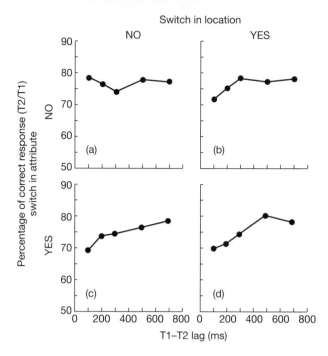

Fig. 4.5 Mean accuracy of second target (T2) identification when location switching (no, yes) and target attribute switching (no, yes) were combined orthogonally. No AB was observed in the no attribute–no location switching condition (upper left cell). From Kawahara, Enns and Di Lollo (in preparation).

in this chapter, and has been discussed extensively by Monsell (1996) and by Visser *et al.* (1999). In brief, what is reconfigured is an intelligent filtering system which operates on the visual input to select and encode task-relevant items, and to distinguish them from other incoming stimuli.

When the task requires an attentional switch between targets, the configuration of the system must be changed from one tuned to the characteristics of the first target to one tuned to those of second. If the second target arrives before the system has been suitably reconfigured, an adequately-encoded representation of the second target cannot be formed, and its identification is correspondingly impaired. Thus, an AB occurs even if the second target is not masked, because an attentional switch prevents a representation of it from being formed at the earliest stage. On this option, the AB shown in the lower two panels in Fig. 4.4 would be mainly attributable to the processing deficit arising from an attentional switch. Under these conditions, the presence or absence of a trailing mask would be largely irrelevant. Our reasoning is that if a representation of the second target was not formed to begin with, there would be nothing of consequence to be overwritten by a trailing mask.

This proposal of attentional reconfiguration leads to an interesting prediction. When an observer views an RSVP stream and is required to make an attentional switch between the first and second target, there should be no priming of a third target by the second target, as has been found by Shapiro, Driver, Ward, and Sorensen (1997) in a task involving no switch between the first and second targets. When there are no switches of this kind, the mechanisms activated by the second target prior to being masked can serve to prime the processing of the third target.

However, when there is a switch, a representation of the second target will not even be formed. Accordingly, the mechanisms for the second target are never activated and so no priming would be expected.

An alternative interpretation to that of attentional switching should be considered and dismissed. It may be suggested that an adequate representation of the second target is formed at an early processing stage even when the task requires an attentional switch. During the time period of the switch, however, that representation undergoes rapid decay and becomes undecipherable by the time the system has been suitably reconfigured. The ensuing AB would then be mediated by the decay of the second target. This option is disconfirmed by two findings in the literature. Reaction-time studies have shown that processing of the second target is substantially delayed while the system is processing the first target (Arnell and Duncan 1997; Jolicœur 1998). Were the representation of the second target to decay during this delay, a lag-dependent deficit should be expected. Yet no such deficit has been revealed in studies in which the second target was not masked, and the level of second-target identification in these cases was well below ceiling (Fig. 4.5a; Brehaut *et al.*, 1999; Giesbrecht and Di Lollo 1998). This all but rules out decay as a factor in the AB.

On the role of second-target masking

It is clear from Fig. 4.4 that backward masking of the second target invariably yields an AB, whether or not the task involves an attentional switch between targets. Despite the consistency of these results, a complete account of the role of masking is not entirely straightforward. When there are no attentional switches (Fig. 4.4, upper panels), the evidence is unambiguous: an AB is obtained only if the second target is followed by a mask. We have seen how this can be explained by the joint actions of backward masking and inattention. Namely, while the first target is being processed, the second target remains unattended and vulnerable to overwriting by the trailing mask.

On the other hand, when there is an attentional switch, an AB is invariably obtained, even if the second target is not followed by a mask (Fig. 4.4, lower panels). Clearly, masking cannot be regarded as necessary under these conditions. But does it still play a role in bringing about an AB? This question cannot be answered on the basis of extant information. There are at least two options. First, the mask may destroy the early representation of the second target, thus rendering the consequences on an attentional switch irrelevant. The second option hinges on the assumption, discussed above, that an attentional switch entails a reconfiguration of the visual system, resulting in failure to encode the second target. This, by itself, would produce a lag-dependent deficit, and would render a trailing mask irrelevant.

Concluding remarks

The view that has come to guide our research on the attentional blink is one in which 'attention' refers to the coordinated effort of a widely distributed group of specialized brain regions. These include lower-level registers of exclusively visual information (LGN and V1), higher-level centres involved in pattern analysis and visual object identification (visual temporal cortex) and supramodal centres involved in the selection, coordination and prioritization of information flow leading to action (thalamus, prefrontal cortex). Described this way, the approach bears a close resemblance to that of several other widely-held views of attention (Colby 1991; Desimone and Duncan 1995; Duncan 1996; Posner and Raichle 1994).

There are two features of our proposal, however, that we believe are unique. The first is a deliberate attempt to incorporate modern anatomical and physiological data concerning recipro- cal and re-entrant connections between specialized brain regions into our accounts and predic- tions regarding behaviour. For example, the idea of intelligent filtering proposed by Visser *et al.* (in press) is premised on the functional importance of re-entrant neural pathways from cortical centres of attentional control (e.g. frontal cortex, parietal cortex) to centres of attentional expression (e.g., LGN, area V1). In other work, we have described how we believe re-entrant neural connections are responsible for object-substitution masking effects, both in the atten- tional blink (Giesbrecht and Di Lollo 1998) and in metacontrast masking (Di Lollo *et al.*, 2000).

The second unique aspect of our proposal concerns its implications for what is being meas- ured in an AB experiment. We have tried to make it clear that the AB, observed in the absence of any masking of the second target (the right side in Fig. 4.4), is an index of the cost incurred by the visual system in reconfiguring itself for a task (the second target) that is different from a recently completed task (the first target). This means that the AB cannot, even in principle, be used to index the time required to perform the task of identifying the first target—the view espoused by the *attentional dwelltime* view of Duncan *et al.* 1994). The view of AB as a measure of dwelltime is premised on the assumption that the task of identifying the first target is indexed by its negative influence on the processing of the later-arriving second target. But as we have seen, whether or not an AB is observed for a given first target is dependent on whether an attentional switch must be made to process the second target. The dwelltime view would therefore force the untenable conclusion that the time required to process the first target is dependent on the processing of the later-arriving target. Our studies of the AB under conditions of no masking (Fig. 4.5) suggest it is much more likely that the AB is an index of the cost of reconfiguring brain mechanisms to process the second target.

The reconfiguration hypothesis is also at odds with the ABs being used as an index of second-target tasks that can be performed without attention (e.g. Joseph *et al.* 1997). This use of the AB is premised on the assumption that the processing of the second target is uninfluenced by the nature of the processing required for the first target. The question is simply whether the second target task can be performed concurrently with an attention-demanding first target task or not. But this assumption also flies in the face of the dependency of the AB on the relationship between the first and second target tasks. The second target task may, as we have shown, either result in an AB or not, depending on whether a switch in target location is involved (upper two panels in Fig. 4.5) or a switch in target attribute is involved (left side of Fig. 4.5). Again, the data are more consistent with the AB as a measure of system reconfigura- tion than as a measure of the attentional requirements of the second target.

It is also worth noting, at least in passing, that our emerging view of the AB as a measure of system reconfiguration has interesting implications in terms of ecological validity. Consider that almost all the AB research to date has required very little by way of reconfiguration for the observer to process each of the two targets. The large bulk of the studies have used target pairs consisting of two letters or two digits in the same spatial location. This can be contrasted with the typical sequence of tasks we encounter in everyday use of the attentional system. Many times during the day we are required to perform serial target identification tasks such as glanc- ing from the road to the speedometer while driving, glancing from our notes to a classroom of students while teaching, and alternating between visual attention to a ball and to opponents in athletic competitions. Considerations such as these encourage us to study the temporal dynamics of attention under a much wider range of task-switching conditions than has hitherto been the norm.

We readily acknowledge that the issues we have outlined in our discussion of the AB will be decided ultimately by subjecting them to the strong light of further empirical research. Our primary goal in this chapter has been to make our assumptions, intuitions, and research strategies as explicit as possible, both to ourselves and others, with the hope of making the empirical search for understanding an efficient one. Regardless of the eventual status of various theories concerning the attentional blink, it is already clear that enormous value has been gained in our practical understanding of the temporal dynamics of visual attention.

Acknowledgements

The preparation of this chapter, and the research reported in it, were sponsored by grants from the Natural Sciences and Engineering Research Council of Canada to the first and last authors.

References

Allport, A., Antonis, B., and Reynolds, P. (1992). On the division of attention: a disproof of the single channel hypothesis. *Quarterly Journal of Experimental Psychology*, **24**, 225–35.

Allport, A., Styles, E. A., and Hsieh, S. (1994). Shifting intentional set: exploring the dynamics of tasks. In C. Umiltà and M. Moscovitch (eds.), *Attention and performance XV*, pp. 421–52. Hillsdale, NJ: Erlbaum.

Arnell, K., and Duncan, J. (1997). Speeded responses or masking can produce an attentional blink. Poster session presented at the 38th Annual Meeting of the Psychonomic Society, Philadelphia, PA.

Bachmann, T., and Allik, J. (1976). Integration and interruption in the masking of form by form. *Perception*, **5**, 79–97.

Brehaut, J. C., Enns, J. T., and Di Lollo, V. (1999). Masking plays two roles in the attentional blink. *Perception and Psychophysics*, **61**, 1436–48.

Breitmeyer, B. G. (1984). *Visual masking: an integrative approach*. New York: Oxford University Press.

Chun, M. M. (1997). Temporal binding errors are redistributed in the attentional blink. *Perception and Psychophysics*, **59**, 1191–9.

Chun, M. M., and Potter, M. C. (1995). A two-stage model for multiple target detection in rapid serial visual presentation. *Journal of Experimental Psychology: Human Perception and Performance*, **21**, 109–27.

Colby, C. L. (1991). The neuroanatomy and neurophysiology of attention. *Journal of Child Neurology*, **6**, S90–S118.

Desimone, R., and Duncan, J. (1995). Neural mechanisms of selective visual attention. *Annual Review of Neuroscience*, **18**, 193–222.

Di Lollo, V., Enns, J. T., and Rensink, R. A. (2000). Competition for consciousness among visual events: the psychophysics of re-entrant visual processes. *Journal of Experimental Psychology: General*, **129**, 481–507.

Duncan, J. (1996). Cooperating brain systems in selective perception and action. In T. Inui and sJ. McClelland (eds.), *Attention and performance, Vol. XVI*, pp. 549–78. Cambridge, MA: MIT Press.

Duncan, J., Ward, R., and Shapiro, K. L. (1994). Direct measurement of attentional dwell time in human vision. *Nature*, **369**, 313–15.

Enns, J. T., and Di Lollo, V. (1997). Object substitution: a new form of visual masking in unattended visual locations. *Psychological Science*, **8**, 135–9.

Felleman, D. J., and Van Essen, D. C. (1991). Distributed hierarchical processing in primate visual cortex. *Cerebral Cortex*, **1**, 1–47.

Ganz, L. (1975). Temporal factors in visual perception. In E. C. Carterette and M. P. Friedman (eds), *Handbook of Perception*, Vol. 5, pp. 169–231. New York: Academic Press.

Giesbrecht, B. L., and Di Lollo, V. (1998). Beyond the attentional blink: visual masking by object substitution. *Journal of Experimental Psychology: Human Perception and Performance*, **24**, 1454–66.

Goldman-Rakic, P. S. (1987). Circuitry of primate prefrontal cortex and regulation of behavior by representational knowledge. In F. Plum and V. B. Mountcastle (eds), *Handbook of physiology*, Section 1, *The nervous system*, Vol. 5, *Higher functions of the brain*, pp. 373–417. Bethesda, MD: American Physiological Society.

Goldman-Rakic, P. S. (1988). Changing concepts of cortical connectivity: parallel distributed cortical networks. In P. Rakic and W. Singer (eds), *Neurobiology of the neocortex*, pp. 177–202. Berlin: Wiley.

Grandison, T. D., Ghirardelli, T. G., and Egeth, H. E. (1997). Beyond similarity: masking of the target is sufficient to cause the attentional blink. *Perception and Psychophysics*, **59**, 266–74.

Jolicœur, P. (1998). Modulation of the attentional blink by on-line response selection: evidence from speeded and unspeeded Task$_1$ decisions. *Memory and Cognition*, **26**, 1014–32.

Joseph, J. S., Chun, M. M., and Nakayama, K. (1997). Attentional requirements in a preattentive feature search task. *Nature*, **387**, 805–7.

Kahneman, D. (1968). Method, findings, and theory in studies of visual masking. *Psychological Bulletin*, **70**, 404–25.

Kawahara, J., Di Lollo, V. and Enns, J. T. (1999). Attentional blink: separating what we see from when we see it. *Investigative Ophthalmology & Visual Science*, **40**, S529 (abstract).

Kawahara, J., Enns, J. T., and Di Lollo, V. (in preparation). Location and attribute switching in the attentional blink.

Martin, J., Isaak, M. I., and Shapiro, K. L. (1995). Probe identification errors support an interference model of the attentional blink in rapid serial visual presentation. Poster presented at the Annual Meeting of the American Psychological Society, New York, NY.

Monsell, S. (1996). Control of mental processes. In V. Bruce (ed.), *Unsolved mysteries of the mind: tutorial essays in cognition*, pp. 93–148. Hove, Sussex: Erlbaum (UK) Taylor & Francis.

Moore, C. M., Egeth, H., Berglan, L. R., and Luck, S. J. (1996). Are attentional dwell times inconsistent with serial visual search? *Psychonomic Bulletin and Review*, **3**, 360–5.

Motter, B. C. (1993). Focal attention produces spatially selective processing in visual cortical areas V1, V2, and V4 in the presence of competing stimuli. *Journal of Neurophysiology*, **70**, 909–19.

Neisser, U. (1967). *Cognitive psychology*. New York: Appleton-Century-Crofts.

Posner, M. I., and Raichle, M. E. (1994). *Images of mind*. New York: Scientific American Library.

Potter, M. C., Chun, M. M., Banks, B. S., and Muckenhoupt, M. (1998). Two attentional deficits in serial target search: The visual attentional blink and an amodal task-switch deficit. *Journal of Experimental Psychology: Learning, Memory, and Cognition*, **24**, 979–92.

Raymond, J. E., Shapiro, K. L., and Arnell, K. M. (1992). Temporary suppression of visual processing in an RSVP task: an attentional blink? *Journal of Experimental Psychology: Human Perception and Performance*, **18**, 849–60.

Rensink, R. A., O Regan, J. K., and Clark, J. J. (1997). To see or not to see: The need for attention to perceive changes in scenes. *Psychological Science*, **8**, 368–73.

Roelfsema, P. R., Lamme, V. A. F., and Spekreijse, H. (1998). Object-based attention in the primary visual cortex of the macaque monkey. *Nature*, **395**, 376–81.

Scheerer, E. (1973). Integration, interruption and processing rate in visual backward masking. *Psychologische Forschung*, **36**, 71–93.

Seiffert, A. E., and Di Lollo, V. (1997). Low-level masking in the attentional blink. *Journal of Experimental Psychology: Human Perception and Performance*, **23**, 1061–73.

Shapiro, K. L. and Raymond, J. E. (1994). Temporal allocation of visual attention: Inhibition or interference? In D. Dagenbach and T. H. Carr (eds), *Inhibitory process in attention, memory, and language*, San Diego: Academic Press.

Shapiro, K. L., Raymond, J. E., and Arnell, K. M. (1994). Attention to visual pattern information produces the attentional blink in RSVP. *Journal of Experimental Psychology: Human Perception and Performance*, **20**, 357–71.

Shapiro, K. L., Driver, J., Ward, R., and Sorensen, R. E. (1997). Priming from the attentional blink: a failure to extract visual tokens but not visual types. *Psychological Science*, **8**, 95–100.

Simons, D. J. (1996). In sight, out of mind: when object representations fail. *Psychological Science*, **7**, 301–5.

Somers, D. C., Dale, A. M., Seiffert, A. E., and Tootell, R. B. H. (1999). Functional MRI reveals spatially specific attentional modulation in human primary visual cortex. *Proceedings of the National Academy of Science*, **96**, 1663–8.

Turvey, M. T. (1973). On peripheral and central processes in vision: inferences from an information-processing analysis of masking with patterned stimuli. *Psychological Review*, **81**, 1–52.

Visser, T. A. W., Bischof, W. F., and Di Lollo (1999). Attentional switching in spatial and non-spatial domains: evidence from the attentional blink. *Psychological Bulletin*, **125**, 458–69.

Ward, R., Duncan, J., and Shapiro, K. (1997). Effects of similarity, difficulty, and nontarget presentation on the time course of visual attention. *Perception and Psychophysics*, **59**, 593–600.

5

The attentional blink bottleneck

Pierre Jolicœur, Roberto Dell'Acqua, and Jacquelyn M. Crebolder

Abstract

Evidence suggesting that the attentional blink (AB) phenomenon is caused by a central processing bottleneck is reviewed. A bottleneck model of the psychological refractory period paradigm is used to motivate four major predictions concerning the patterns of results expected in dual-task experiments. This model, which was designed to explain results from experiments in which both tasks are speeded is modified to make predictions for experiments in which one task, or the other, is not speeded. This extention of the bottleneck model allows us to test four major predictions in the context of the AB paradigm. The bulk of the evidence is consistent with these predictions, and so we conclude that the AB phenomenon is caused by a processing bottleneck. The evidence also provides some contraints on the possible locus of the bottleneck in the information-processing stream.

The AB phenomenon can be succinctly defined as an increase in the difficulty of reporting a second (masked) target that follows (after a short delay) a first target that required immediate processing. In this chapter we review evidence supporting the view that the attentional blink (AB) phenomenon results from a bottleneck in the information-processing stream required to perform tasks designed to reveal the AB phenomenon.

The chapter is organized into six sections. In the first four we describe a major prediction of bottleneck models of dual-task interference (Pashler and Johnston 1989), followed by relevant evidence from AB and related paradigms. In the fifth section we review evidence pertinent to the issue of the locus of the bottleneck in the AB phenomenon. The last section presents some conclusions.

Carry-over to Task$_2$ of pre-bottleneck and bottleneck Task$_1$ manipulations

Prediction of postponement model

Figure 5.1 shows two sets of stage diagrams that illustrate one of the major predictions of one class of bottleneck models of dual-task interference. These diagrams are meant to capture task interactions in experiments designed to investigate the psychological refractory period (PRP). In these experiments, both tasks are speeded. In this class of models it is assumed that one stage of processing constitutes a processing bottleneck. This stage is labelled B in Fig. 5.1. This stage can only process information from one information-processing stream at a time. If the bottleneck stage is occupied by the processing required for one task, then it is unavailable for the processing in another task. Stages before the bottleneck (labelled A in the figure) or after it (labelled C) can go on in parallel with other stages of processing in a concurrent information-processing stream. For example, as shown in the figure, pre-bottleneck processing required for Task$_2$ (A$_2$) can overlap pre-bottleneck (A$_1$) and bottleneck (B$_1$) processing required for Task$_1$.

Processing at the bottleneck stage in Task$_2$ (B$_2$), however, cannot begin while the bottleneck is busy with Task$_1$. A period of waiting in the information-processing stream for Task$_2$ will result if pre-bottleneck processing finishes before the bottleneck is available. The probability and duration of such a period of waiting, called slack in scheduling theory, increases as the

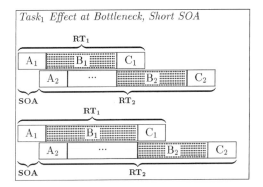

Fig. 5.1 Stage diagrams showing the predicted task interactions in the PRP paradigm for a manipulation in Task$_1$ that affects the bottleneck stage. At short SOA the onset of processing at the bottleneck stage in Task$_2$ is delayed by processing at the bottleneck in Task$_1$. This delay is longer when Task$_1$ processing at the bottleneck is longer.

stimulus onset asynchrony (SOA) between the target stimulus for Task$_1$ and the target stimulus for Task$_2$ is reduced. We call the first target T$_1$ and the second target T$_2$.

As can be seen in Fig. 5.1 across the top and bottom pair of stage diagrams, a longer duration of processing at stage B$_1$ will increase reaction times in Task$_2$. We refer to this dual-task interaction as a carry-over from Task$_1$ to Task$_2$. Clearly, the model predicts that this carry-over will only occur at short SOAs. Furthermore, carry-over would also be expected if the duration of pre-bottleneck processing in Task$_1$ were varied.

This class of models can be applied to the AB paradigm in the following way. Rather than measuring RT in Task$_2$, experiments investigating the AB effect measure accuracy in Task$_2$ to a briefly-presented and masked target stimulus (T$_2$). Figure 5.2 illustrates how the model in Fig. 5.1 can be extended to the AB paradigm. We suppose that the representation of T$_2$ decays (or is degraded by the mask) during the period of waiting. This is represented in the figure by the falling sequence of dots during the period of slack. We also assume that a longer period of waiting allows more decay to take place before the representation of T$_2$ can be stabilized by processing at the bottleneck stage (Arnell and Jolicœur 1999; Chun and Potter 1995; Dell'Acqua and Jolicœur 2000; Jolicœur 1998, 1999*b*,*c*; Jolicœur and Dell'Acqua 1998, 1999; Ross and Jolicœur 1999). With these assumptions, it is clear that a carry-over from a manipulation affecting the duration of bottleneck processing in Task$_1$ to accuracy in Task$_2$ is predicted by the model. More waiting, and hence more decay of T$_2$, would occur as the duration of bottleneck or pre-bottleneck processing in Task$_1$ is increased. As in the PRP paradigm, this carry-over is predicted to occur only at short SOAs.

Tests of the carry-over prediction

Several tests have confirmed the Task$_1$ carry-over to Task$_2$ predicted by postponement models in the AB paradigm. We consider first the experiments of Crebolder, Jolicœur, and Mclwaine, (2002). They presented upper-case letters using rapid serial visual presentation (RSVP). T$_1$ was an H, O, or S, presented in red; T$_2$ was an X or a Y, presented in white; the other (filler) letters were all white; the background was black. T$_2$ was presented in every trial, but T$_1$ was presented only on some trials. Task$_1$ was to report the identity of T$_1$, when T$_1$ was presented, or to press the space bar, when T$_1$ was not presented. Task$_2$ was to report the identity of T$_2$.

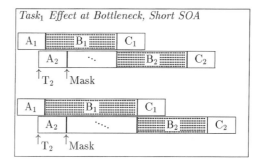

Fig. 5.2   Stage diagrams showing the predicted task interactions in the attentional blink paradigm for a manipulation in Task$_1$ that affects the bottleneck stage. At short SOA the onset of processing at the bottleneck stage in Task$_2$ is delayed by processing at the bottleneck in Task$_1$. This delay is longer when Task$_1$ processing at the bottleneck is longer. The representation of T$_2$ decays or is overwritten during the period of waiting if T$_2$ is masked; more decay results from a longer waiting period.

The manipulation of greatest interest in these experiments concerned the relative frequency of occurrence of the three possible letters used as targets in Task$_1$. For any given subject, one letter was shown 9 times in every 14 T$_1$-present trials, one was shown 4 times, and one was shown once. The assignment of letters to probability conditions was counterbalanced across subjects.

The results from two AB experiments are shown in Fig. 5.3. In both cases, the magnitude of the AB effect was largest for the least frequent T$_1$, intermediate for the intermediate-frequency T$_1$, and smallest for the most frequent T$_1$. The pattern of results was found both when Task$_1$ required a deferred unspeeded response (middle panel), as in the usual AB paradigm, and also when Task$_1$ was performed with a speeded response (top panel). Task$_2$ was always unspeeded. Results from the experiment in which Task$_1$ was speeded confirmed the expectation that RT$_1$ should increase as the relative frequency of T$_1$ decreased (Kornblum 1968). The results shown in the bottom panel are discussed in a subsequent section.

Jolicœur, Dell'Acqua, and Crebolder (2000) found clear-cut evidence for carry-over in a cross-modal AB experiment in which Task$_1$ required either a speeded two-alternative or four-alternative discrimination based on the pitch of a tone presented concurrently with an RSVP stream that contained T$_2$ (X vs. Y). As can be seen in Fig. 5.4, a larger AB effect was found when the duration of central processing in Task$_1$ was lengthened.

Jolicœur (1999c) also manipulated Task$_1$ variables in AB paradigms. In Experiment 1, Task$_1$ required either a two-alternative discrimination or a four-alternative discrimination, with a deferred unspeeded response. In Experiment 2, the same discriminations were made, but with a speeded response. Task$_2$ was the same X–Y two-alternative discrimination as in the Crebolder *et al.* (2002) experiments. An AB effect was found in both experiments. A larger AB effect was found when Task$_1$ required a four-alternative discrimination than when it required a two-alternative discrimination, but only when Task$_1$ was speeded. Equivalent AB effects in the two-alternative discrimination and four-alternative discrimination versions of Task$_1$ were found when Task$_1$ was unspeeded, as reported by Ward, Duncan, and Shapiro (1996), who had also found equivalent interference on Task$_2$ from a two-alternative and a four-alternative Task$_1$ in a modified AB paradigm. Jolicœur (1999c) argued that response selection in Task$_1$ was deferred when the response was deferred, but that it could not be deferred when a speeded response was

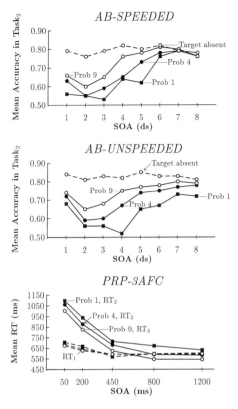

Fig. 5.3 Results from Crebolder *et al.* (2002). Top panel: Mean accuracy in Task$_2$ when Task$_1$ was speeded; for control trials (dashed line) and experimental trials (solid lines) for each level of relative signal frequency of T$_1$, as a function of T$_1$–T$_2$ SOA. Middle panel: results with Task$_1$ not speeded. Bottom panel: Mean response time in Task$_1$ and Task$_2$ of a PRP experiment in which relative signal probability of T$_2$ was manipulated in Task$_2$.

required in Task$_1$. Interference between Task$_1$ and Task$_2$ is only expected if the capacity-demanding processing required to perform each of these two tasks have the potential to overlap temporally. For these tasks and stimuli, it appears that subjects could defer one of the capacity-demanding operations (response selection) when Task$_1$ was not speeded.

The effects of requiring a speeded response in Task$_1$ on the carry-over effect highlights the importance of ensuring that manipulations in Task$_1$ affect processing taking place concurrently with Task$_2$ processing. It is possible that some previous failures to find carry-over effects were due to an ability to schedule capacity-demanding operations after the time-critical processing in Task$_2$ had run to completion (cf. e.g. Raymond *et al.* 1995; Ward *et al.* 1996, 1997).

In a third experiment, Task$_1$ was either a simple RT task or a speeded two-alternative discrimination based on letter identity (H vs. S; Jolicœur 1999c). A much larger AB effect was caused by the two-alternative discrimination task than by the simple RT task, providing another demonstration of the carry-over of a Task$_1$ manipulation on Task$_2$ performance in the AB paradigm. These results are shown in Fig. 5.5.

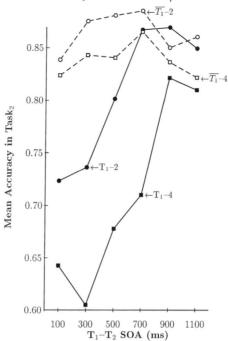

Fig. 5.4 Results from Jolicœur, Dell'Acqua, and Crebolder (2000). Mean accuracy in $Task_2$ for control trials (unfilled symbols, dashed lines) vs. experimental trials (filled symbols, solid lines) as a function of T_1–T_2 SOA, for two levels of $Task_1$ difficulty ($Task_1$ had either two response alternatives (T_1–2) or four response alternatives (T_1–4).

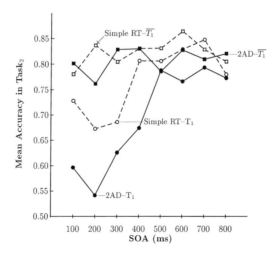

Fig. 5.5 Results from Jolicœur (1999c). Mean accuracy in $Task_2$ for control trials (square symbols) vs. experimental trials (round symbols) as a function of T_1–T_2 SOA, for two levels of $Task_1$ difficulty ($Task_1$ required either a simple RT response, or a speeded two-alternative discrimination (2AD).

Jolicœur (1998) provided additional converging evidence simply by comparing the magnitude of the AB effect across conditions in which Task$_1$ was unspeeded or speeded. At short SOAs, larger deficits were found in Task$_2$ when Task$_1$ was speeded than when it was unspeeded. If we assume that a longer period of central processing was required to perform Task$_1$ when it was speeded, then the results provide additional evidence of carry-over.

In summary, several experiments in which a longer period of central processing was required to process T$_1$ while T$_2$ was encoded have confirmed the carry-over prediction of postponement models in the context of the AB paradigm (Crebolder *et al.* 2002) as well as in AB paradigms modified by requiring a speeded response in Task$_1$ (Jolicœur 1998, 1999*b*, *c*; Jolicœur *et al.* 2000).

Correlation between Task$_1$ and Task$_2$ performance at short SOA

Prediction of postponement model

Figure 5.1 can also be used to understand a major prediction of bottleneck models of dual-task interference. As in previous sections, we begin with an application of bottleneck models to the PRP paradigm. At short SOAs between T$_1$ and T$_2$, variations in the duration of processing in the bottleneck stage (or at earlier stages) in Task$_1$ should cause corresponding variations in the duration of processing in Task$_2$. This is illustrated in the figure by supposing that the top and bottom pairs of stage diagrams represent different trials in which the duration of Task$_1$ bottleneck processing was either short (top pair) or long (bottom pair) due to randomly-occurring variance in the duration of processing. On the assumption that a significant proportion of the variance in Task$_1$ processing duration is due to variance in the bottleneck stage (Pashler 1994), the model leads to the prediction of a positive correlation between Task$_1$ RTs and Task$_2$ RTs. This correlation should be larger at shorter SOAs than at longer SOAs. This prediction has been confirmed in several experiments for the PRP paradigm (Pashler 1994).

Tests of correlation prediction

Jolicœur and his colleagues have examined whether a similar prediction holds for the dual-task interference observed in the AB paradigm. In order to estimate the duration of processing in Task$_1$, the paradigm was modified slightly by requiring an immediate and speeded response in Task$_1$, rather than a deferred and unspeeded response, as had typically been used in previous work. As was stated in the foregoing section, we suppose that the representation of T$_2$ decays or is overwritten by subsequent stimulation (Chun and Potter 1995; Giesbrecht and Di Lollo 1998) if it is not consolidated. The model in Fig. 5.2 leads to the prediction that a longer period of processing in Task$_1$ should be associated with a longer period of decay, and hence with lower accuracy, in Task$_2$.

The most direct way to estimate the duration of processing in Task$_1$ is to require an immediate and speeded response and to measure RT in Task$_1$ (RT$_1$). Jolicœur (1998) was the first to examine the effects of requiring a speeded response in Task$_1$ of the AB paradigm, which made it possible to test the prediction of postponement models that RT$_1$ and Task$_2$ accuracy should be negatively correlated at short SOAs, and that this relationship should attenuate at longer SOAs. Results from Jolicœur (1998, Experiments 2 and 3) are shown in Fig. 5.6. Task$_1$ was to decide whether a red letter, presented in an RSVP stream of white letters, was an H or an S. The response in Task$_1$ was a button press, and it was to be made as quickly as possible, while keeping errors to a minimum. Task$_2$ was to decide whether the RSVP stream contained an X or a Y, and the response in Task$_2$ was unspeeded and deferred until the end of the trial. Response times in Task$_1$ were first sorted based on RT$_1$, for each subject and each SOA, and aggregated into four bins defined by RT$_1$ quartiles. For each bin, accuracy in Task$_2$ was computed, and the resulting means are shown in Fig. 5.6. Accuracy in Task$_2$ was highest for trials that were in the first RT$_1$ quartile (labelled Q$_1$) and lowest for trials in the fourth RT$_1$ quartile (Q$_4$), as would be expected from a postponement process, and this relationship between RT$_1$ and Task$_2$ accuracy was not found at the longest SOA, where we expect the relationship to be weaker.

This relationship between speed of processing in Task$_1$ and the magnitude of the AB effect has been found in numerous experiments in which a direct measure of processing time in Task$_1$ has been available (Jolicœur 1998, Experiments 2 and 3; Jolicœur 1999a, Experiments 1–2;

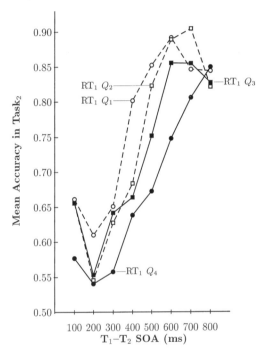

Fig. 5.6   Results from Jolicœur (1998). Mean accuracy in Task$_2$ depending on the speed of the response in Task$_1$ and on T$_1$–T$_2$ SOA. First quartile (Q$_1$) RT$_1$s were the shortest.

Jolicœur 1999c, Experiments 2 and 3; Jolicœur et $al.$ 2000, Experiment 1; Ross and Jolicœur 1999, Experiments 1–5). A similar relationship was also found in related experiments designed to provide converging evidence with experiments using the AB paradigm (see Dell'Acqua and Jolicœur 2000; Jolicœur and Dell'Acqua 1999).

The relationship between RT$_1$ and Task$_2$ accuracy is correlational in nature, and as such it is open to alternative interpretations. However, this correlation was predicted by postponement models of dual-task interference, and the absence of this correlation would have disconfirmed this class of models. The available published studies, however, provide consistent and strong confirmations of the prediction.

Absorption of pre-bottleneck Task$_2$ effect

Prediction of postponement model

Next we consider what is probably the most striking prediction of postponement models of dual-task interference: the absorption into slack of a Task$_2$ pre-bottleneck effect. Figure 5.7 illustrates the key concepts behind this prediction. It is assumed that a variable affecting the duration of stage A$_2$ (a pre-bottleneck stage in Task$_2$ processing) has been manipulated. As is shown in the figure, although the duration of stage A$_2$ has been affected by the manipulation, this has no effect on RT$_2$. The additional processing at stage A$_2$ has been absorbed into the period of slack. At long SOAs, however, there is no period of slack (no postponement), and so the effect of the manipulation is observed in the RT$_2$s. Thus, postponement models predict that effects of a variable altering the duration of a pre-bottleneck stage in Task$_2$ should be largest at longer SOAs and should become increasing smaller as SOA is reduced. This interaction is often referred to as an underadditive interaction, as SOA is reduced. This prediction of postponement models has been verified several times in the context of the PRP paradigm (cf. e.g. Pashler and Johnston 1989; Pashler 1994; Van Selst and Jolicœur 1997).

Tests of absorption into slack prediction

Jolicœur (1999d) performed an experiment to determine whether the type of encoding task that triggers an AB effect in the AB paradigm would also trigger a form of dual-task interference that could be characterized as a bottleneck. The logic of the experiment was to begin with a

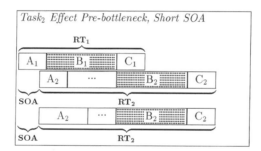

Fig. 5.7 Stage diagrams showing the predicted task interactions in the PRP paradigm for a manipulation in Task$_2$ that affects a pre-bottleneck stage. At short SOA the onset of processing at the bottleneck stage in Task$_2$ is determined by when the bottleneck is freed by Task$_1$ rather than by when Task$_2$ pre-bottleneck processing runs to completion. The Task$_2$ manipulation has no effect on RT$_2$.

classic AB paradigm, with a typical Task$_1$, but with Task$_2$ modified in such a way as to allow for a test of the absorption into slack prediction. The first task, not speeded, required visual encoding, and had been used in earlier experiments to investigate the AB phenomenon. T$_1$ was either an H, an O, or an S, presented in red in a stream of white letters shown using RSVP, and was reported after the end of the RSVP stream. Task$_2$ was a speeded discrimination between two possible T$_2$s (X vs. Y). T$_2$ was always the last item in the RSVP stream. In addition, the contrast of T$_2$ was manipulated (two levels: normal contrast vs. low contrast). Response times in Task$_2$ were measured. The most important results concern the pattern of interaction of this contrast manipulation in Task$_2$ with SOA. This manipulation, in the context of the PRP paradigm, has produced underadditive interactions with decreasing SOA in numerous earlier experiments (e.g. Van Selst and Jolicœur 1997).

The results are shown in Fig. 5.8. The contrast effect was strongly underadditive with decreasing SOA. At longer SOAs, low-contrast T$_2$s caused longer RTs than high-contrast T$_2$s. As SOA was reduced the contrast effect decreased, and became very small at the shortest SOA. These results provide strong evidence that the interference in the AB paradigm takes the form of a processing bottleneck and that the bottleneck is after a stage of processing affected by stimulus contrast. The additional processing time caused by the low-contrast stimulus was not strongly reflected in RT$_2$ at short SOA, presumably because it was absorbed into a period of slack in the processing stream for Task$_2$.

The underadditive interaction between contrast and SOA is consistent with a bottleneck interpretation of the interference observed in the AB paradigm, although other forms of interference, such as capacity sharing, can also predict this form of interaction. Figure 5.8 also shows results from a companion PRP experiment that was in all ways identical to the AB exper-

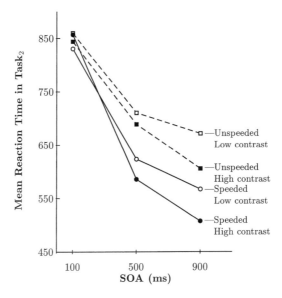

Fig. 5.8   Results from Jolicœur (1999*d*). Mean reaction time, in milliseconds, for Task$_2$ responses depending on the SOA between T$_1$ and T$_2$, on whether the response in Task$_1$ was speeded or unspeeded, and on the contrast of T$_2$.

iment described in the foregoing paragraphs, except that a speeded response was also required in Task$_1$. The two experiments generally produced strikingly similar results, with both sets of results exhibiting clear and similar underadditive interactions between contrast and SOA. Interestingly, RT$_2$s were actually longer when Task$_1$ was not speeded relative to when Task$_1$ was speeded, a result that has been replicated in other similar experiments. We discuss there results further in the last paragraphs of this chapter.

Converging evidence for a bottleneck account of the interference in the AB paradigm was provided by Ruthruff, Johnston, and McCann (1998). T_1 was an S or a T presented in an RSVP stream (100 ms/item). A speeded response to T_2 was required, and T_2 was presented either just above or below an RSVP stream that contained T_1 and continued after the presentation of T_2. Rather than manipulating the contrast of T_2, the letters that could be used as T_2 characters were distorted, as illustrated in Fig. 5.9. T_2 could be either an A or an H, shown in a normal format, as in the left pair of letters in Fig. 5.9, or distorted, as in the right pair in Fig. 5.9. Task$_2$ was to decide, as quickly as possible, whether T_2 was an A or an H. As expected, distorted letters produced longer RT_2s than normal letters. The most important finding, however, was that the effect of letter distortion was largest at longer SOAs and that it decreased as SOA was reduced. That is, an underadditive interaction was found between letter distortion and decreasing SOA.

Ruthruff *et al.* (1998) interpreted their results as evidence for a bottleneck after the stage at which visual patterns are classified into letter categories, on the assumption that the letter-distortion effect occurred at a stage where letters are categorized.

The results of Ruthruff *et al.* (1998) and of Jolicœur (1999*d*) both confirm the prediction of absorption into slack made by bottleneck models of dual-task interference in the context of the

Fig. 5.9 Illustration of the normal (left) and distorted (right) As and Hs used by Ruthruff, Johnston, and McCann (1998).

AB paradigm. The results also provide evidence concerning the locus of the bottleneck, which will be discussed in more detail in a subsequent section.

Additivity of bottleneck and post-bottleneck Task$_2$ effect

Prediction of postponement model

Another prediction of bottleneck models is that changes in the duration of processing in Task$_2$ at the bottleneck stage, or after the bottleneck stage, should produce additive effects with SOA on response times in Task$_2$. Unlike effects taking place before the bottleneck (Fig. 5.7), effects at the bottleneck stage in Task$_2$ are expected to be fully reflected in RT_2 even at short SOAs, as is shown in Fig. 5.10. Because the effect size of bottleneck or post-bottleneck manipulations is expected to be the same at all SOAs, the model predicts additive effects with SOA.

Tests of additivity prediction for bottleneck and post-bottleneck effects

Additivity of Task$_2$ stimulus–response compatibility and SOA in a PRP paradigm was reported by McCann and Johnston (1992). Jolicœur (1999*d*) examined the additivity prediction in the AB paradigm by manipulating stimulus–response compatibility in a speeded second task that

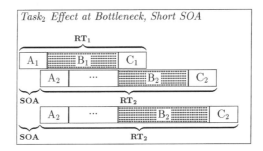

Fig. 5.10 Stage diagrams showing the predicted task interactions in the PRP paradigm for a manipulation in Task$_2$ that affects the bottleneck stage. Even at short SOA RT$_2$s reflect the full magnitude of the manipulation.

followed a typical unspeeded first task known to produce a robust AB effect in a typical AB paradigm (e.g., in Crebolder *et al.* 2002).

T$_1$ was an H, O, or S, presented in red in an RSVP stream of white letters. Task$_1$ was to report the identity of T$_1$, at the end of the trial, without speed pressure. The RSVP stream always ended with T$_2$, which was an L or an R. Task$_2$ was a speeded two-alternative discrimination based on the identity of T$_2$. The subjects were instructed to think of the L and R as meaning 'left' and 'right,' respectively. Responses in Task$_2$ were performed with the index and middle fingers of the left hand, using response buttons on the bottom row of a computer keyboard. Every subject performed two blocks of trials. In one block they responded to the L with the left response button and to the R with the right response button. This was the compatible stimulus–response condition. In the other block they responded to the L with the right button and to the R with the left button. This was the incompatible stimulus–response condition. The incompatible condition was expected to produce a longer mean response time than the compatible condition.

If we assume that the stimulus–response compatibility manipulation has an effect at response selection (McCann and Johnston 1992), and if response selection is at or after a processing bottleneck that causes the AB effect, then the compatibility manipulation should have additive effects with SOA in this paradigm. That is, the difference between the mean RT across the incompatible and compatible conditions should be the same at all SOAs. The contrast of T$_2$ was also manipulated (low vs. high). This manipulation should produce an underadditive interaction with decreasing SOA, replicating the results shown in Fig. 5.8 for the unspeeded Task$_1$ condition.

The logic of this experiment demanded that subjects perform Task$_1$ (encoding and later reporting a single letter) with high accuracy, otherwise subjects could trade off accuracy in Task$_1$ in order to process T$_2$. For this reason, only subjects who maintained an accuracy of $87\frac{1}{2}$ per cent or better in Task$_1$ were included in the final analyses. An accuracy level of 85 per cent was also required in Task$_2$. The results of 14 subjects who met these requirements are shown in Fig. 5.11. As expected, the incompatible stimulus–response condition produced a longer mean RT than the compatible condition. Although there was a tendency for the compatibility effect to become smaller with decreasing SOA, the interaction between SOA and compatibility was not statistically significant ($p > 0.13$). The contrast effect, on the other hand, was clearly underadditive with decreasing SOA: the contrast effect was largest at the longest SOA and smaller at two

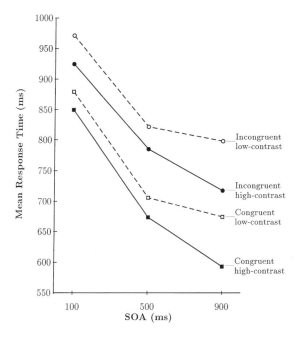

Fig. 5.11 Results from Jolicœur (1999*d*). Mean reaction time, in milliseconds, for Task$_2$ responses depending on the SOA between T$_1$ and T$_2$, on the contrast of T$_2$, and on the stimulus–response compatibility in Task$_2$.

shorter SOAs (p <0.012). There was a large change in the measured contrast effect across the middle and longest SOAs, and no change across the middle and shortest SOAs. The residual contrast effect at the two shorter SOAs was significant (p < 0.002).

The stimulus–response compatibility results were not as clear as I would have wished. Although the interaction between compatibility and SOA was not significant, there was a tendency for the compatibility effect to decrease as SOA was reduced; there was a 124 ms congruency effect at 900 ms SOA, a 114 ms effect at 500 ms SOA, and an 84 ms effect at 100 ms SOA. Such a trend, even if not significant, should probably not be ignored (the trend was significant if subjects with lower accuracy in Task$_1$ were included in the analysis). Interpreted in the context of bottleneck models, this trend suggests that some of the stimulus-compatibility effect is at a stage that is before the bottleneck. The large residual effect at the shortest SOA, however, suggests that much of the effect is at or after the bottleneck. Generally, the results are consistent with the view that the AB bottleneck coincides with a stage affected by stimulus–response compatibility, but that the bottleneck may extend to an even later stage.

The absorption into slack of the contrast effect was expected given the results shown in Fig. 5.8 and the results of Ruthruff *et al.* (1998). This suggests that a large component of the contrast effect was in a stage that was before the AB bottleneck. Interestingly, not all of the contrast effect was absorbed, leaving a significant residual effect of about 36 ms at the two shorter SOAs (also evident in Fig. 5.8 for the unspeeded condition, but the effect there was 15–20 ms). According to postponement models, this would suggest that the contrast manipulation also had an effect in or after the AB bottleneck.

Another line of evidence that appears consistent with the additivity prediction has been provided by Maki, Frigen, and Paulson (1997). T_2 stimuli (words) were preceded by neutral or semantically-related prime words at various points in the RSVP sequence. The results of several experiments revealed additive effects of priming and SOA on accuracy in Task$_2$. Although there are scaling issues when we interpret additivity or interactions for accuracy results, because accuracy scores may not be on an interval scale, the results of Maki *et al.* provide some evidence that appears to confirm the additivity prediction. Given that the variable manipulated involved semantic relationships, which presumably occur fairly late in the information-processing stream, the results suggest that the AB bottleneck occurs at a stage at which semantics is extracted, or at a later stage, but not at an earlier stage.

Locus of the bottleneck

Several converging lines of evidence suggest that the locus of dual-task interference in the AB paradigm (i.e., the locus of the bottleneck) is relatively late in the information-processing sequence. Luck and his colleagues have used evoked response potential (ERP) techniques to constrain the locus of the effect (cf. e.g. Luck *et al.* 1996). They focused on the N400 ERP response, which is generally evoked when a stimulus does not match the present semantic context. Each trial began with the presentation of a word. The purpose of this word was to create a semantic context for the remainder of the trial. Then an RSVP stream was presented. Each frame of the RSVP stream contained several characters. One of the frames, T_1, was a repeated string of digits (e.g., 22222). Subjects were asked to report the digit string shown in T_1 as 'even' or 'odd'. At different lags following T_1 a word, T_2, was also presented. This word could either match or mismatch the semantic context created by the word shown at the beginning of the trial. As expected, at long SOAs, when subjects could report the identity of T_2, T_2 words that did not match the semantic context of the initial word elicited a robust N400 ERP response relative to words that matched the semantic context. More importantly, at a lag where a large proportion of T_2 words could not be overtly reported, because the behavioural data at this lag exhibited a robust AB effect, the N400 response was just as large as at longer lags. The results suggest that the stimuli were, in some sense, deeply processed, despite the poor performance in the overt responses. The suggestion is that the AB bottleneck is relatively late in the information-processing stream, occurring after stimuli such as words have activated representations in semantic memory.

Jolicœur and his colleagues (e.g., Arnell and Jolicœur 1999; Jolicœur 1998, 1999*a,b,c*; Jolicœur and Dell'Acqua 1998, 1999; Ross and Jolicœur 1999), as had Chun and Potter (1995), proposed that the interference occurs at a stage of processing where perceptually identified representations are transferred to a form of durable storage (Coltheart 1984). They called the transfer process short-term consolidation (Jolicœur and Dell'Acqua 1998). According to Chun and Potter (1995) the consolidation of T_2 is postponed by the consolidation of T_1. Jolicœur and his colleagues made the same supposition, but in addition they hypothesized that other processes can interfere with, and be disrupted by, short-term consolidation. For example, Jolicœur and Dell'Acqua (1998, 1999) hypothesized that response selection in a speeded pitch-discrimination task is postponed by the short-term consolidation process required to encode a visually-presented letter. Indeed, response times to a tone, presented after a visual character to be encoded into memory, increased as the SOA between the character and the tone was reduced, but only if the character had to be remembered (Jolicœur and Dell'Acqua 1998, Experiment 7).

The results of Ruthruff *et al.* (1998) are also consistent with a locus of AB interference after letter identification, given that the letter-distortion manipulation, which presumably affects the duration of letter identification, produced an underadditive interaction with decreasing SOA. Thus, letter identification had to occur before the bottleneck.

The results of Jolicœur *et al.* (2000) and Crebolder *et al.* (2002) both suggest that some portion of the AB effect may occur at the same stage as the locus of the PRP effect. Recall that Jolicœur *et al.* (1999c) manipulated the number of stimulus and response alternatives in Task$_1$ of a speeded AB experiment and found that this manipulation carried over into the accuracy results in Task$_2$ (Fig. 5.4). Crebolder *et al.* (2002) found Task$_1$ carry-over for a stimulus probability manipulation (Fig. 5.3). These results suggest that effects of the number of S–R alternatives and of relative signal probability occur either at or before the bottleneck producing the AB effect.

Jolicœur *et al.* (2000) and Crebolder *et al.* (2002) also manipulated these variables (probability and number of S–R alternatives) in Task$_2$ of a standard PRP paradigm, in order to determine the locus of probability and of the number of S–R alternatives relative to the PRP bottleneck. Additive effects of probability and SOA (see Fig. 5.3, bottom panel) and additive effects of number of S–R alternatives and SOA were found, indicating that these variables had effects that were either at or after the bottleneck stage in the PRP paradigm. If we assume that the stage of processing affected by the manipulations was the same in the AB and PRP paradigms (an assumption that seems especially sound for the effects of S–R quantity), the results imply that there is a bottleneck in the AB paradigm that is either at the same stage as in the PRP paradigm, or later. Although it is logically possible, it seems unlikely that the AB bottleneck would come after the PRP bottleneck. Thus, the most likely interpretation of these results is that the two bottlenecks coincide, at least under some conditions.

The results shown in Fig. 5.11, involving the joint effects of SOA and stimulus–response compatibility, also suggest a late locus of interference in the AB paradigm. The large residual stimulus–response compatibility effect at the shortest SOA suggests that the bottleneck is at least as late as the stage affected by stimulus–response compatibility. This stage is probably response selection. The trend for a decreasing compatibility effect as SOA was reduced suggests the possibility that the AB bottleneck might extend beyond response selection.

Summary and conclusions

In this chapter we reviewed findings relevant to the issue of the nature of the dual-task interference observed as the AB phenomenon. The review was organized around four major predictions of bottleneck models of dual-task interference. We began with predictions for response-time results in the context of the PRP paradigm (Figs 5.1 and 5.7; Pashler and Johnston 1989) and extended the model to generate predictions for results from the AB paradigm or close variants (Fig. 5.2). For each of the four major predictions:

- carry-over to Task$_2$ of pre-bottleneck and bottleneck Task$_1$ manipulations,
- correlation between Task$_1$ and Task$_2$ performance at short SOA,
- absorption of pre-bottleneck Task$_2$ effect, and
- additivity of bottleneck and post-bottleneck Task$_2$ effect,

we found significant converging evidence, often from different labs, confirming the prediction either in the AB paradigm or in a close variant in which $Task_2$ required a speeded response, but in which $Task_1$ was unchanged from that used in the usual AB paradigm. We assume that paradigmatic variations in which $Task_2$ requires a speeded response provide critical evidence for the processing consequences of the encoding operations required to perform $Task_1$, which are the trigger for the AB phenomenon.

The tests and confirmation of predictions of bottleneck models suggest strongly that the interference in the AB paradigm takes the form of a processing bottleneck. One important issue concerns the nature of that bottleneck. Is the bottleneck all-or-none, or is the bottleneck a functional bottleneck, but one that allows for the sharing of a restricted pool of resources? Theoretical work by Tombu and Jolicœur (2001) shows that there is a class of central capacity sharing models that makes the same four key predictions as the all-or-none bottleneck model. When sharing takes place, however, $Task_1$ suffers at short SOAs. We examine this issue next by examining $Task_1$ performance in a typical AB paradigm.

The results from $Task_1$ are generally unaffected by the time at which T_2 is presented relative to T_1 (i.e., by SOA). One might think to argue that there is a channel that has enough capacity for just a little more than one item, such that $Task_1$ processing can be 'protected' from the allocation of some resources to the processing of T_2. This line of argument runs into difficulty, however. Accuracy in $Task_1$ is often well below ceiling, suggesting that all the capacity in the channel is required to process T_1. Therefore, devoting some resources to the processing of T_2 should take away capacity for T_1, which should produce a drop in accuracy in $Task_1$.

Representative results examining effects of SOA in $Task_1$ and $Task_2$ are shown in Fig. 5.12, in which results of Jolicœur (1998), Experiment 1, are replotted to show $Task_1$ accuracy. T_1 was a red H or S embedded in an RSVP steam (100 ms/item) of white letters. T_2 was an X or a Y. T_1 was presented on half the trials (experimental trails) and not shown on the other half (control trials). $Task_1$ was to report whether T_1 was an H or an S, or to report that no red letter had been presented. $Task_2$ was to report the identity of T_2. $Task_1$ and $Task_2$ were unspeeded, and responses were deferred to the end of the trial. Clearly, the appearance of T_2 had no effect on mean accuracy in $Task_1$. Furthermore, as can be seen from the error bars, which represent 95 per cent confidence intervals for within-subject designs (Loftus and Masson 1994), accuracy in $Task_1$ was significantly below 100 per cent.

The paradigmatic AB task can be thought of as a dual-task situation that places high demands on the subjects' abilities to attend, encode, and remember information presented under challenging perceptual conditions. It is not surprising that a complete understanding of the paradigm will be likely to involve elements from many different levels of analysis and many subsystems. It seems likely, for example, that subjects have some ability to trade off accuracy in $Task_1$ for accuracy in $Task_2$, at a relatively macro level, perhaps by preparing more for one task than the other (De Jong and Sweet 1994; Pashler 1994). At this level of analysis, it is possible that capacity-sharing models might capture significant aspects of the results, although no formal attempts to do so have been proposed to date. We tried to show, for a different level of analysis, at a more micro level, that the processing required to perform $Task_1$ of the AB paradigm is sharply capacity-demanding, and appears to create a processing bottleneck for other concurrent cognitive operations. This bottleneck is particularly apparent when $Task_2$ is similar to $Task_1$, and presumably requires similar encoding and mnemonic processes. Evidence for a bottleneck can also be found, however, even when $Task_1$ and $Task_2$ are quite different, as when $Task_1$ requires a speeded response to a tone and $Task_2$ an unspeeded deferred report of a letter embedded in an RSVP stream (Jolicœur 1998; Jolicœur et al. 2000).

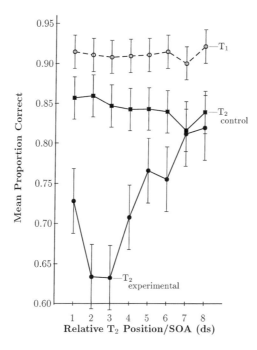

Fig. 5.12 Results from Jolicœur (1998), Experiment 1. Mean accuracy for Task$_2$ control trials (filled squares, solid line), Task$_2$ experimental trials (filled circles, solid line), and Task$_1$ accuracy (unfilled circles, dashed line) as a function of T_1–T_2 SOA. The error bars are 95 per cent confidence intervals for within subject designs.

Some of the evidence that we reviewed suggests that the AB bottleneck is similar to the bottleneck implicated in the PRP paradigm (Pashler 1994) in terms of the degree of interference caused by the bottleneck and by its general location in the information-processing stream (Crebolder *et al.* 2002); Jolicœur and Dell'Acqua 1998; Jolicœur *et al.* 2000). For example, the results shown in Fig. 5.8 show that the interference on Task$_2$ caused by encoding T_1 for a deferred unspeeded report (AB paradigm) is quite similar in magnitude to that found when both tasks are speeded (i.e., in the PRP paradigm).

There were also obvious differences, however. First, although similar, the slope of the SOA effect was somewhat larger in the PRP paradigm than in the AB paradigm. Second, at longer SOAs, RT$_2$s were longer when Task$_1$ was unspeeded than when Task$_1$ was speeded. However, at short SOAs, there was no difference in RT$_2$s across the two Task$_1$ conditions. One interpretation of these results is that, at short SOA, there was equivalent interference on Task$_2$ produced by the two Task$_1$ conditions. As SOA was increased, the need to maintain a durable representation of T_1 for later report in the unspeeded Task$_1$ condition may have created additional interference on Task$_2$. A durable memory for T_1 is not required after the response in the speeded version of Task$_1$. Other interpretations of the results are possible. Our main point is that the different Task$_1$ requirements in the speeded condition versus the unspeeded condition very likely recruit different processing mechanisms (for example, on-line response selection in the speeded condition, short-term consolidation in the unspeeded condition). Thus, although there are some very interesting similarities between the interference observed in the AB and PRP paradigms, it is clear that there are also some differences. More work will be required to understand fully these similarities and differences, and the mechanisms that underlie both of these paradigms. The evidence reviewed in this chapter highlights one important similarity, namely that the interference in both paradigms appears to be caused by a central processing bottleneck.

Acknowledgements

We thank Marg Ingleton for technical support. This work was supported by research grants from NSERC and HFSP.

References

Arnell, K. M., and Jolicœur, P. (1999). The attentional blink across stimulus modalities: evidence for central processing limitations. *Journal of Experimental Psychology: Human Perception and Performance*, **25**, 630–48.

Chun, M. M., and Potter, M. C. (1995). A two-stage model for multiple target detection in rapid serial visual presentation. *Journal of Experimental Psychology: Human Perception and Performance*, **21**, 109–27.

Coltheart, M. (1984). Sensory memory: a tutorial review. In H. Bouma and D. G. Bouwhuis (eds), *Attention and performance X: Control of language processes*, pp. 259–85. Hillsdale, NJ: Erlbaum.

Crebolder, J. M., Jolicœur, P., and McIlwaine, J. (2002). Loci of signal probability effects and of the attentional blink bottleneck. Unpublished manuscript, Defence and Civil Institute of Environmental Medicine, Toronto, Ontario, Canada.

De Jong, R., and Sweet, J. B. (1994). Preparatory strategies in overlapping-task performance. *Perception and Psychophysics*, **55**, 142–51.

Dell'Acqua, R., and Jolicœur, P. (2000). Visual encoding of patterns is subject to dual-task interference. *Memory and Cognition*, **28**, 184–91.

Giesbrecht, B. L., and Di Lollo, V. (1998). Beyond the attentional blink: visual masking by object substitution. *Journal of Experimental Psychology: Human Perception and Performance*, **24**, 1454–66.

Jolicœur, P. (1998). Modulation of the attentional blink by on-line response selection: evidence from speeded and unspeeded $Task_1$ decisions. *Memory and Cognition*, **26**, 1014–32.

Jolicœur, P. (1999a). Restricted attentional capacity between sensory modalities. *Psychonomic Bulletin and Review*, **6**, 87–92.

Jolicœur, P. (1999b). Dual-task interference and visual encoding. *Journal of Experimental Psychology: Human Perception and Performance*, **25**, 596–616.

Jolicœur, P. (1999c). Concurrent response selection demands modulate the attentional blink. *Journal of Experimental Psychology: Human Perception and Performance*, **25**, 1097–113.

Jolicœur, P. (1999d). A processing bottleneck causes the attentional blink. Unpublished manuscript, University of Waterloo, Waterloo, Ontario, Canada.

Jolicœur, P., and Dell'Acqua, R. (1998) The demonstration of short-term consolidation. *Cognitive Psychology*, **36**, 138–202.

Jolicœur, P., and Dell'Acqua, R. (1999). Attentional and structural constraints on visual encoding. *Psychological Research*, **62**, 154–64.

Jolicœur, P., Dell'Acqua, R., and Crebolder, J. M. (2000). Multitasking performance deficits: forging some links between the attentional blink and the psychological refractory period. In S. Monsell and J. Driver (eds), *Attention and performance XVIII: Control of cognitive processes*, pp. 309–30. Cambridge, MA: MIT Press.

Kornblum, S. (1968). Serial-choice reaction time: inadequacies of the information hypothesis. *Science*, **159**, 432–4.

Loftus, G. F., and Masson, M. E. J. (1994). Using confidence intervals in within-subject designs. *Psychonomic Bulletin and Review*, **1**, 476–90.

Luck, S. J., Vogel, E. K., and Shapiro, K. L. (1996). Word meaning can be accessed but not reported during the attentional blink. *Nature*, **382**, 616–18.

McCann, R. S., and Johnston, J. C. (1992). Locus of the single-channel bottleneck in dual-task interference. *Journal of Experimental Psychology: Human Perception and Performance*, **18**, 471–84.

Maki, W. S., Frigen, K., and Paulson, K. (1997). Associative priming by targets and distractors during rapid serial visual presentation: Does word meaning survive the attentional blink? *Journal of Experimental Psychology: Human Perception and Performance*, **23**, 1014–34.

Pashler, H. (1994). Dual-task interference in simple tasks: data and theory. *Psychological Bulletin*, **116**, 220–44.

Pashler, H., and Johnston, J. C. (1989). Chronometric evidence for central postponement in temporally overlapping tasks. *Quarterly Journal of Experimental Psychology*, **41A**, 19–46.

Raymond, J. E., Shapiro, K. L., and Arnell, K. M. (1995). Similarity determines the attentional blink. *Journal of Experimental Psychology: Human Perception and Performance*, **21**, 653–62.

Ross, N. E., and Jolicœur, P. (1999). Attentional blink for color. *Journal of Experimental Psychology: Human Perception and Performance*, **25**, 1483–94.

Ruthruff, E., Johnston, J. C., and McCann, R. S. (1998). The attentional blink does not prevent character identification. Poster presented at the 39th Annual Meeting of the Psychonomic Society, Dallas Texas. Published abstract in *Abstracts of the Psychonomic Society*, **3**, 60.

Tombu, M. and Jolicœur, P. (2001). A central capacity sharing model of dual-task performance. Manuscript submitted for publication, Department of Psychology, University of Waterloo, Waterloo, Ontario, Canada.

Van Selst, M., and Jolicœur, P. (1997). Decision and response in dual-task interference. *Cognitive Psychology*, **33**, 266–307.

Ward, R., Duncan, J., and Shapiro, K. (1996). The slow time-course of visual attention. *Cognitive Psychology*, **30**, 79–109.

Ward, R., Duncan, J., and Shapiro, K. (1997). Effects of similarity, difficulty, and nontarget presentation on the time course of visual attention. *Perception and Psychophysics*, **59**, 593–600.

6

Perceptual and central interference in dual-task performance

Eric Ruthruff and Harold E. Pashler

Abstract

Humans often experience difficulty when asked to perform multiple tasks at the same time. Two of the better-known forms of dual-task interference are the attentional blink (AB) effect and the Psychological Refractory Period (PRP) effect. These phenomena have traditionally been studied independently, using divergent methodologies and different dependent measures. The purpose of this chapter is to explore the possibility that these dual-task phenomena might reflect the same underlying processing limitation—a central bottleneck. We also discuss how AB and PRP effects are related to other phenomena such as repetition blindness and movements of spatial attention across visual space.

Most observers can reliably find and report a single target object (e.g. a digit) embedded within a rapid serial visual presentation (RSVP) display, even when each object is presented for only a tenth of a second. After detecting one target, however, observers frequently have difficulty detecting targets that appear within the next several hundred ms (cf. e.g. Broadbent and Broadbent 1987; Chun and Potter 1995; Raymond *et al.* 1992; Shapiro *et al.* 1994). Because this dual-task interference is not attributable to sensory masking, it has been dubbed the 'attentional blink' (AB) effect.

This phenomenon at least superficially resembles a type of dual-task interference known as the Psychological Refractory Period (PRP) effect. The PRP effect occurs when subjects are asked to make speeded responses to two different stimuli. Whereas subjects usually respond quickly to the first stimulus, they tend to respond slowly to the second stimulus when it appears shortly after the first (cf. e.g., Vince 1948; Welford 1952; see Pashler and Johnston 1998 for a review).

The PRP and AB designs differ in several respects. The typical PRP design requires simple perceptual discriminations involving stimuli presented well above threshold, often in different sensory modalities. The stimulus–response mappings are usually arbitrary, however, and subjects are pressured to produce their responses very quickly. In contrast, the typical AB design involves perceptual discriminations that are made difficult by visual backward masking, but without any pressure to select and produce a response quickly. Thus, whereas interference in the PRP design appears to arise from response selection and other decision-making processes, interference in the AB design appears to arise in perceptual processing. One might suspect, therefore, that these phenomena have different causes. On the other hand, AB and PRP effects are similar in that performance of the first task is relatively unaffected by task overlap, but performance of the second task is impaired when presented within a few hundred ms of the first. This surface similarity raises the possibility that the two phenomena might be related. It has been recently suggested, in fact, that they might be two different manifestations of a single underlying processing limitation (for example, a central bottleneck: see Jolicœur and Dell'Aqua 1998, 1999, this volume, Chapter 6).

A unified account of AB and PRP effects would obviously be very attractive by virtue of its parsimony. The purpose of this chapter, therefore, will be to consider whether such an account is viable. We begin by briefly reviewing the AB and PRP literatures. We then present several experiments using hybrid AB/PRP designs to evaluate the unified account of these phenomena. Finally, we will relate our work to the closely related studies conducted by Pierre Jolicœur and his colleagues (of which we were unaware of when we began the present experiments) and to work on other attentional phenomena, such as the effects of spatial attention and repetition blindness.

The Psychological Refractory period (PRP) effect

In the PRP design, subjects are presented with two tasks (Task 1 and Task 2), each requiring a separate speeded response. The key independent variable is the time between the stimulus onsets, better known as the stimulus onset asynchrony (SOA). Although the response time (RT) to Task 1 usually does not depend much on the SOA, the RT to Task 2 can be elevated by

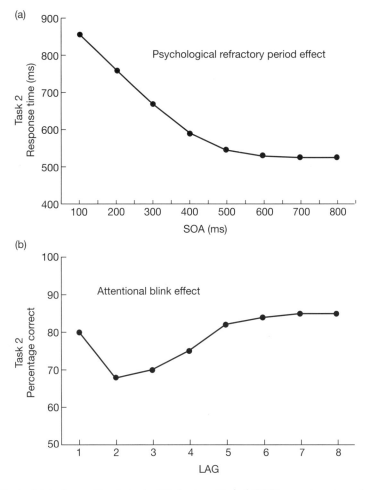

Fig. 6.1   (a) Typical data from a Psychological Refractory Period (PRP) experiment as a function of the stimulus onset asynchrony (SOA). (b) Typical data from an attentional blink (AB) experiment as a function of the lag between the first and second targets.

300 ms or more at short SOAs (i.e. when task overlap is high). This phenomenon, shown in Fig. 6.1a, was labelled the 'Psychological Refractory Period' (or PRP) effect on the mistaken assumption that it is a type of recovery phenomenon, similar to the refractory period of a neuron.

The PRP effect has been found using a very wide range of speeded tasks, including some very simple ones.[1] The effect has been found even when the responses are made with different output modalities (e.g. vocal versus manual) and even when the stimulus are presented in different input modalities (e.g. auditory versus visual). In fact, the PRP effect occurs even when subjects respond to two different attributes of the same visual object (Fagot and Pashler 1992), which presumably serves to minimize input interference (see Duncan 1984).

The fact that the PRP effect does not depend on any obvious input or output conflicts led Welford (1952) to propose the Central Bottleneck Model. This model asserts that central mental operations (for example response selection, planning, decision-making) can proceed on only one task at a time. As is shown in Fig. 6.2, while the Task 1 central stage is under way, the Task 2 central stage must wait. This waiting time (a.k.a. postponement, or 'bottleneck delay') is the primary cause of the PRP effect.[2]

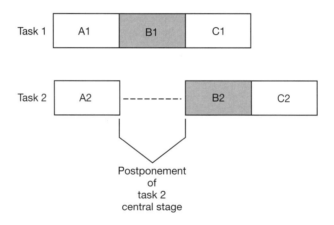

Fig. 6.2   The central bottleneck model.

The Central Bottleneck Model makes a number of distinctive predictions that have been confirmed in a wide range of PRP experiments (see Pashler and Johnston 1998 for a review). For example, the model correctly predicts that the effects of prolonging prebottleneck stages of Task 2 (for example by reducing stimulus contrast) on Task 2 RT should be greatly reduced at

---

[1] The PRP effect sometimes fails to occur for speeded tasks with extremely high stimulus– response compatibility, such as fixating visual stimuli (Pashler *et al.* 1993) or zero-order stick-tracking (Johnston and Delgado 1993; see also Greenwald 1972; Greenwald and Shulman 1973). In addition, the PRP effect can in some cases be greatly reduced with extensive task practice (Ruthruff *et al.*, 2001; Van Selst *et al.*, 1999).

[2] Dual-task slowing at short SOAs might also be due in part to reduced task preparation or perceptual interference.

short SOAs (cf. e.g. Pashler and Johnston 1989). Consequently, it seems clear that a central bottleneck, or something very close to it, is chiefly responsible for the PRP effect.

The attentional blink (AB) effect

In the standard AB design subjects view a stream of visual items (usually at a rate of about 10 per second), each presented in the same location on a computer monitor; this presentation technique is known as rapid serial visual presentation (or just 'RSVP'). The subjects' task is to report two targets (T1 and T2) within the RSVP stream. For example, subjects might be asked to identify two digits embedded within a stream consisting mostly of letters. Subjects are allowed to report these identities at their leisure; the primary dependent measure in this design is accuracy, not response time. As in the PRP design, task overlap is controlled by varying the SOA between T1 and T2. However, rather than express the task overlap variable in terms of the SOA, many investigators refer instead to the stimulus 'lag', defined as the number of frames between the onsets of T1 and T2.

Typically, subjects can accurately report T1, but tend to miss T2 when it appears within 300–400 ms of T1. This phenomenon, known as the attentional blink, is shown in Fig. 6.1b. The AB effect does not occur when subjects are told to ignore T1; because the mere presence of T1 in the RSVP stream is not sufficient to elicit an AB effect, the effect cannot be due to sensory masking (Raymond *et al.* 1992). The AB effect also does not occur reliably at lag 1. This 'lag-1 sparing' effect provides one potentially important clue to the cause of the AB effect.

Several different theoretical interpretations of the AB effect have been proposed. Raymond *et al.* (1992) postulated an 'attentional gate' that closes following detection of T1, preventing subsequent items in the RSVP stream (except perhaps for the item immediately following T1) from entering visual short-term memory (VSTM). Once T1 has been encoded from VSTM into a more durable form of short-term memory, the gate can then be reopened. This putative attentional gate would help preserve the clarity of T1's representation in VSTM at the cost of missing T2 when it is presented shortly after T1.

The attentional gate model can explain lag-1 sparing by adding the additional assumption that the gate does not close immediately following detection of T1. However, this model is inconsistent with several converging lines of evidence that, despite being unavailable for report, T2 is identified on most trials. Shapiro, Driver, Ward, and Sorensen (1997*b*), for example, showed that the identity of T2 primed the processing of a later target (T3) even when subjects could not reliably report T2 (see also Maki *et al.* 1997; Shapiro *et al.* 1997*a*). Further, the electrophysiological consequences of stimulus identification (e.g. the 'N400' peak of the event-related potential waveform) are largely preserved at short lags, even when T2 is missed (Luck *et al.* 1996; Vogel *et al.* 1998). In addition, Johnston, Ruthruff, and McCann (2001) provided chronometric evidence using locus-of-slack logic (see McCann and Johnston 1992) that T2 identification is not suppressed during the attentional blink. It seems likely, therefore, that the failure to report T2 arises primarily in postperceptual processing stages.

Several different accounts of the AB effect are consistent with the evidence that T2 can be identified but not reported. Chun and Potter (1995), for example, hypothesized that all RSVP items pass through a categorization stage (i.e. are identified), but only a select few then pass through a limited-capacity stage that forms a reportable perception (e.g. that consolidates categorical representations into short-term memory [STM]). At short lags, the consolidation of T2 will be delayed by the consolidation of T1. Because of this delay, the categorical representation of T2 is likely to become unavailable owing to rapid decay and/or overwriting by subsequent RSVP items before it can be consolidated. Hence, T2 will often be missed at short lags. This 'two-stage' model can account for lag-1 sparing by adding the auxiliary assumption that T1 and T2 can be concurrently subjected to consolidation if they arrive in close temporal proximity.

Shapiro *et al.* (1994) proposed a somewhat different account of the AB effect based on interference in visual short-term memory (VSTM). According to this model, items that match the template for T1 and/or T2 are granted entry into VSTM. T1 and T2 are therefore likely to enter VSTM, and so are the items immediately following T1 and T2, owing to the sluggishness of the selection mechanism. Upon entering VSTM the items are assigned a weighting that determines the order in which their representations (which are assumed to decay rapidly) will be passed on to a limited-capacity reporting stage. These weightings are based jointly on (a) how well the item matches the target templates and (b) the extent of the remaining 'resources'. T1, and perhaps the item immediately following T1 in the RSVP stream, will use up much of the available resources. Thus if T1 has not cleared VSTM by the time T2 arrives (i.e. at short lags), T2 will be assigned a low weighting and will probably be missed.

The VSTM interference model of Shapiro *et al.* (1994) differs in a few details from the two-stage model of Chun and Potter (1995). However, both models agree that consolidation of T1 into a reportable percept interferes (directly or indirectly) with the consolidation of T2 into a reportable percept.

A potential unified central bottleneck model of AB and PRP effects

The primary purpose of this chapter is to ask whether the AB and PRP effects can reasonably be attributed to the same central bottleneck. According to the unified central bottleneck (UCB) model considered in this chapter, there is some constraint that prevents the simultaneous occurrence of any two operations that belong to a set of demanding central operations. By hypothesis, at least one of these central operations is required to select and produce a speeded response (i.e. the typical PRP task); hence a processing bottleneck occurs when two such tasks are presented in close temporal synchrony, causing the PRP effect. Likewise, finding and remembering a target embedded in an RSVP stream requires at least one operation that belongs to the set of demanding central operations; thus, a processing bottleneck occurs when two RSVP tasks are presented in close temporal synchrony, causing the AB effect.

Note that the demanding central operation(s) responsible for the AB effect need not be the exact same operation(s) responsible for the PRP effect. In fact, the central bottleneck in the PRP design appears to be due in large part to the process of response selection (see Pashler and Johnston 1998), whereas the AB design allows subjects to respond at their leisure, and thus does not even appear to require on-line response selection.[3] Although it is premature to specify

---

[3] It is conceivable that subjects do select an immediate on-line response to the Task 1 stimulus, even though response selection could in principle be deferred until after the RSVP stream has ended. However, it seems unlikely that Task-1 response selection is responsible for the AB effect. Johnston *et al.* (2001) found robust AB effects even in a condition where the response could not be selected on-line because the response mapping was not revealed until well after the RSVP stream had ended.

definitively which central operation(s) are responsible for the bottleneck in the AB design, one plausible candidate is the consolidation of categorical representations into STM (see Chun and Potter 1995; Jolicœur 1999*a*, *b*; Jolicœur and Dell'Aqua 1998).

There is reason for scepticism about the prospects for the UCB model. As noted above, the AB effect appears to reflect a difficulty in determining what stimuli were presented, whereas the PRP effect appears to reflect a difficulty in determining how to respond to stimuli once they have been identified. Further, Pashler (1989) has provided evidence that the cause of the PRP effect is different from the cause of interference between simultaneous visual discriminations. Pashler presented subjects with a tone requiring an immediate speeded response, followed 50, 150 or 650 ms later by a masked visual array of characters. In one experiment, subjects looked for the highest number in an array of eight digits; in another they searched for a green T among green Os and red Ts. In both cases, performance on the visual task depended very little on the degree of overlap with the speeded tone task. In contrast, a related set of experiments showed that performance on the visual task was severely impaired by overlap with another visual task (whether speeded or unspeeded). On the basis of this evidence, Pashler (1989) argued that dual-task interference has (at least) two different causes: (1) there is a constraint requiring that demanding central processes (including but not limited to response selection) be performed one at a time, and (2) visual processes may proceed simultaneously, but suffer mutual interference dependent upon the complexity of the visual operations required. On this view, the PRP effect would be due to the first component, whereas the AB effect might be attributed to the second component.

Testing a unified account: experimental logic

According to the UCB model, speeded-response tasks (used in the PRP design) and RSVP tasks (used in the AB design) both require demanding central operations that can take place on only one task at a time. Thus, this model predicts that one speeded task should interfere with another speeded-response task at short task lags (causing the PRP effect). Likewise, one unspeeded RSVP task should interfere with another unspeeded RSVP task at short task lags (causing the AB effect). Furthermore, the UCB model predicts that an unspeeded RSVP task should interfere with a speeded-response task. In other words, while demanding central operations are under way on an RSVP Task 1, the critical central operations on a speeded-response Task 2 will be postponed. Experiment 1 was designed to test this prediction. Experiment 2 was designed to test the converse prediction, namely that a speeded-response task should interfere with an unspeeded RSVP task.

Table 6.1 summarizes the relations between the AB design, the PRP design, and the designs of Experiments 1 and 2. Each of the four cells represents an experimental design resulting from pairing together speeded-response and/or unspeeded RSVP tasks as Task 1 and Task 2. When Task 1 and Task 2 are both speeded-response tasks, the experiment is a pure PRP design. When Task 1 and Task 2 are both unspeeded RSVP tasks, the experiment is a pure AB design. When Task 1 is an unspeeded RSVP task and Task 2 is a speeded-response task (Experiment 1), the experiment is a 'PRP/AB' design (where the first set of initials refer to the type of Task 1 and the second set of initials refer to the type of Task 2). When Task 1 is a speeded-response task and Task 2 is an unspeeded RSVP task (Experiment 2), the experiment is an 'AB/PRP' design. If the UCB model is correct, then we should observe substantial dual-task interference in the hybrid designs.

Table 6.1 Dual-task designs created by pairing speeded-response and/or unspeeded RSVP tasks. PRP = Psychological Refractory Period design; AB = Attentional Blink design.

		TASK 1	
		Speeded-Response	Unspeeded
RSVP TASK 2 AB/PRP	Speeded-Response	PRP	Hybrid
	Unspeeded RSVP	Hybrid PRP/AB	AB

As will be seen, the hybrid designs of Experiments 1 and 2 do in fact produce substantial interference effects, consistent with the UCB model. The purpose of Experiments 3 and 4 was to provide a more direct test for the involvement of central interference in the AB effect. Finally, Experiment 5 looked for evidence that the AB effect is due to both central interference and perceptual interference.

Experiment 1 (AB/PRP design)

Experiment 1 paired a typical AB Task 1 (i.e. an unspeeded report of a target in an RSVP stream) with a typical PRP Task 2 (i.e. a speeded report of an unmasked letter). We refer to this as 'the AB/PRP design'. The goal is to see if central operations engendered by a typical AB Task 1 prevent the simultaneous performance of central operations on a typical PRP Task 2, as predicted by the UCB model. If so, then we should observe substantial dual-task interference, in the form of Task 2 slowing at short lags (similar to the PRP effect).

Task 1 (unspeeded) was to report a target digit (1, 2, 3, or 4) presented within an RSVP stream of letter distractors. Task 2 was to respond rapidly to a letter (E or F) by pressing the corresponding key on the keyboard. The E or F was always the last item in the RSVP stream (and therefore unmasked). Further, it was slightly brighter than the other RSVP items and remained present until the subject responded to it. Thus as with most PRP experiments, the Task 2 stimulus was fairly easy to perceive.

Method

Subjects. Twenty-one students at the University of California – San Diego participated in return for a partial course credit. All reported normal or corrected-to-normal vision. No subject participated in more than one experiment reported in this chapter.

Stimuli. Stimuli were alphanumeric characters, which subtended approximately 0.5 degrees horizontally by 0.7 degrees vertically from a typical viewing distance of 60 cm. They were grey (IBM VGA color code #8) against a black background, displayed on NEC Multisynch monitors connected to IBM PC-compatible computers. Characters were displayed using rapid serial visual presentation (RSVP). Each RSVP display consisted of fifteen characters in the middle of the screen, presented strictly sequentially for 85 ms each. The non-target characters in the RSVP sequence were letters drawn randomly from the alphabet, excluding the letters E, F, I, and O.

Procedure. The Task 1 stimulus (S1) was a digit (1, 2, 3, or 4) presented in the fourth frame of the RSVP stream. Subjects were asked to identify the digit and report it at the end of the trial (i.e. some time after they had made their speeded response to Task 2). The digits 1–4 were mapped in numerical order on to the 'N', 'M', '<', and '>' keys. The Task 2 stimulus (E or F) appeared at a lag of two to eight frames after S1. We avoided lag 1 because this condition does not reliably produce an AB effect (cf. e.g. Raymond *et al.* 1992; Chun and Potter 1995). Each lag occurred equally often. Subjects were asked to press the key ('E' or 'F') corresponding to the Task 2 stimulus as fast as possible, without making too many mistakes.

Subjects completed 6 blocks of 66 trials. The first two blocks were considered practice and therefore not analysed. Similarly, the first two trials within a block were considered warm-ups. Feedback on Task 1 accuracy, Task 2 accuracy and RT2 was provided at the end of each block.

The sequence of events within a trial was as follows. A fixation point was presented in the middle of the screen for 500 ms followed by a blank screen for 250 ms. Next, the RSVP stream was presented at a rate of 85 ms per character. Once both responses had been made, they were echoed to the screen. The word 'WRONG' was displayed adjacent to any incorrect response(s) for 1 second. The next trial began 1 second later.

Analyses. We excluded any subject who did not achieve an average of at least 65 per cent correct on Task 1 and 85 per cent correct on Task 2. Further, we excluded from RT2 analyses all trials on which the Task 1 response was incorrect.

Results and discussion

The results are shown in Fig. 6.3 as a function of lag. The Task 1 percentage correct, which did not depend on lag, is shown in Table 6.2. The main purpose of this experiment was to determine if an unspeeded RSVP Task 1 would interfere with a speeded Task 2, as predicted by the UCB model. Task 2 accuracy (mean = 98 per cent) was affected relatively little by lag; however, mean Task 2 RT was 124 ms longer at short lags than at long lags, $F(6,120) = 54.7$, $p < 0.001$. Similar findings have been obtained by Johnston, Ruthruff, and McCann (2001) and Arnell and Duncan (1999) using other variants of the AB/PRP design.

Second-task slowing in this experiment occurred even though the Task 2 stimulus was unmasked and therefore relatively easy to perceive. In fact, we have informally replicated this

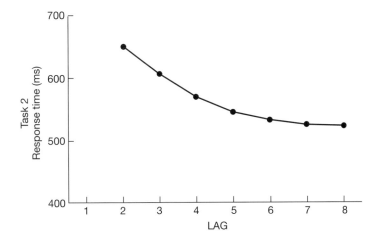

Fig. 6.3   Task 2 response time in Experiment 1 as a function of lag.

Table 6.2 Task 1: percentage correct as a function of lag for each experiment. S = Speeded; U = Unspeeded; V = Visual; A = Auditory; C = Compatible; I = Incompatible

		LAG							
		1	2	3	4	5	6	7	8
Exp. 1	UV	—	97.4	88.6	88.6	89.3	87.7	90.6	87.0
Exp. 2	SA	95.1	97.0	95.1	94.8	95.3	96.0	93.8	95.3
Exp. 3	SV	75.8	79.5	78.9	81.3	81.9	81.3	79.7	81.8
	UV	76.5	79.2	80.9	80.5	79.5	80.0	80.6	80.1
Exp. 4	SV-C	93.5	95.2	95.0	93.5	92.9	94.4	95.0	90.2
	SV-I	90.2	89.9	90.2	88.1	89.0	91.4	88.7	89.3
Exp. 5	SV	93.5	93.8	92.9	94.4	95.0	93.8	95.9	94.7
	SA	89.9	90.5	92.6	93.2	91.4	89.9	90.5	89.3

finding even when the Task 2 stimulus was Auditory (see also Arnell and Duncan 1999; Jolicœur and Dell'Aqua 1998, 1999). This pattern of results is consistent with a central locus of interference.

How does the size of this interference effect compare to that typically found in PRP experiments? First, it is important to note that the present experiment used a somewhat restricted range of lags (2 to 8) compared to that used in most PRP experiments (the equivalent of lags 0 to 9). Even after accounting for the difference in lags, the slowing observed here (124 ms) is somewhat less than the 170+ ms PRP effects observed across comparable task lags (cf. e.g. McCann and Johnston 1992; Pashler and Johnston 1989; Ruthruff *et al.* 1995). According to the UCB model, this means that the Task 1 central operations required by a typical AB task finish sooner than the central operations required by a typical PRP task. This conclusion is plausible given that the AB task requires only that subjects remember the target identity (presumably a highly-practised operation), whereas PRP tasks usually require subjects to negotiate a novel and arbitrary stimulus-response mapping.

Experiment 2 (PRP/AB design)

In Experiment 1 an unspeeded RSVP Task 1 (like that used in the AB design) interfered with the processing of a speeded, unmasked Task 2 (like that used in the PRP design) at short lags. This outcome is consistent with the UCB model, which says that central operations on an unspeeded RSVP Task 1 can postpone central operations on a speeded Task 2. In Experiment 2 we reversed the order of these two tasks (the 'PRP/AB' design) to test the converse claim: that central operations on a speeded Task 1 can postpone the central operations critical to performing an unspeeded RSVP Task 2. If so, then Task 2 accuracy should be reduced at short lags relative to long lags (as in the AB effect).

Task 1 required a speeded response to the pitch of a pure tone, which was presented in half the trials and absent in the other half (selected at random). The target-absent trials provide a baseline against which we can measure performance in the target-present condition. The tone was presented for 85 ms (for the same duration as the characters in the RSVP stream) and was not masked. Subjects were instructed to press the 'N' key if they heard the low-frequency tone (300 Hz) and press the 'M' key if they heard the high-frequency tone (900 Hz). Task 2 was to make an unspeeded response indicating whether the 'E' or the 'F' was present in the RSVP stream.

Method

Except where noted, the method was identical to that of Experiment 1.

Subjects. Fifty-four students at the University of California—San Diego participated in exchange for a partial course credit.

Procedure. The Task 1 tone, when present, coincided with the fourth frame of the RSVP stream, which contained a random distractor letter. Subjects were instructed to respond immediately to the tone. After a lag of 1 to 8 frames, the Task 2 stimulus (an 'E' or an 'F') was presented. Subjects were allowed to respond at their leisure to this stimulus.

Analyses. We excluded any subject who did not achieve an average of at least 65 per cent correct overall on Task 1 and 65 per cent correct on Task 2 at the two longest lags. Further, we excluded from analyses of Task 2 RT and Task 2 accuracy any trials on which the Task 1 response was incorrect.

Results and discussion

The results are shown in Fig. 6.4. Mean Task 1 RT (620 ms) did not depend on lag, $F(1,31) < 1$; this result indicates that subjects did not interrupt Task 1 in order to perform Task 2. The main purpose of this experiment was to see if a speeded Task 1 (like that used in the PRP design) would interfere with an unspeeded RSVP Task 2 (like that used in the AB design), as predicted by the UCB model. As is shown in Fig. 6.4, the Task 2 accuracy was in fact much lower overall when the Task 1 stimulus was present than when it was absent, $F(1,31) = 3.66$, $p < 0.01$. Further, Task 2 accuracy declined substantially at short lags when the Task 1 stimulus was present, $F(7,217) = 5.78$, $p < 0.001$.

Using methods similar to those of the present experiment, Jolicœur (1999b) found essentially the same pattern of results and arrived at a similar conclusion. We have also replicated this result in several different pilot experiments. One key finding is that PRP/AB interference occurs even when the Task 1 tone judgement is very easy. For example, we have found this effect when the high and low tone frequencies are even more distinct (200 and 2000 Hz) and the tone is presented for three times as long (i.e. 255 ms). The fact that interference in the PRP/AB design can be obtained even when the stimuli are cross-modal and relatively easy to perceive is certainly consistent with models that propose a central (rather than perceptual) locus for the AB effect.

Experiment 3 (task-1 carryover)

It appears that dual-task interference occurs regardless of how speeded-response tasks and RSVP tasks are combined (i.e. occurs in all four cells of Table 6.1). Performance of two speeded-response tasks produces the robust PRP effect and performance of two unspeeded RSVP tasks produces the well-documented AB effect. Experiment 1 (the AB/PRP design) showed that dual-task interference also occurs when an RSVP Task 1 is paired with a speeded-response Task 2. Further, Experiment 2 (the PRP/AB design) showed that interference occurs when a speeded-response Task 1 is paired with an RSVP Task 2. These results are consistent the UCB model, which says that both of these types of tasks (speeded-response and unspeeded RSVP) require central operations that can proceed on only one task at a time.

Although the results from Experiments 1 and 2 are consistent with the UCB model, they are consistent with a few other models as well. For example, the results are consistent with a model in which the PRP effect is due primarily to a central bottleneck, whereas the AB effect is due to an earlier bottleneck that is completely separate from the central bottleneck. The locus of this

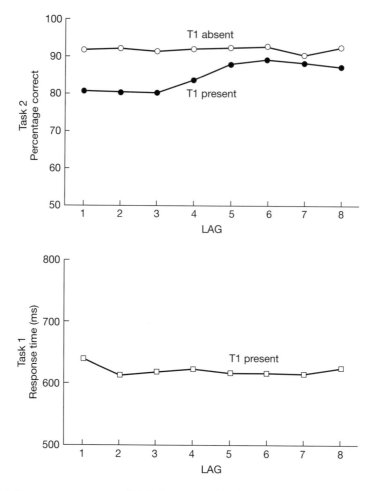

Fig. 6.4 Task 2 percentage correct and Task 1 response time in Experiment 2 as a function of lag. Data from trials in which the Task 1 stimulus was present are represented by squares. Data from trials in which the Task 1 stimulus was absent are represented by triangles.

early bottleneck need not be perceptual processing *per se*. It might be STM consolidation, though there are other plausible candidates as well, such as stimulus classification.[4] To explain the interference observed in both Experiment 1 (the AB/PRP design) and Experiment 2 (the PRP/AB design), this dual-bottleneck account need only assume that both speeded tasks and RSVP tasks require the same early stage that causes the AB bottleneck.

---

[4] One obvious way to test the early bottleneck model would be to find a Task 1 judgement that does not contain this early stage and then see if such a Task 1 interferes with an unspeeded RSVP Task 2. Unfortunately this approach is not feasible. We cannot specify what stage causes the putative early bottleneck, without sacrificing generality; it is, of course, impossible to find a task that lacks an unspecified stage.

If the AB effect is due to an early bottleneck rather than a central bottleneck, then the amount of interference in the AB design should not depend on the time required to complete the central stages of Task 1. Instead, the size of the AB effect should depend only on the time required to complete the early stages of Task 1 (i.e. up to and including the stage that causes the early bottleneck). According to the UCB model, on the other hand, dual-task interference occurs primarily because the Task 1 central stage postpones the onset of the Task 2 central stage. At short lags, any manipulation that delays the completion of the Task 1 central stage will cause an additional postponement in the onset of the Task 2 central stage. In the AB design, where Task 2 is an RSVP task, the extra postponement of the Task-2 central stage might cause the categorical representation of T2 to decay even further before it can be consolidated. Consequently, T2 would be especially unlikely to be reported when the Task 1 central stage is prolonged. More precisely, when the Task 1 central stage is prolonged the AB effect should be deeper and more prolonged (i.e. recovery from dual-task interference should occur at a later lag). Henceforth we refer to this as the 'Task-1 carryover prediction' because the effects of prolonging the Task 1 central stage carry over into Task 2.

To test between the UCB model and the dual-bottleneck model, Experiment 3 evaluated the Task-1 carryover prediction. We manipulated the duration of the Task 1 central processing in an AB design by varying whether the response to the Task 1 stimulus was speeded or unspeeded. We reasoned that the requirement to perform an immediate response selection should increase the duration of the central stage relative to a condition where subjects could defer response selection until after T2 had been processed. If the UCB model is correct, the decrement in Task-2 accuracy should be deeper and longer-lasting when Task 1 is speeded. If the early-bottleneck account is correct, however, the decrement in Task-2 accuracy should be roughly the same for both the speeded and the unspeeded conditions.

Method

Except where noted, the method was identical to that of Experiment 2.

Subjects. Forty-two students at the University of California—San Diego participated in exchange for a partial course credit.

Procedure. Task 1 was to identify a digit (1, 2, 3, or 4) embedded in an RSVP stream. Half the subjects were instructed to respond immediately to Task 1 (the speeded condition), and the other half were instructed to respond only after all 15 characters had been displayed (the unspeeded condition). Task 2 (unspeeded) was to indicate whether the E or the F had been presented.

Results and discussion

The results are shown in Fig. 6.5. Task 2 performance was worse overall when Task 1 was speeded than when Task 1 was unspeeded, $F(1,40) = 18.3$, $p < 0.001$, although it is debatable whether the recovery from dual-task interference was also prolonged. These results are broadly consistent with the UCB model, but appear to contradict the dual-bottleneck model.

Jolicœur (1998) has also compared speeded and unspeeded Task-1 judgements and found a very similar pattern of results. Further, Jolicœur (1999*b*) manipulated the number of stimulus–response alternatives on Task 1. In some blocks of trials subjects performed a two-alternative forced-choice Task 1 based either on letter size (small or large) or letter identity (H versus S), while in other blocks of trials subjects performed a four-alternative forced-choice Task 1 based on *both* letter identity and letter size. This manipulation also produced the same basic Task-1 carryover effect observed in the present experiment and in Jolicœur (1998).

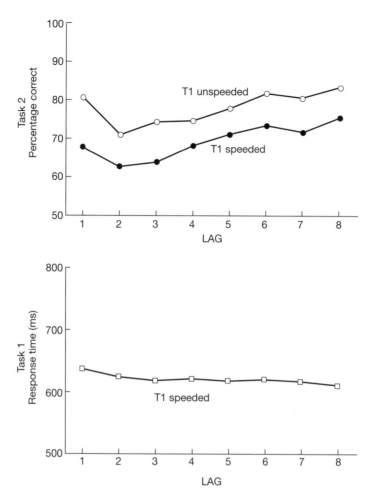

Fig. 6.5 Task 2 percentage correct and Task 1 response time in Experiment 3 as a function of lag. The speeded Task 1 data are represented by squares. The unspeeded Task 1 data are represented by diamonds.

Experiment 4 (Task-1 carryover using a within-block manipulation)

Varying whether subjects need to select an on-line response to Task 1 (as in Experiment 3) would seem to be a natural way to control the duration of the Task 1 central stage; however, this manipulation has several drawbacks. First, although it seems likely that Task 1 central stage is prolonged in the speeded condition relative to the unspeeded condition, this is not necessarily the case. It is conceivable, for example, that it takes about as long to consolidate T1 into memory when the task is unspeeded as it does to select an immediate response to T1 when the task is speeded. Because the unspeeded Task-1 condition does not produce an observable response time, there is no easy way to confirm that central operations were in fact prolonged in the speeded condition relative to the unspeeded condition.

An even more serious problem with Experiment 3 (and with the studies reported by Jolicœur 1998, 1999*b* as well) is that the speeded and unspeeded Task-1 conditions were not run within the same block of trials. Consequently, subjects might have adopted very different response criteria or task strategies in the speeded and unspeeded blocks. Furthermore, subjects might have experienced higher levels of arousal, or simply exerted more effort, in one condition or the other. In addition, more pre-trial preparation might be required for a speeded Task 1 than an unspeeded Task 1, leaving subjects less prepared for Task 2 (see Gottsdanker 1980). This preparation effect alone could account for the poorer Task 2 performance observed in Experiment 3 when the Task 1 response was speeded.

In Experiment 4 we addressed these problems by varying the stimulus–response compatibility of a speeded Task 1 *within* blocks of trials. The Task 1 stimulus was a digit drawn randomly from the set {1, 2, 3, 4, 5, 6, 7, 8}. As in Experiment 3, the digits 1–4 were mapped in numerical order (i.e. compatibly) on to four response keys. However, the digits 5–8 were mapped in reverse order (i.e. incompatibly) on to the same response keys. These compatible and incompatible Task-1 conditions were mixed randomly within a block of trials to ensure that these conditions did not differ in arousal, response criteria, the degree of preparation for Task 2, etc. Furthermore, note that both the compatible and incompatible conditions produce an observable RT, allowing us to determine whether we were in fact successful in prolonging the duration of the Task 1 response-selection stage and to estimate the amount of prolongation. An additional advantage of this design is that it seems especially clear that the Task 1 stage affected by the present response-compatibility manipulation is central rather than perceptual.

Method

Except where noted, the method was identical to that of Experiment 3.

Subjects. Twenty-two students at the University of California—San Diego participated in exchange for a partial course credit.

Procedure. The Task 1 stimulus was chosen at random (with replacement) from the set {1, 2, 3, 4, 5, 6, 7, 8}, with the exception that the same stimulus was never presented twice in a row. Subjects responded by pressing either the 'N', 'M', '<', or '>' key. The digits 1–4 were mapped in numerical order on to the four response keys from left to right (i.e. compatibly), while the digits 5–8 were mapped in reverse numerical order from left to right (i.e. incompatibly). Subjects were instructed to respond to the digit as fast as possible without making too many errors. As in the previous experiment, Task 2 (unspeeded) was to indicate whether an E or an F had been presented in the RSVP stream.

Results and discussion

The results are shown in Fig. 6.6. The compatible Task 1 condition (M = 720 ms) produced much faster responses than the incompatible Task 1 condition (M = 846 ms), $F(1,21) = 7.6$, $p < 0.01$. This result suggests that we were successful in increasing the duration of Task 1 central operations, making it possible to evaluate meaningfully the Task-1 carryover prediction. Relative to the compatible condition, the incompatible Task 1 condition produced a deeper decrement in Task 2 accuracy, $F(1,20) = 13.2$, $p < 0.01$. The incompatible condition also appears to have produced a more prolonged decrement in Task 2 accuracy. These observed Task-1 carryover effects contradict the dual-bottleneck account. However, they provide compelling evidence for the involvement of central operations in the AB effect, as proposed by the UCB model.

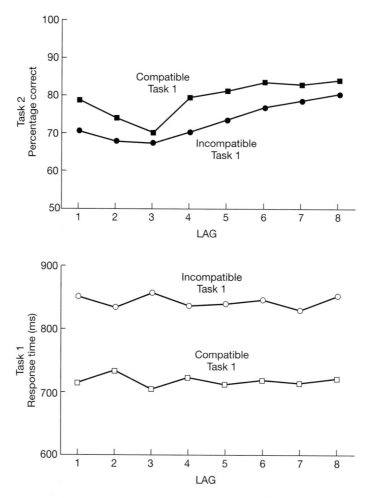

Fig. 6.6   Task 2 percentage correct and Task 1 response time in Experiment 4 as a function of lag. Data from the compatible Task 1 condition are represented by squares. Data from the incompatible Task 1 condition are represented by circles.

Experiment 5 (within-modal versus cross-modal interference)

The interference observed in the hybrid designs of Experiments 1 and 2 along with the Task-1 carryover effects observed in Experiments 3 and 4 clearly suggest that central interference plays a role in the AB effect. As discussed next, however, there is reason to suspect that a central bottleneck is not the only cause of interference in the AB design. The hybrid AB/PRP design of Experiment 1 produced only 124 ms of dual-task slowing between lags 2 and 8, whereas comparable PRP experiments typically show 170 ms or more of Task-2 slowing between the same lags (cf. e.g. McCann and Johnston 1992; Pashler and Johnston 1989; Ruthruff *et al.* 1995). In other words, a speeded-response Task 2 appears to suffer less interference when Task 1 is an RSVP task (as in the AB/PRP design) than when Task 1 is another speeded-response task.

According to the UCB model this difference occurred because central operations simply finish earlier when Task 1 is an RSVP task than when Task 1 is a speeded-response task, resulting in less postponement. This explanation is plausible given that an RSVP task merely requires that the subject remember a single target character, whereas most speeded-response tasks require subjects rapidly to carry out a novel and arbitrary mapping of stimulus on to response. However, the UCB must then predict that an RSVP Task 1 should also produce less interference than a speeded-response Task 1 when Task 2 is an RSVP task. In other words, the PRP/AB design should produce more interference than the pure AB design. Contrary to this prediction, the PRP/AB design appeared to produce somewhat *less* interference than is typically found in pure AB experiments.

One explanation for this pattern of results is that a central bottleneck is not the only cause of interference in these designs. It is possible, for example, that the AB effect is due in part to a central bottleneck and in part to perceptual interference. Experiment 5 was designed to evaluate this possibility. Specifically, the goal was to determine if the AB effect is due to perceptual interference in addition to any central interference.

Our approach was to see if an unspeeded RSVP Task 2 suffers more interference when Task 1 is another RSVP task (the within-modal condition) than when Task 1 requires a speeded response to a pure, unmasked tone (the cross-modal condition). The within-modal condition was essentially a replication of the standard AB effect. The cross-modal condition was explicitly designed to require the same amount of Task 1 central processing as the within-modal condition, but with greatly reduced potential for perceptual interference (both because the inputs were cross-modal and because the tone was unmasked). If the AB effect is due in part to perceptual interference, then the AB effect should be substantially reduced in the cross-modal condition relative to the within-modal condition.

To ensure that the within- and cross-modal conditions did not differ with respect to arousal, effort, or task preparation, these conditions were randomly mixed within blocks of trials. Furthermore, to estimate the duration of central processing on Task 1, we required a speeded response to the Task-1 stimulus in both the within- and cross-modal conditions. The task parameters were chosen (on the basis of pilot studies) so that the auditory and visual Task 1 judgements would produce roughly the same RT. Because these judgements required exactly the same responses (pressing the 'L' or 'H' key), we can then infer that they required roughly the same amount of central processing. Therefore the UCB model predicts that the two conditions should produce roughly equal amounts of Task 2 postponement, which in turn should lead to roughly equal amounts of dual-task interference. However, if the AB effect is due in part to perceptual interference then we should observe greater amounts of interference in the within-modal condition.

Method

Except where noted, the method was identical to that of Experiment 3.

Subjects. Twenty-one students at the University of California–San Diego participated in exchange for a partial course credit.

Procedure. S1 on each trial was either a tone (300 Hz or 900 Hz) or a green letter (L or H). When S1 was a letter, it occupied the fourth frame of the RSVP stream. When it was a tone, it was presented simultaneously with the fourth frame of the RSVP stream, which was a grey distractor character selected at random. Subjects responded to the low-pitched tone and to the letter L by pressing the key marked 'L'; subjects responded to the high-pitched tone and to the letter H by pressing the key marked 'H'. Subjects were instructed to respond quickly without

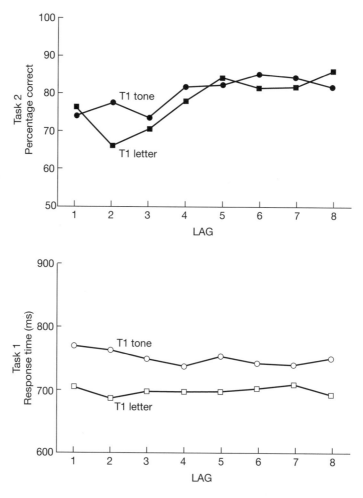

Fig. 6.7 Task 2 percentage correct and Task 1 response time in Experiment 5 as a function of lag. Data from the letter Task 1 condition are represented by squares. Data from the tone Task 1 condition are represented by the 'X' symbols.

making too many errors. Task 2 (unspeeded) was to determine whether an E or an F had been presented in the RSVP stream.

Results and discussion

The results are shown in Fig. 6.7. Subjects responded somewhat more slowly to Task 1 when S1 was auditory (551 ms) than when it was visual (491 ms), $F(1,20) = 9.5, p < 0.01$. Because subjects made exactly same physical responses to both Task 1 judgements, the duration of response processes should have been similar. It therefore seems likely that the central stage of the tone task tended to finish at least as late as the central stage of the letter task (if not later). Consequently, the UCB model predicts at least as much interference when the Task 1 stimulus was a tone (i.e. in the cross-model condition) as when the Task 1 stimulus was a letter (i.e. the within-modal condition). Contrary to this prediction, we actually found *less* interference in the cross-modal condition than in the within-modal condition. More precisely, we observed less effect of lag in the cross-modal condition than in the within-modal condition, $F(2,40) = 2.2$, $p < 0.05$. Furthermore, Task 2 performance at short lags (2, 3 and 4) was significantly better in the cross-modal condition than in the within-modal condition, $F(1,20) = 5.7$, $p < 0.05$.

Similar results were obtained in a recent study by Arnell and Duncan (1999). Each task in their experiment (i.e. Task 1 and Task 2) could be either a speeded-response to an auditory stimulus or a unspeeded response to a visual stimulus. As in the present Experiment 5, these conditions were mixed together randomly within a block of trials. Consistent with the present results, Arnell and Duncan found greater interference in both the pure conditions (i.e. speeded tone task followed by another speeded tone task or unspeeded digit task followed by another unspeeded digit task) than in the hybrid conditions (i.e. speeded tone task followed by unspeeded digit task or unspeeded digit task followed by speeded tone task).

In summary, the present results and those of Arnell and Duncan (1999) suggest that the AB effect is due not only to central interference (see Experiments 1–4) but also to perceptual interference resulting from the requirement to perform two demanding visual discriminations simultaneously. One specific possibility is that perceptual interference in the within-modal conditions results in relatively weak categorical representations of T2 relative to the cross-modal condition. Because the representation of T2 is especially weak to begin with, it is unlikely to survive the period of Task 2 postponement.

General discussion

The purpose of this chapter is to explore the relationship between two well-studied forms of dual-task interference: the Psychological Refractory Period (PRP) effect and the Attentional Blink (AB) effect. The PRP effect is a tendency to respond slowly to one speeded-response task when it is presented shortly after another speeded-response task. The AB effect is a tendency to miss one target embedded in an RSVP stream when it is presented shortly after another target.

One might suspect that these two phenomena are caused by distinct attentional limitations. The PRP effect arises with two speeded tasks even when input conflicts are minimized. In fact, it occurs even with unmasked supra-threshold stimuli presented in different modalities (e.g. one auditory and one visual stimulus). The AB effect, in contrast, would seem to reflect a sort of input interference limited to rapid presentation rates. It is observed even in the complete absence of speed pressure on either response. The experiments reported in this chapter, however, suggest that AB and PRP effects may nonetheless be due, at least in part, to the same attentional limitation, commonly termed the 'central bottleneck'.

We conducted several tests of the unified central bottleneck (UCB) model, which says that (a) there is some constraint that prevents the simultaneous occurrence of any two operations that belong to a set of demanding central operations and (b) speeded-response tasks and RSVP tasks require at least one of these demanding central operations (though not necessarily the same ones). Thus, both AB and PRP effects occur because the demanding central operations on Task 2 are postponed until the demanding central operations on Task 1 have been completed. In the PRP design, this postponement manifests itself as a slowing of the Task-2 response; in the AB design, the postponement manifests itself as a tendency to miss T2.

If the UCB model is correct, then a central bottleneck should occur not only in the pure AB and PRP design, but also in hybrid designs that pair a speeded-response task with an RSVP task (see Table 6.1). Experiment 1 verified that an unspeeded RSVP Task 1 does in fact slow processing on a speeded-response Task 2 (see also Johnston *et al.* 2001). Similarly, Experiment 2 verified that a speeded-response Task 1 impaired performance of an unspeeded RSVP Task 2. A very similar finding was also reported by Jolicœur (1999*b*).

Experiments 3 and 4 tested an alternative account of Experiments 1 and 2, which says that the PRP effect is due to a central bottleneck, whereas the AB effect is entirely due to a separate bottleneck that occurs before the central bottleneck. If this dual-bottleneck account is correct, then the accuracy of an RSVP Task 2 should depend on the duration of early Task-1 stages, but should not depend on the duration of Task-1 central stages. In contrast, the UCB model says that Task-2 central operations cannot begin until Task-1 central operations have finished; hence the accuracy of T2 at short lags should depend very much on the duration of the Task 1 central stage. Specifically, the decrement in Task 2 accuracy should be deeper and longer-lasting when Task-1 central operations are prolonged. We refer to this as the 'Task-1 carryover' prediction of the UCB model.

To evaluate the Task-1 carryover prediction, we manipulated whether or not Task 1 was speeded (Experiment 3). We reasoned that the requirement to make an immediate speeded response to Task 1 should add to the duration of the central operations relative to a condition where subjects could defer response selection. As was predicted by the UCB model, Task 2 performance was indeed much worse when Task 1 was speeded than when Task 1 was unspeeded. Similar results have also been demonstrated by Jolicœur (1998, 1999*a*).

The results of Experiment 3 clearly support the UCB model, but are open to several objections. Because the speeded and unspeeded Task 1 conditions of Experiment 3 were run in different blocks of trials, they may have differed in terms of arousal, effort, response criteria, or task preparation. The same criticism applies to Jolicœur (1998, 1999*a*) as well. Differences in task preparation are especially serious, since greater preparation for a speeded Task 1 could result in less preparation (and thus worse performance) on Task 2. To eliminate these possible confounds, Experiment 4 replicated the Task-1 carryover effect using a within-block manipulation of Task-1 difficulty. The digits 1–4 were mapped in numerical order (i.e. compatibly) on to the four response keys, but the digits 5–8 were mapped in reverse numerical order (i.e. incompatibly) on to the same four response keys. Relative to the compatible condition, the incompatible Task 1 condition caused both a deeper and a more prolonged decrement in Task-2 performance (i.e. a substantial Task-1 carryover effect). These results clearly contradict the dual-bottleneck account, which predicted that the duration of Task-1 central stages should not affect Task 2 accuracy. These Task-1 carryover effects, however, provide direct support for the UCB model.

Locus of the central bottleneck

The UCB implies the existence of a set of demanding central processes that cannot operate concurrently. Although it clearly suggests that very early input processes and very late output processes do not belong to this set, the model does not specify exactly which operations in-between do belong to this set. This issue is therefore left as an empirical matter. In the PRP design there is clear evidence that character identification does not belong to the set of demanding central operations (cf. e.g. Pashler and Johnston 1989), but that later processes such as response-selection (cf. e.g. Pashler 1984; McCann and Johnston 1992), memory retrieval (Carrier and Pashler 1995), and mental rotation (Ruthruff *et al.* 1995) do belong to this set.

It is less clear what mental operations in the AB paradigm belong to the set of demanding central operations. As was noted earlier, however, there are several lines of evidence indicating that character identification is not suppressed during the attentional blink (Johnston *et al.* 2001; Luck *et al.* 1996; Maki *et al.* 1997; Shapiro *et al.* 1997*a*; Shapiro *et al.* 1997*b*; Vogel *et al.* 1998). Therefore, the earliest operation that does belong to the set of demanding central operations must come after the process of character identification. Consistently with this inference, Chun and Potter (1995) have proposed that the AB effect is due to interference between the operations that make the categorical representations of the targets available for subsequent report (for example by consolidating those representation into short-term memory [STM]).

Jolicœur and Dell'Acqua (1998, 1999) have provided particularly clear-cut support for the conjecture that consolidation into STM belongs to the set of demanding central processes. In one experiment (not unlike the AB/PRP design of Experiment 1), they presented a masked display of 1–3 characters, which subjects were to remember for a later report, followed after a variable SOA by a pure tone requiring an immediate speeded response. Jolicœur and Dell'Acqua found that processing of the visual stimulus slowed the response to the tone at short SOAs. More importantly, the magnitude of the tone-task slowing was greater when three letters needed to be consolidated than when only one letter needed to be consolidated. This result is consistent with the claim that central operations required to make a speeded response to the tone are postponed until short-term consolidation of the visual stimulus has been completed. As noted by Jolicœur and Dell'Acqua, the hypothesis that STM consolidation cannot proceed on two tasks at the same time provides a ready explanation for interference in the AB design as well (see also Jolicœur 1998, 1999*a*,b).

Multiple sources of interference

The fact that we observed substantial interference in a hybrid AB/PRP design (Experiment 1) and a hybrid PRP/AB design (Experiment 2) supports the UCB model. However, the interference effects observed in these hybrid designs appears to be somewhat smaller than interference effects typically found in the standard PRP and AB designs. This finding raises the suspicion that a central bottleneck cannot fully account for interference in both designs. Consistent with this conjecture, Experiment 5 showed that performance of an RSVP Task 2 was impaired more when the Task 1 stimulus was presented in the same input modality. This effect cannot easily be explained by the UCB model, because the two tasks were explicitly designed so that they would require roughly equal amounts of central interference. Results similar to those of the present Experiment 5 have also been obtained by Arnell and Duncan (1999).

The fact that interference effects were greater in the within-modal condition than in the cross-modal condition of Experiment 5 suggests that RSVP tasks interfere at a central stage *and* at some earlier perceptual stage. For example, perceptual interference at short lags might result in categorical representations that are relatively weak. These representations might then decay even further because short-term consolidation of T2 must wait until short-term consolidation of T1 is complete. Thus, perceptual interference reduces the initial activation of the categorical representation of T2 and central interference then causes the activation to decay even further.

Further work is needed to specify better the nature of the non-central sources of interference in the AB paradigm and how they interact with the sources of central interference. For the present, however, it seems safe to conclude that although a central bottleneck plays an important role in both the AB and PRP paradigms, it is probably not the only source of interference.

Relation of the central bottleneck to other attentional limitations

The present data and those of Jolicœur and his colleagues suggest that AB and PRP effects are both due in large part to a central bottleneck. In this section we briefly discuss whether a central bottleneck might play a role in other well-known phenomena.

Covert movements of spatial attention. People are capable of choosing particular regions of a visual display and processing stimuli in these regions more intensively than other stimuli. Furthermore, they are capable of rapidly altering which stimuli are attended and which are unattended without moving the eyes (Helmholtz 1924, p. 455; James 1950 [1890]). Two lines of evidence suggest that these covert movements of visual attention are not subject to the central bottleneck constraint.

Pashler (1991), for example, presented a tone stimulus followed after a variable SOA by a masked visual stimulus consisting of an array of eight letters and a bar probe. Subjects made a speeded response to the tone, and then indicated (at their leisure) the identity of the letter located next to the bar probe. If subjects were unable to allocate covert attention to the probed item until after they had selected a response to the tone then, at short SOAs, attentional allocation would not have occurred until well after the display had been masked. This delay in the allocation of attention should have catastrophic effects on performance, as demonstrated by a control condition in which the *presentation* of the bar probe was delayed until after the display had been masked. Contrary to this prediction, letter task accuracy was very high at all SOAs, indicating that spatial attention was in fact allocated to the probed item well before tone-task response selection had been completed. Hence, Pashler concluded that movements of spatial attention are not limited by the central bottleneck.

Johnston, McCann, and Remington (1995) arrived at a similar conclusion regarding movements of spatial attention. These authors used locus-of-slack logic to show that the processing stage(s) affected by a spatial cuing manipulation come *at or before* stimulus identification, whereas the processing stage(s) that cause interference in the PRP paradigm come *after* stimulus identification. Thus it appears that the locus of the spatial-attention effect is earlier than the locus of the central bottleneck.

Repetition blindness. When the same target is presented twice in an RSVP stream, subjects often miss the second presentation of the target (cf. e.g. Kanwisher 1987). This effect, known as repetition blindness (or 'RB'), occurs even when the RSVP stream consists of words, and failure to detect a repeated word makes a sentence ungrammatical. Some have attributed RB to a failure of token individuation; in other words, subjects fail to establish a separate token for the repeated target (e.g. Kanwisher 1987). This view can be incorporated into the UCB model by assuming that token individuation belongs to the set of demanding central operations; hence, token individuation of the second instance of a target is postponed until token individuation of the first instance of that target has been completed.

On the other hand, a variety of evidence suggests that failures in retrieval may play a key role in RB (Armstrong and Mewhort 1995; Fagot and Pashler 1995; Whittlesea *et al.* 1995; Whittlesea and Wai 1997). Fagot and Pashler (1995), for instance, argued that repeated items are perceived just as clearly (or unclearly) as non-repeated items, but that subjects are biased against reporting the repeated items. In favour of this account, Fagot and Pashler showed that RB effects are greatly reduced or eliminated in designs that minimized the subject's memory load (but see also Hochhaus and Johnston 1996). Further, they were able to show that RB is sensitive to the type of retrieval cue presented *after* the RSVP stream has ended, directly implicating retrieval failure rather than perceptual failure as the cause of RB. These data argue against a central bottleneck account of the RB effect. In addition, Chun (1997) has shown that

(a) AB and RB have different time courses, (b) RB is found in cases where AB is absent, and (c) AB is found in cases where RB is absent. Given the evidence in this chapter that the AB effect is due in part to a central bottleneck, Chun's results would appear to indicate directly that RB is not due to a central bottleneck.

Summary

This chapter examined the possibility that the AB effect (interference between two unspeeded RSVP tasks) and the PRP effect (interference between two speeded-response tasks) are attributable to the same central bottleneck. If this unified central bottleneck model is correct, then substantial dual-task interference should occur in hybrid AB/PRP designs where Task 1 is an unspeeded RSVP task and Task 2 is a speeded-response task, and vice versa. Experiments 1 and 2 confirmed this prediction. Furthermore, Experiments 3 and 4 showed that prolongation of the Task 1 central stage caused a deeper and longer-lasting interference effect on an unspeeded RSVP Task 2. This 'Task-1 carryover' effect directly supports the assertion that central stages on Task 1 postpone the critical operations needed to form a reportable percept of the Task 2 RSVP target. Although the AB effect appears to be due in part to a central bottleneck, it appears to be due to specifically visual interference as well. Experiment 5 demonstrated substantially greater impairment of an RSVP Task 2 when the Task 1 stimulus was visual than when the Task 1 stimulus was auditory, even though both Task 1 judgements required similar amounts of central processing. Whereas the present chapter showed that a central bottleneck model can account (at least in part) for interference in two very different types of dual-task designs (AB and PRP), the central bottleneck appears not to provide a satisfactory account of certain other dual-task phenomena, such as repetition blindness. Furthermore, the central bottleneck does not appear to play a major role in covert movement of spatial attention across the visual field.

Acknowledgements

This work was supported by a National Research Council postdoctoral fellowship to Eric Ruthruff and by grants from the National Institute of Mental Health (1-R01-MH45584) and the National Science Foundation (SBR #9729778). We thank James C. Johnston, Robert McCann, Roger Remington, Valerie Lewis, Beth Siegel, Julie Creagh, and Mark Van Selst for useful discussions.

References

Armstrong, I. T. and Mewhort, D. J. (1995). Repetition deficit in rapid-serial-visual-presentation displays: encoding failure or retrieval failure? *Journal of Experimental Psychology: Human Perception and Performance*, **21**, 1044–52.

Arnell, K. M. and Duncan, J. (in press). Specific and general sources of dual-task cost in stimulus identification and response selection. *Cognitive Psychology*.

Broadbent, D. E., and Broadbent, M. H. P. (1987). From detection to identification: response to multiple targets in rapid serial visual presentation. *Perception and Psychophysics*, **42**, 105–13.

Carrier, M. and Pashler, H. E. (1995). Attentional limitations in memory retrieval. *Journal of Experimental Psychology: Learning, Memory and Cognition*, **21**, 1339–48.

Chun, M. M. (1997). Types and tokens in visual processing: a double dissociation between the attentional blink and repetition blindness. *Journal of Experimental Psychology: Human Perception and Performance*, **23**, 738–55.

Chun, M. M. and Potter, M. C. (1995). A two-stage model for multiple target detection in rapid serial visual presentation. *Journal of Experimental Psychology: Human Perception and Performance*, **21**, 109–27.

Duncan, J. (1984). Selective attention and the organization of visual information. *Journal of Experimental Psychology: General*, **113**, 501–17.

Fagot, C. and Pashler, H. E. (1992). Making two responses to a single object: implications for the central attentional bottleneck. *Journal of Experimental Psychology: Human Perception and Performance*, **18**, 1058–79.

Fagot, C. and Pashler, H. E. (1995). Repetition blindness: perception or memory failure? *Journal of Experimental Psychology: Human Perception and Performance*, **21**, 275–92.

Gottsdanker, R. (1980). The ubiquitous role of preparation. In G. E. Stelmach and J. Requin (eds), *Tutorials in motor behavior*, pp. 315–71. Amsterdam: North-Holland.

Greenwald, A. (1972). On doing two things at once: time sharing as a function of ideomotor compatibility. *Journal of Experimental Psychology*, **94**, 52–7.

Greenwald, A. and Shulman, H. G. (1973). On doing two things at once: II. Elimination of the psychological refractory period effect. *Journal of Experimental Psychology*, **11**, 72–89.

Helmholtz, H. (1924). *Helmholtz's Treatise on Physiological Optics*, translated from the 3rd German edn, ed. James P. C. Southall. Rochester, NY: The Optical Society of America.

Hochhaus, L., and Johnston, J. C. (1996). Perceptual repetition blindness effects. *Journal of Experimental Psychology: Human Perception and Performance*, **22**, 355–66.

James, W. (1950 [1890]). *The principles of psychology*, Vol 1. New York: Dover.

Johnston, J. C. and Delgado, D. F. (1993). Bypassing the single-channel bottleneck in dual-task performance. Paper presented at the annual meeting of the Psychonomic Society, Washington, DC.

Johnston, J. C., McCann, R. S., and Remington, R. W. (1995). Chronometric evidence for two types of attention. *Psychological Science*, **5**, 365–9.

Johnston, J. C., Ruthruff, E., and McCann, R. S. (2001). Chronometric evidence for a late selection locus to the attentional blink. Unpublished manuscript.

Jolicœur, P. (1998). Modulation of the attentional blink by on-line response selection: evidence from speeded and unspeeded Task$_1$ decisions. *Memory and Cognition*, **26**, 1014–32.

Jolicœur, P. (1999a). Concurrent response selection demands modulate the attentional blink. *Journal of Experimental Psychology: Human Perception and Performance*, **25**, 1097–113.

Jolicœur, P. (1999b). Restricted attentional capacity between sensory modalities. *Psychonomic Bulletin and Review*, **6**, 87–92.

Jolicœur, P., and Dell'Acqua, R. (1998). The demonstration of short-term consolidation. *Cognitive Psychology*, **36**, 138–202.

Jolicœur, P., and Dell'Acqua, R. (1999). Attentional and structural constraints on visual encoding. *Psychological Research*, **62**, 154–164.

Kanwisher, N. G. (1987). Repetition blindness: type recognition without token individuation. *Cognition*, **27**, 117–43.

Luck, S. J., Vogel, E. K., and Shapiro, K. L. (1996). Word meanings can be accessed but not reported during the attentional blink. *Nature*, **383**, 616–18.

McCann, R. S. and Johnston, J. C. (1992). Locus of the single-channel bottleneck in dual-task interference. *Journal of Experimental Psychology: Human Perception and Performance*, **18**, 471–84.

Maki, W. S., Frigen, K., and Paulson, K. (1997). Associative priming by targets and distractors during rapid serial visual presentation: Does word meaning survive the attentional blink? *Journal of Experimental Psychology: Human Perception and Performance*, **23**, 1014–34.

Pashler, H. E. (1984). Processing stages in overlapping tasks: evidence for a central bottleneck. *Journal of Experimental Psychology: Human Perception and Performance*, **10**, 358–77.

Pashler, H. E. (1989). Dissociation and dependencies between speed and accuracy: evidence for a two-component theory of divided attention in simple tasks. *Cognitive Psychology*, **21**, 469–514.

Pashler, H. E. (1991). Shifting visual attention and selecting motor responses: distinct attentional mechanisms. *Journal of Experimental Psychology: Human Perception and Performance*, **17**, 1023–40.

Pashler, H. E. and Johnston, J. C. (1989). Chronometric evidence for central postponement in temporally overlapping tasks. *Quarterly Journal of Experimental Psychology, Section A: Human Experimental Psychology*, **41**, 19–45.

Pashler, H. E. and Johnston, J. C. (1998). Attentional limitations in dual-task performance. In H. Pashler (ed.), *Attention*, pp. 155–89. Hove, UK: Psychology Press.

Pashler, H. E., Carrier, M., and Hoffman, J. (1993). Saccadic eye movements and dual-task interference. *Quarterly Journal of Experimental Psychology*, **46A**, 51–82.

Raymond, J. E., Shapiro, K., and Arnell, K. M. (1992). Temporary suppression of visual processing in an RSVP task: an attentional blink? *Journal of Experimental Psychology: Human Perception and Performance*, **18**, 849–60.

Ruthruff, E., Miller, J. O., and Lachmann, T. (1995). Does mental rotation require central mechanisms? *Journal of Experimental Psychology: Human Perception and Performance*, **21**, 552–70.

Ruthruff, E., Johnston, J. C. and Van Selst, M. A. (2001). Why practice reduces the dual-task interference. *Journal of Experimental Psychology: Human Perception and Performance*, **27**, 3–21.

Shapiro, K., Raymond, J. E. and Arnell, K. M. (1994). Attention to visual pattern information produces the attentional blink in rapid serial visual presentation. *Journal of Experimental Psychology: Human Perception and Performance*, **20**, 357–71.

Shapiro, K. L., Caldwell, J., and Sorensen, R. E. (1992). Personal names and the attentional blink: a visual 'cocktail party' effect. *Journal of Experimental Psychology: Human Perception and Performance*, **23**, 504–14.

Shapiro, K., Driver, J., Ward, R. and Sorensen, R. E. (1997*b*). Priming from the attentional blink: a failure to extract visual tokens not visual types. *Psychological Science*, **8**, 95–100.

Van Selst, M. A., Ruthruff, E., and Johnston, J. C. (1999). Can practice eliminate the Psychological Refractory Period effect? *Journal of Experimental Psychology: Human Perception and Performance*, **25**, 1268–83.

Vince, M. A. (1948). The intermittency of control movements and the psychological refractory period. *British Journal of Psychology*, **38**, 149–57.

Vogel, E. K., Luck, S. J., and Shapiro, K. L. (1998). Electrophysiological evidence for a postperceptual locus of suppression during the attentional blink. *Journal of Experimental Psychology: Human Perception and Performance*, **24**, 1656–74.

Welford, A. T. (1952). The 'psychological refractory period' and the timing of high-speed performance— a review and a theory. *British Journal of Psychology*, **43**, 2–19.

Whittlesea, B. W. A., and Wai, K. H. (1997). Reverse 'repetition blindness' and release from 'repetition blindness': constructive variations on the 'repetition blindness' effect. *Psychological Research*, **60**, 173–82.

Whittlesea, B. W. A., Dorken, M. D., and Podrouzek, K. W. (1995). Repeated events in rapid lists: Part I. Encoding and representation. *Journal of Experimental Psychology: Learning, Memory, and Cognition*, **21**, 1670–88.

7

Multiple sources of interference in dual-task performance: the cases of the attentional blink and the psychological refractory period

Steven J. Luck and Edward K. Vogel

Abstract

Limitations in dual-task performance are frequently studied by presenting subjects with two targets on each trial and varying the delay between them. Under many conditions, subjects are either slower or less accurate at detecting the second target when it occurs shortly after the first target, presumably because the subject cannot efficiently respond to the second target while the first target is still being processed. In this chapter, we explore two such dual-task paradigms, the psychological refractory period paradigm and the attentional blink paradigm. In the psychological refractory period paradigm, two highly discriminable stimuli are presented on each trial and the subject makes a speeded response to both stimuli. Interference in this paradigm takes the form of increased reaction times for the second target when it appears shortly after the first target. Two targets are also used in the attentional blink paradigm, but they are embedded in a rapid stream of distractor items; interference in this paradigm takes the form of decreased accuracy for the second target when it appears shortly after the first target. In this chapter, we describe a series of electrophysiological experiments that indicate that the impairments for the second target in these paradigms reflect two different forms of interference. Specifically, the impaired accuracy for the second target in the attentional blink paradigm appears to reflect input-level interference, whereas the slowed reaction times for the second target in the psychological refractory period paradigm appear to reflect output-level interference. However, the input and output interference effects are not completely independent, and we therefore propose that both may reflect limitations in accessing a central working memory system.

Overview

Human performance of both simple and complex tasks is often impaired when multiple tasks must be performed concurrently, especially when the individual tasks are not highly practised. An inexperienced driver, for example, cannot easily drive while carrying on a conversation, and even an experienced driver prefers silence when driving under difficult conditions. Interference can occur between two tasks with heavy input requirements, such as listening to two simultaneous conversations, and it can also occur between two tasks with heavy output requirements, such as writing with one hand and drawing with the other. Input and output interference have been intensively studied in recent years with two experimental paradigms, the attentional blink paradigm (Broadbent and Broadbent 1987; Raymond *et al.* 1992; Reeves and Sperling 1986) and the psychological refractory period paradigm (De Jong 1993; Pashler 1994; Welford 1952). These paradigms are illustrated in Fig. 7.1.

In the attentional blink paradigm (see Fig. 7.1A), a very rapid stream of approximately 20 stimuli is presented at fixation on each trial, and the subject is required to detect two targets (called T1 and T2) from this stream. Subjects frequently fail to detect T2 if it occurs shortly after T1, and this temporary impairment in T2 accuracy is called the attentional blink because it

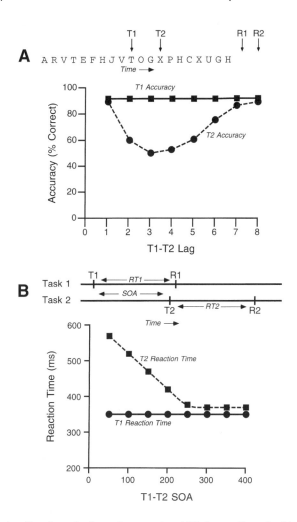

Fig. 7.1 A. Example stimuli and results from the attentional blink paradigm. In this version (based on Raymond *et al.* 1992), the letters are presented in black on a grey background except for the first target (T1), which is white. At the end of the trial, subjects make two responses (R1 and R2), one to indicate the identity of the white letter and one to indicate whether the letter X (T2) was present or absent. B. Example stimuli and results from the psychological refractory period paradigm. In this paradigm, two targets are presented (T1 and T2), and subjects are required to make speeded responses (R1 and R2) as soon as possible after each target. The amount of time between the onset of T1 and the onset of T2 is the called the stimulus onset asynchrony (SOA).

is analogous to the impairment in accuracy that would result from a T1-triggered eyeblink. In this paradigm, the subjects are not required to respond until the end of the trial, and so the errors in T2 detection appear to reflect interference in input processing rather than output processing.

  The psychological refractory period paradigm is similar to the attentional blink paradigm, but stresses output processing rather than input processing. As in the attentional blink paradigm, two targets are presented on each trial in the psychological refractory period paradigm (see Fig. 7.1B), but there are no other distractor items, and each target requires an immediate, speeded response. When the amount of time between the onset of T1 and the onset of T2 (the stimulus onset asynchrony or SOA) is reduced, reaction times (RTs) for T2 become progressively longer. Thus both the attentional blink and psychological refractory period paradigms manipulate the interference between two independent tasks by varying the delay between two targets, but the attentional blink paradigm stresses input processing by presenting a very rapid stream of stimuli, whereas the psychological refractory period paradigm stresses output processing by requiring speeded responses.

Although the attentional blink and psychological refractory period paradigms appear to stress input and output systems, respectively, recent research indicates that they actually reflect relatively central limitations in information processing rather than low-level sensory or motor interference. In fact, Jolicœur (1998, 1999) and Arnell (this volume, Chapter 8) have proposed that the two paradigms tap a common limited resource. The purpose of the present chapter is to describe electrophysiological evidence indicating that interference in the attentional blink paradigm occurs at an earlier stage than interference in the psychological refractory period paradigm, as might be expected on the basis of the task descriptions. However, the chapter will conclude with an explanation of how these different stages of interference can be reconciled with the proposal of a common central limitation, namely working memory access. That is, both input and output processes may need to access working memory, and interference may occur when multiple processes attempt to access working memory simultaneously.

Locus of the attentional blink

There is a growing consensus that the impairment in T2 accuracy during the attentional blink does not reflect a failure to identify T2, but instead reflects a failure to store T2 in a durable form in working memory. That is, when T2 is presented while T1 is still being processed, T2 is identified but is then overwritten by the next item in the input stream. There are several pieces of evidence supporting the conclusion that T2 is fully identified (Maki et al. 1997; Shapiro et al. 1997a; Shapiro et al. 1997b); but we will focus on a study that used event-related potential (ERP) recordings to covertly monitor the processing of T2 (Luck et al. 1996; Vogel et al. 1998).

ERPs are voltage deflections that can be recorded non-invasively from normal human subjects while they perform cognitive tasks. ERPs are extracted from electroencephalogram (EEG) recordings by means of signal-averaging procedures, in which the EEG segments following each of several stimuli are averaged together. When multiple stimulus-locked EEG segments are averaged together, the random EEG activity averages to near zero and the stimulus-related neural response can be observed. As is illustrated in Fig. 7.2, the resulting averaged ERP waveform consists of a sequence of positive and negative voltage deflections that are called peaks, waves, or components. Each peak is named according to its polarity and either its latency (for example, 'P220' for a positive peak at 220 ms post-stimulus) or its ordinal position in the waveform (for example 'N2' for the second major negative peak). The initial peaks reflect sensory processes, and later peaks reflect progressively higher-level cognitive processes. ERP recordings have two distinct advantages for assessing the stage of processing at which a given experimental effect occurs. First, the ERP waveform provides a continuous measure of processing following a stimulus, allowing the timing of an effect to be measured with millisecond accuracy. Second, ERPs are elicited even for stimuli that do not elicit an overt response, making

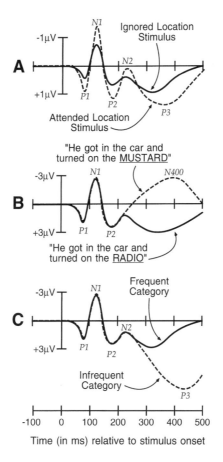

Fig. 7.2 Example event-related potential (ERP) waveforms. Note that negative is plotted upward and that time zero represents the onset of the ERP-eliciting stimulus. A. Example of the effects of spatial attention on the P1 and N1 components. In this example, stimuli are presented in random order at two locations, one of which is attended, and stimuli presented at the attended location are found to elicit larger P1 and N1 waves than stimuli presented at the ignored location (see, for example, Mangun and Hillyard 1990). B. Example of the effects of semantic mismatch on the N400 component. In this example (based on Kutas and Hillyard 1980), subjects view sentences presented one word at a time on a computer monitor. The waveforms shown here illustrate the response elicited by the final word in the sentence. C. Example of the effects of stimulus probability on the P3 component. If, for example, subjects categorize names as male and female and the male names are less frequent than the female names, the male names will elicit a larger P3 wave even if any individual name is presented only once (see, for example, McCarthy and Donchin 1981). Reprinted with permission from Vogel *et al.* (1998).

them ideal for measuring the processing of stimuli that subjects fail to detect (for more details, see Hillyard and Picton 1987; Luck and Girelli 1998).

Early sensory processes during the attentional blink—the P1 and N1 waves

Earlier studies of spatial attention have demonstrated that the sensory-evoked P1 and N1 components are larger for stimuli presented at attended locations than for stimuli presented at ignored locations (see, for example, Luck *et al.* 1994; Mangun and Hillyard 1987; Van Voorhis and Hillyard 1977). As is illustrated in Fig. 7.2A, these effects begin within 100 ms of stimulus onset and appear to consist of a relatively pure change in the amplitude of the sensory response. Moreover, the same effects are found for targets, for non-targets, and for completely task-irrelevant stimuli, suggesting that the P1 and N1 modulations reflect a preset sensory gain control mechanism (for more details, see Hillyard *et al.* 1998). Functional neuroimaging studies suggest that the P1 attention effect arises in extrastriate areas of visual cortex (Heinze *et al.* 1994; Mangun *et al.* 1997), and single-unit recordings from monkeys have shown similar modulations of sensory activity in visual area V4 (Luck *et al.* 1997). Thus the ERP technique has been very useful in demonstrating that spatial attention influences early sensory processes.

We used a similar approach to determine whether the attentional blink reflects a suppression of sensory processes. In our first experiment, we measured the P1 and N1 components to determine whether they are suppressed for stimuli presented during the attentional blink period, just as they are for stimuli presented at ignored locations in spatial attention experiments. As is illustrated in Fig. 7.3A, the rapid presentation of stimuli in the attentional blink paradigm leads to an overlap problem. That is, the response to a given stimulus lasts for hundreds of milliseconds, overlapping the responses to several of the subsequent stimuli. This makes it difficult to isolate the response to any given stimulus, which is necessary for using ERPs to assess sensory processing during the attentional blink period. To overcome this problem, we used a 'probe' technique, in which a task-irrelevant probe stimulus was presented simultaneously with T2 on some of the trials. The ERP waveform on these trials reflects the response to the stimulus stream plus the response to the probe, and the response to the probe can therefore be isolated by subtracting no-probe trials from probe trials. The isolated probe-elicited ERP response can then be used to assess sensory processing at the time of the probe.

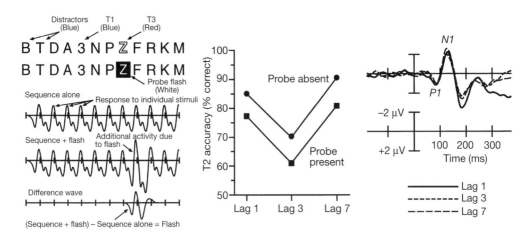

Fig. 7.3 Paradigm and results for the P1/N1 experiment. A. Example of stimulus sequences and overlap problem. All items were successively presented at fixation. At the end of each trial, subjects reported whether T1 was an odd or even digit and whether T2 was a consonant or a vowel. A task-irrelevant probe flash appeared around T2 on 50 per cent of trials. Each stimulus in the sequence was expected to elicit a response that overlapped with the other stimuli, but an additional response was also expected on probe-present trials; this response was isolated by subtracting probe-absent trials from probe-present trials. B. Mean accuracy for identifying T2 as a function of lag and probe presence. C. ERP difference waveforms from lateral occipital sites, averaged across subjects and the left and right hemispheres. Adapted with permission from Vogel *et al.* (1998).

This approach is similar to subtraction techniques that have been used both in previous ERP studies and in neuroimaging studies (see, for example, Luck *et al.* 1993; Petersen *et al.* 1992). For example, Luck *et al.* (1993) found that probes presented at the location of a visual search target elicited larger P1 and N1 waves than probes presented at the location of a non-target item. If sensory processing is suppressed during the attentional blink, then the P1 and N1 waves should similarly be suppressed for probes presented during the attentional blink period relative to probes presented before or after this period.

The results of this experiment are shown in Fig. 7.3 (panels B and C). A substantial decrement in T2 accuracy was observed at lag 3 relative to lags 1 and 7, which is the typical attentional blink pattern. In contrast, the probe-elicited P1 and N1 components were not suppressed during the attentional blink. Thus, there was no evidence of sensory suppression during the attentional blink period, which contrasts with the sensory suppression observed for stimuli presented at unattended locations in studies of spatial attention.

Although these results suggest that the impaired performance observed during the attentional blink arises at a later stage than the effects of spatial attention, there are two limitations on this conclusion. First, this conclusion is based on a null result, that is, a lack of a significant difference between the sensory responses elicited during the attentional blink period and the sensory responses elicited before and after this period. This is not a major problem, because the statistical power of this experiment was similar to that of spatial attention experiments in which significant P1 and N1 modulations were observed. A more substantial problem, however, is that it is possible that a later perceptual process was suppressed during the attentional blink. That is, these results provide no means of assessing whether T2 was *fully* identified. We therefore conducted a follow-up experiment to provide more direct evidence that T2 is identified during the attentional blink period, even though subjects cannot accurately report it.

Word identification during the attentional blink—the N400 component

To determine whether T2 is identified during the attentional blink period, this experiment focused on the language-related N400 component. As is illustrated in Fig. 7.1B, the N400 component is typically elicited by words that mismatch a previously established semantic context. For example, a large N400 would be elicited by the last word of the sentence, 'John sat at the kitchen table and drank a large glass of aluminium,' but a small N400 would be elicited if this same sentence ended with the word 'milk'. The N400 can also be elicited with simple word pairs, such that the second word in CAT–VASE will elicit a large N400, whereas the second word in TABLE–CHAIR will elicit a small N400. The N400 component is well-suited for determining whether a word has been identified, because a semantically mismatching word cannot elicit a larger response unless that word has been identified to the point of semantic (or at least lexical) access. Thus, if a given word elicits a larger N400 when it mismatches the semantic context than when it matches the semantic context, this can be taken as strong evidence that the word was identified to a fairly high level. We therefore designed an experiment to determine whether the N400 component would be suppressed for words presented during the attentional blink period (as it is for words presented at unattended locations, as demonstrated by McCarthy and Nobre 1993).

The stimuli and task for this experiment are illustrated in Fig. 7.4A. Each trial began with the presentation of a 1000-ms 'context word' that established a semantic context for that trial. After a 1000-ms delay, a rapid stream of stimuli was presented at fixation. Each stimulus was 7 characters long. Distractor stimuli were 7-letter sequences of randomly-selected consonants. T1 was

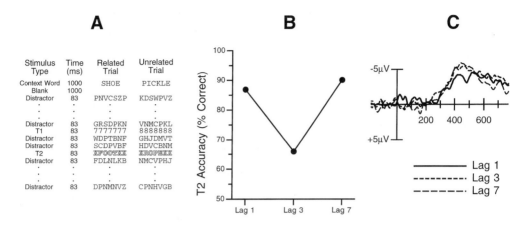

Fig. 7.4 Paradigm and results for the N400 experiment. A. Example stimuli. B. Mean discrimination accuracy for T2 as a function of lag. C. ERP difference waveforms, formed by subtracting related T2 trials from unrelated T2 trials. These waveforms were recorded from a central midline site (Cz) and were averaged across participants. Adapted with permission from Vogel *et al.* (1998).

a digit that was repeated 7 times to form a 7-character stimulus. T2 was a word, presented in red, that was either semantically related or semantically unrelated to the context word that had been presented at the beginning of the trial. The T2 word was flanked by Xs, if necessary, to ensure that it was 7 characters long. At the end of each trial, the subjects made two responses, one to indicate whether T1 was an odd digit or an even digit, and another to indicate whether T2 was semantically related or unrelated to the context word. Related and unrelated T2 words occurred equally often (as did odd and even T1 digits), and related words were very highly related to the context word (e.g., TREE–LEAF). Each T2 word was presented twice for each subject (in widely separated trial blocks), appearing once as a related word and once as an unrelated word. This made it possible to be certain that any differences in the ERPs elicited by related and unrelated words were due to their semantic relationship with the context word rather than any peculiarities of the words themselves.

As in the previous experiment, overlapping ERP responses were a concern in this experiment. To overcome this problem, we computed difference waves in which we subtracted the response elicited by T2 when it was a related word from the response elicited by T2 when it was an unrelated word. The responses to the other items in the stimulus stream should be essentially identical on related-T2 and unrelated-T2 trials, and this difference wave therefore provides a relatively pure measure of the brain's differential response as a function of the semantic relatedness of T2 relative to the context word. A large difference between related-T2 and unrelated-T2 trials can therefore be used as evidence that the T2 word was identified sufficiently to determine its semantic relationship to the context word.

The results of this experiment are shown in Fig. 7.4 (panels B and C). As in the previous experiment, T2 accuracy was highly suppressed at lag 3 relative to lags 1 and 7, but the N400 component was equally large at all three lags. Thus, although the subjects could not accurately report whether T2 was related or unrelated to the context word at lag 3, the N400 wave differentiated between related and unrelated trials, indicating that the brain made this discrimination quite accurately. This result provides strong evidence that stimuli are fully identified during the attentional blink, but are not reported accurately because they are not stored in a durable form in working memory. These results also suggest that a great deal of processing can occur in the absence of awareness (Greenwald *et al.* 1996; He *et al.* 1996; Jacoby and Whitehouse 1989; Kunst-Wilson and Zajonc 1980), although it is possible that subjects were briefly aware of the T2 word even though they could not report its semantic relationship moments later.

Like the P1/N1 experiment, the conclusions of this experiment rely on a null finding, the lack of an N400 suppression during the attentional blink period. Thus, before concluding that the presence of an unsuppressed N400 at lag 3 indicates a lack of perceptual interference, it is necessary to demonstrate that the N400 is sufficiently sensitive to perceptual degradation. To demonstrate this, we conducted an experiment in which subjects ignored T1 and performed only the T2 task. We then added varying amounts of simultaneous masking noise to T2 so that we could assess the effects of perceptual degradation on N400 amplitude. As is shown in Fig. 7.5, the addition of simultaneous masking noise led to a large and statistically significant decline in N400 amplitude that paralleled a decline in T2 discrimination accuracy. Importantly, this experiment used the same number of trials and subjects as the attentional blink experiment, and it therefore had a comparable level of statistical power. Moreover, the difference in accu-

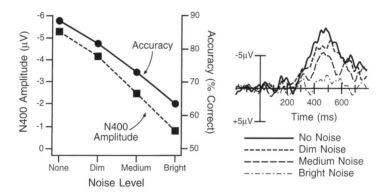

Fig. 7.5 Comparison of N400 amplitude (averaged across scalp sites) and behavioural accuracy as a function of visual noise intensity. The grand average ERP waveforms at the right of the figure were recorded at the central midline site (Cz). Reprinted with permission from Vogel *et al.* (1998)

racy between the no-noise condition and the maximum noise condition was comparable to the difference in accuracy between lag 3 and lags 1 and 7, and this difference was accompanied by a very large decrement in N400 amplitude. Thus, the finding of no N400 suppression at lag 3 provides strong evidence that perceptual processing is unimpaired during the attentional blink.

Working memory encoding during the attentional blink—the P3 component

Because perceptual processing appears to be unimpaired during the attentional blink, and because the use of unspeeded responses implies that the impaired accuracy during the attentional blink is not caused by errors in output processes, it seems likely that the attentional blink reflects an impairment in working memory encoding (for similar conclusions, see Chun and Potter 1995; Maki *et al.* 1997; Shapiro *et al.* 1997*a*). To test this proposal more directly, we conducted an additional experiment focusing on the P3 component, which appears to reflect the updating of working memory (Donchin 1981; Donchin and Coles 1988). Unlike the P1, N1, and N400 components, we expected that the P3 wave would be suppressed during the attentional blink.

As is illustrated in Fig. 7.1C, the hallmark of the P3 wave is that it is much larger for improbable stimulus categories than for probable stimulus categories. In a typical P3-eliciting experiment, subjects might view a sequence of stimuli presented at one per second, with 90 per cent of the stimuli being the letter O and 10 per cent being the letter X. The subjects would typically be required to respond differentially to the Xs and Os, perhaps counting the number of Xs or pressing one button for Xs and another for Os. In this sort of 'oddball' experiment, the improbable Xs would elicit a much larger P3 wave than the probable Os (for reviews, see Donchin 1981; Johnson 1986). P3 amplitude is actually a function of the probability of the task-defined category rather than the probability of the physical stimulus. For example, consider an experiment in which subjects are presented with a sequence of names and asked to press one button for male names and another button for female names. If male names occur less frequently than female names, the male names will elicit a larger P3 than the female names, even if any given name is only presented once. On the basis of these and other findings, Donchin (1981; Donchin and Coles 1988) has proposed that the P3 wave reflects a cognitive process related to the updating of working memory.

To examine the P3 wave in the attentional blink paradigm, we conducted an experiment in which one of the two T2 alternatives was infrequent and therefore had the potential to elicit a P3 wave. As in the experiment that examined the P1 and N1 waves, T1 in this experiment was a digit, and T2 was a white letter; the distractors were black letters. At the end of each trial, subjects made two button-press responses, one to indicate whether T1 was an odd or even digit and another to indicate whether T2 was an E or some other letter. T2 was an E on 15 per cent of trials and a randomly selected non-E letter on the remaining 85 per cent of trials. We therefore expected a larger P3 wave when T2 was an E than when it was a non-E, even though the letter E was more common than any other individual letter. To eliminate overlapping ERPs from the other stimuli in the stream, we computed difference waves in which the ERP waveform elicited when T2 was a non-E letter was subtracted from the ERP elicited when T2 was an E. This difference wave provides a relatively pure measure of the brain's differential processing of probable and improbable T2 stimuli (i.e., the P3 wave).

The results of this experiment are shown in Fig. 7.6. As in the previous experiments, T2 accuracy was greatly reduced at lag 3. Unlike what occurred in the previous experiments, however, the P3 wave was also completely suppressed at lag 3. Thus, as is summarized in Fig. 7.7, the P1, N1, and N400 components are unsuppressed at lag 3, but the P3 wave is completely eliminated during the attentional blink period. These results are consistent with the hypothesis that the attentional blink reflects a failure to store T2 in a durable form in working memory while T1 is being processed (Chun and Potter 1995; Shapiro et al. 1994).

Overwriting of T2 during the attentional blink—delayed P3 latency

Giesbrecht and Di Lollo (1998) have provided converging behavioural evidence for this proposal, showing that T2 is missed because it is overwritten by the subsequent items in the stimulus stream (see also Chun and Potter 1995; Isaak et al. 1999). Specifically, they demonstrated that the attentional blink is eliminated if T2 is the last item in the stimulus stream, but a normal attentional blink is observed if T2 is followed by at least one item. Moreover, this is not simply a ceiling effect, because the attentional blink is still eliminated if T2 is perceptually degraded

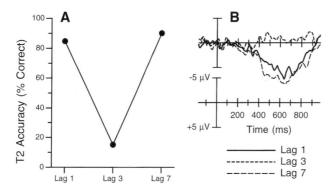

Fig. 7.6   A. Mean discrimination accuracy for T2 as a function of lag for the P3 attentional blink experiment. B. ERP difference waveforms, formed by subtracting frequent-T2 trials from infrequent-T2 trials. These waveforms were recorded at a midline parietal site (Pz) and were averaged across subjects. Adapted with permission from Vogel *et al.* (1998).

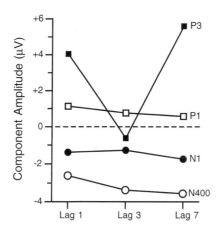

Fig. 7.7   Mean amplitudes for the P1, N1, N400, and P3 components as a function of T1–T2 lag. Reprinted with permission from Vogel *et al.* (1998).

by presenting it simultaneously with another letter. These findings imply that subjects are able to encode T2 into working memory if it occurs during the attentional blink and is not immediately overwritten by another stimulus, presumably because enough information about T2 is still available when the processing of T1 has completed. In other words, the entry of T2 into working memory is merely delayed when it is the last item in the stream, leading to no deficit in accuracy when T2 is reported at the end of the trial. To test this, we have recently conducted a new experiment to assess whether P3 latency is delayed under these conditions (Vogel and Luck, unpublished observations).

Previous studies have shown that P3 latency is a sensitive measure of the amount of time required to perceive and categorize a stimulus (e.g., Kutas *et al.* 1977; McCarthy and Donchin 1981). Because the P3 wave is sensitive to the task-defined category of a stimulus, it cannot be elicited until the stimulus has been perceived and categorized. As a result, P3 latency is increased when the amount of time required to discriminate and categorize a target is increased. For example, P3 latency perfectly parallels RT in difficult visual search tasks (Luck and Hillyard 1990). If working memory encoding is merely delayed when T2 occurs during the usual attentional blink period but is the last item in the stream, then the P3 wave should be delayed on such trials. In other words, even though T2 accuracy exhibits no attentional blink when T2 is the last item in the stream, the P3 wave should exhibit evidence of delayed working memory encoding.

We examined P3 latency during the attentional blink using essentially the same procedures as described above (i.e. Vogel *et al.* 1998). The primary difference, however, was that T2 was either the last item in the stream (unmasked) or the next-to-last item in the stream (masked). Only lags 3 and 7 were tested. When T2 was unmasked, accuracy was near-perfect both during and after the typical attentional blink period (i.e., at lags 3 and 7, respectively). P3 amplitude was equivalent for unmasked T2 items presented for both lags; however, the onset latency of the P3 elicited at lag 3 was significantly later than at lag 7. Thus, P3 onset latency during the normal attentional blink period was delayed compared to that elicited for T2 items presented after the attentional blink period, despite equivalent behavioural performance for both conditions. These results suggest that working memory encoding is indeed delayed for unmasked T2 items when they are presented during the attentional blink period.

Together, the results of these ERP experiments indicate that an object presented during the attentional blink period is fully identified, but will be overwritten in working memory by the next item in the stream unless it is the last item in the stream, in which case it will be encoded into working memory after a delay of approximately 100 ms.

Locus of the psychological refractory period

Unlike theories of the attentional blink, the leading theories of the psychological refractory period propose that the slowed RTs observed in this paradigm reflect interference in output-related processes, such as response selection (Pashler 1994) and response initiation (De Jong 1993). In support of these proposals, Osman and Moore (1993) demonstrated that the lateralized readiness potential (LRP)—a response-related ERP component—is delayed when the interval between T1 and T2 is shortened. However, it is still possible that the interference occurs at an earlier stage and that these early delays are simply propagated forward to response-related processes. Thus, the finding of a delayed LRP provides only an upper bound on the locus of interference in the psychological refractory period paradigm.

To provide a lower bound, we examined the P3 wave in this paradigm (Luck 1998). As was discussed above, P3 latency provides a measure of the amount of time required to identify and categorize a stimulus. Moreover, several studies have shown that the P3 is insensitive to manipulations of response-related processes (see, for example, Magliero *et al.* 1984; McCarthy and Donchin 1981). Thus, if the delayed RTs observed at short T1–T2 SOAs reflect delays in response-related processes, P3 latency should not be delayed even though RT is delayed.

The design of this experiment is illustrated in Fig. 7.8A. On each trial, two stimuli were presented, separated by an SOA of 50, 150, or 350 ms. The first stimulus was a red or green outline square, and subjects were required to respond immediately to this stimulus by pressing one of two buttons with one hand to indicate the colour of the square. The second stimulus was an X or an O, and subjects were required to respond immediately to this stimulus by pressing one of two buttons with the other had to indicate whether it was an X or an O. On

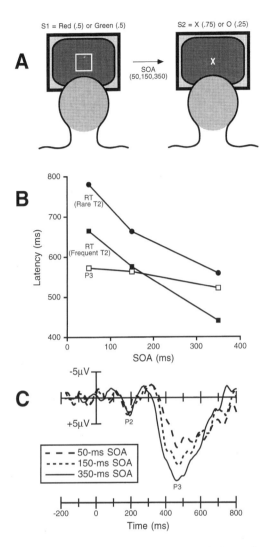

Fig. 7.8 Paradigm and results for the psychological refractory period experiment. A. Example stimulus sequence. Two stimuli were presented on each trial, a red or green outline box followed by an X or an O. For half the subjects, X occurred on 75 per cent of trials and O on the remaining 25 per cent; this was reversed for the other half. Subjects made a speeded two-alternative forced-choice response for each of the two targets. B. Mean reaction times and P3 latencies as a function of the T1–T2 SOA. C. ERP difference waveforms, formed by subtracting frequent-T2 trials from infrequent-T2 trials. These waveforms were recorded at a midline parietal site (Pz) and were averaged across subjects. Adapted with permission from Luck (1998).

some trial blocks, X appeared on 25 per cent of trials and O appeared on 75 per cent of trials, and on other trial blocks these probabilities were reversed. Because the ERP elicited by T1 overlapped the ERP elicited by T2, a subtraction procedure was used, just as in the above-described attentional blink experiments. Specifically, the response to the 75 per cent-probable T2 stimulus was subtracted from the response to the 25 per cent-probable T2 stimulus, yielding a waveform that reflected the brain's differential response to the improbable T2 versus the probable T2.

The results of this experiment are summarized in Fig. 7.8B. As is typically found, T2 reaction times were more than 200 milliseconds longer at the shortest SOA than at the longest SOA. T2 reaction time was longer for the improbable T2 alternative than for the probable T2 alternative, but this effect did not interact with SOA. In contrast, P3 latency was only slightly increased at the shortest SOA compared to the longest SOA. The P3 latency effect was statistically significant, but it was less than 25 per cent of the size of the RT effect. Thus, the delayed RTs observed at short T1–T2 SOAs in the psychological refractory period paradigm primarily reflect delays in a process that follows the P3 wave (and, hence, follows the identification and categorization of T2).

Before accepting this conclusion, it is necessary to demonstrate that the P3 latency and RT are equally sensitive to manipulations that delay target identification and categorization. Care was taken in this experiment to ensure that comparable methods were used to measure P3 latency and RT, but it was still important to provide an empirical verification of the sensitivity of P3 latency in this experimental paradigm. We therefore conducted an additional experiment in which T2 was very dim on half the trials. Reductions in stimulus contrast are a common means of manipulating the time required to perceive a stimulus, and we predicted that P3 latency and RT would be increased by similar amounts for dim relative to bright stimuli. In fact, the contrast manipulation had an even larger effect on P3 latency than on RT in this experiment, indicating that the small increase in P3 latency at short SOAs does not reflect a lack of sensitivity. In addition, a complex pattern of interactions with SOA was also observed that provided even more support for the proposal that the delayed RTs at short T1–T2 SOAs reflect delays in a process that follows the P3 wave (for more details, see Luck 1998).

It should be noted that P3 amplitude was somewhat suppressed at short SOAs, as is shown in Fig. 7.8C. This suggests that the processing of T1 did interfere at least slightly with the encoding of T2 in working memory. However, given that T2 was highly discriminable and required an immediate response, this modest impairment in working memory encoding probably had little effect on the response to T2. Pashler (1989) also found evidence for a relatively minor impairment in processes that preceded response selection at short T1–T2 SOAs, and he similarly concluded that these earlier effects are not responsible for the large delays in T2 reaction time.

Comparison of the two paradigms

The results of these ERP experiments provide a clear dissociation between the locus of interference in the attentional blink paradigm and the locus of interference in the psychological refractory period paradigm. The interference in both paradigms occurs after the stimuli have been identified, but the attentional blink primarily reflects an impairment in working memory, whereas the psychological refractory period primarily reflects an impairment in a later, response-related process. Thus, when an attentional blink occurs, the P3 wave is completely eliminated, and if the attentional blink is eliminated by making T2 the last item in the stream, P3 latency is highly delayed. In contrast, the P3 wave is only slightly reduced in amplitude and slightly delayed in latency during the psychological refractory period, with the majority of the RT effect being due to interference at a later stage.

These results seem to suggest that the attentional blink and the psychological refractory period reflect completely unrelated and independent forms of dual-task interference. However, Jolicœur (1999) and Arnell (this volume, Chapter 8) have provided compelling evidence indicating that these two phenomena are related. Specifically, these investigators have demonstrated that an attentional blink occurs in a hybrid paradigm in which T1 is an unmasked simple target, as in the psychological refractory period paradigm, and T2 is masked, as in the attentional blink paradigm. Moreover, Jolicœur (1998) found that the attentional blink for T2 is larger if a speeded response rather than an unspeeded response is required for T1. This suggests that the processes involved in making an immediate response to T1 interfere with the storage of T2 in working memory, which leads to a conundrum: Why should response-related processing (i.e., the speeded response to T1) interfere with working memory storage (as indexed by T2 accuracy)? Many investigators over the past several decades have proposed the existence of a general-purpose, limited-capacity resource that is involved in many tasks; but few have speculated on the nature of this resource. Without theories of the nature of this resource, we are left with some sort of mystical 'brain juice' that is important for a variety of apparently unrelated cognitive processes.

One possible solution to this puzzle is that both the storage of a new item in working memory and the selection of a response may require access to the same working memory system. Just as most conventional computers cannot simultaneously store and retrieve information from memory, it may not be possible for the brain to store a new item in working memory while a set of stimulus–response translation rules are being accessed from working memory. This proposal is supported by a recent study by Jolicœur and Dell' Acqua (1998), who studied a dual-task combination of a memory encoding task and a speeded response task. On each trial, subjects were presented with a set of letters to store in working memory, and this memory set was followed by a high- or low-pitched tone that required an immediate button-press response. Increasing the number of items in the memory set was found to yield longer RTs to the tone, and this effect became larger as the SOA between the memory set and the tone decreased. These results suggest that subjects could not select the tone response while they were busy storing the digits in memory. This parallels our proposal that subjects cannot store T2 in memory while they are selecting the T1 response. Thus, even though working memory storage and response selection are, in some sense, discrete stages of processing, they may both rely on access to a common working memory system.

This proposal would appear to conflict with one of the ERP results described above. Specifically, if making a speeded response to T1 interferes with the storage of T2 in working memory, then why was P3 latency not delayed at short SOAs in the psychological refractory period study? In fact, P3 *amplitude* was reduced at short SOAs in this experiment, suggesting that the selection of the T1 response did indeed interfere with the storage of T2 in working memory. However, because T1 and T2 were unmasked and required speeded responses, a modest impairment in working memory presumably had little impact on task performance.

Conclusions

Studies of the attentional blink and the psychological refractory period have generally assumed that processing in these paradigms consists of a sequence of feedforward stages, leading from the perception of a stimulus to the execution of a response. This has been a useful simplifying assumption and has led to substantial evidence that the dual-task interference observed in these paradigms occurs within qualitatively different cognitive processes. However, the finding of an interaction between these processes suggests that it may be time to move beyond sequential stage models and to consider more dynamic models of information processing.

Acknowledgements

Preparation of this chapter was supported by grants from the National Science Foundation (SBR 98–09126), the National Institute of Mental Health (MH56877), and the Human Frontier Science Program (RG0136).

References

Broadbent, D. E., and Broadbent, M. H. P. (1987). From detection to identification: response to multiple targets in rapid serial visual presentation. *Perception and Psychophysics*, **42**, 105–13.

Chun, M. M., and Potter, M. C. (1995). A two-stage model for multiple target detection in rapid serial visual presentation. *Journal of Experimental Psychology: Human Perception and Performance*, **21**, 109–27.

De Jong, R. (1993). Multiple bottlenecks in overlapping task performance. *Journal of Experimental Psychology: Human Perception and Performance*, **19**, 965–80.

Donchin, E. (1981). Surprise! . . . Surprise? *Psychophysiology*, **18**, 493–513.

Donchin, E., and Coles, M. G. H. (1988). Is the P300 component a manifestation of context updating. *Behavioral Brain Science*, **11**, 357–74.

Giesbrecht, B. L., and Di Lollo, V. (1998). Beyond the attentional blink: visual masking by object substitution. *Journal of Experimental Psychology: Human Perception and Performance*, **24**, 1454–66.

Greenwald, A. G., Draine, S. C., and Abrams, R. L. (1996). Three cognitive markers of unconscious semantic activation. *Science*, **273**, 1699–1702.

He, S., Cavanagh, P., and Intriligator, J. (1996). Attentional resolution and the locus of visual awareness. *Nature*, **383**, 334–7.

Heinze, H. J., Mangun, G. R., Burchert, W., Hinrichs, H., Scholz, M., Münte, T. F., Gös, A., Scherg, M., Johannes, S., Hundeshagen, H., Gazzaniga, M. S., and Hillyard, S. A. (1994). Combined spatial and temporal imaging of brain activity during visual selective attention in humans. *Nature*, **372**, 543–6.

Hillyard, S. A., and Picton, T. W. (1987). Electrophysiology of cognition. In F. Plum (ed.), *Handbook of physiology: Section 1. The nervous system: Volume 5. Higher functions of the brain. Part 2*, pp. 519–84). Bethesda, MD: Waverly Press.

Hillyard, S. A., Vogel, E. K., and Luck, S. J. (1998). Sensory gain control (amplification) as a mechanism of selective attention: electrophysiological and neuroimaging evidence. *Philosophical Transactions of the Royal Society: Biological Sciences*, **353**, 1257–70.

Isaak, M. I., Shapiro, K. L., and Martin, M. J. (1999). The attentional blink reflects retrieval competition among multiple RSVP items: tests of the interference model. *Journal of Experimental Psychology: Human Perception and Performance*, **25**, 1774–92.

Jacoby, L. L., and Whitehouse, K. (1989). An illusion of memory: false recognition influenced by unconscious perception. *Journal of Experimental Psychology: General*, **118**, 126–35.

Johnson, R., Jr. (1986). A triarchic model of P300 amplitude. *Psychophysiology*, **23**, 367–84.

Jolicœur, P. (1998). Modulation of the attentional blink by on-line response selection: evidence from speeded and unspeeded Task-sub-1 decisions. *Memory and Cognition*, **26**(5), 1014–32.

Jolicœur, P. (1999). Restricted attentional capacity between sensory modalities. *Psychonomic Bulletin and Review*, **6**, 87–92.

Jolicœur, P., and Dell' Acqua, R. (1998). The demonstration of short-term consolidation. *Cognitive Psychology*, **36**(2), 138–202.

Kunst-Wilson, W. R., and Zajonc, R. B. (1980). Affective discrimination of stimuli that cannot be recognized. *Science*, **207**, 557–8.

Kutas, M., and Hillyard, S. A. (1980). Reading senseless sentences: brain potentials reflect semantic incongruity. *Science*, **207**, 203–5.

Kutas, M., McCarthy, G., and Donchin, E. (1977). Augmenting mental chronometry: the P300 as a measure of stimulus evaluation time. *Science*, **197**, 792–5.

Luck, S. J. (1998). Sources of dual-task interference: evidence from human electrophysiology. *Psychological Science*, **9**, 223–7.

Luck, S. J., and Girelli, M. (1998). Electrophysiological approaches to the study of selective attention in the human brain. In R. Parasuraman (ed.), *The attentive brain*, pp. 71–94. Cambridge, MA: MIT Press.

Luck, S. J., and Hillyard, S. A. (1990). Electrophysiological evidence for parallel and serial processing during visual search. *Perception and Psychophysics*, **48**, 603–17.

Luck, S. J., Fan, S., and Hillyard, S. A. (1993). Attention-related modulation of sensory-evoked brain activity in a visual search task. *Journal of Cognitive Neuroscience*, **5**, 188–95.

Luck, S. J., Hillyard, S. A., Mouloua, M., Woldorff, M. G., Clark, V. P., and Hawkins, H. L. (1994). Effects of spatial cuing on luminance detectability: psychophysical and electrophysiological evidence for early selection. *Journal of Experimental Psychology: Human Perception and Performance*, **20**, 887–904.

Luck, S. J., Vogel, E. K., and Shapiro, K. L. (1996). Word meanings can be accessed but not reported during the attentional blink. *Nature*, **382**, 616–18.

Luck, S. J., Chelazzi, L., Hillyard, S. A., and Desimone, R. (1997). Neural mechanisms of spatial selective attention in areas V1, V2, and V4 of macaque visual cortex. *Journal of Neurophysiology*, **77**, 24–42.

McCarthy, G., and Donchin, E. (1981). A metric for thought: a comparison of P300 latency and reaction time. *Science*, **211**, 77–80.

McCarthy, G., and Nobre, A. C. (1993). Modulation of semantic processing by spatial selective attention. *Electroencephalography and Clinical Neurophysiology*, **88**, 210–19.

Magliero, A., Bashore, T. R., Coles, M. G. H., and Donchin, E. (1984). On the dependence of P300 latency on stimulus evaluation processes. *Psychophysiology*, **21**, 171–86.

Maki, W. S., Frigen, K., and Paulson, K. (1997). Associative priming by targets and distractors during rapid serial visual presentation: Does word meaning survive the attentional blink? *Journal of Experimental Psychology: Human Perception and Performance*, **23**, 1014–34.

Mangun, G. R., and Hillyard, S. A. (1987). The spatial allocation of visual attention as indexed by event-related brain potentials. *Human Factors*, **29**, 195–211.

Mangun, G. R., and Hillyard, S. A. (1990). Allocation of visual attention to spatial location: event-related brain potentials and detection performance. *Perception and Psychophysics*, **47**, 532–50.

Mangun, G. R., Hopfinger, J. B., Kussmaul, C. L., Fletcher, E. M., and Heinze, H.-J. (1997). Covariations in ERP and PET measures of spatial selective attention in human extrastriate visual cortex. *Human Brain Mapping*, **5**, 273–9.

Osman, A., and Moore, C. M. (1993). The locus of dual-task interference: psychological refractory effects on movement-related brain potentials. *Journal of Experimental Psychology: Human Perception and Performance*, **19**, 1292–1312.

Pashler, H. (1989). Dissociations and dependencies between speed and accuracy: evidence for a two-component theory of divided attention in simple tasks. *Cognitive Psychology*, **21**, 469–514.

Pashler, H. (1994). Dual-task interference in simple tasks: data and theory. *Psychological Bulletin*, **116**, 220–44.

Petersen, S. E., Fiez, J. A., and Corbetta, M. (1992). Neuroimaging. *Current Opinion in Neurobiology*, **2**, 217–22.

Raymond, J. E., Shapiro, K. L., and Arnell, K. M. (1992). Temporary suppression of visual processing in an RSVP task: an attentional blink? *Journal of Experimental Psychology: Human Perception and Performance*, **18**, 849–60.

Reeves, A., and Sperling, G. (1986). Attention gating in short term visual memory. *Psychological Review*, **93**, 180–206.

Shapiro, K. L., Raymond, J. E., and Arnell, K. M. (1994). Attention to visual pattern information produces the attentional blink in rapid serial visual presentation. *Journal of Experimental Psychology: Human Perception and Performance*, **20**, 357–71.

Shapiro, K., Driver, J., Ward, R., and Sorensen, R. E. (1997*a*). Priming from the attentional blink: a failure to extract visual tokens but not visual types. *Psychological Science*, **8**, 95–100.

Shapiro, K. L., Caldwell, J. I., and Sorensen, R. E. (1997*b*). Personal names and the attentional blink: a visual 'cocktail party' effect. *Journal of Experimental Psychology: Human Perception and Performance*, **23**, 504–14.

Van Voorhis, S. T., and Hillyard, S. A. (1977). Visual evoked potentials and selective attention to points in space. *Perception and Psychophysics*, **22**, 54–62.

Vogel, E. K., Luck, S. J., and Shapiro, K. L. (1998). Electrophysiological evidence for a postperceptual locus of suppression during the attentional blink. *Journal of Experimental Psychology: Human Perception and Performance*, **24**, 1656–74.

Welford, A. T. (1952). The 'psychological refractory period' and the timing of high speed performance—a review and a theory. *British Journal of Psychology*, **43**, 2–19.

8

Cross-modal interactions in dual-task paradigms

Karen M. Arnell

Abstract

Researchers have employed various combinations of visual, auditory and tactile stimuli in a number of different dual-task situations. Experiments using two stimuli presented in different modalities have provided compelling evidence for cross-modal dual-task costs and benefits. For example, researchers using the spatial cuing paradigm have demonstrated response-time benefits for targets when the preceding cue is presented in a different modality than the target. Researchers employing the Psychological Refractory Period paradigm have repeatedly demonstrated response slowing to the second target when a speeded response is required to an earlier cross-modal target. Further, using several paradigms, costs and benefits to cross-modal targets have also been found in the accuracy of unspeeded responses. For example, some neglect patients extinguish the leftmost of two bilaterally presented simultaneous targets, even if the two targets are presented in different modalities (touch and vision). In addition, attentional blinks have been demonstrated when one of two targets is visual and the other is auditory. Also, research with both speeded and unspeeded responses has suggested asymmetries in both cross-modal interference and facilitation. In some studies auditory stimuli have been shown to have greater influence on the response times (or accuracy) for subsequent visual stimuli than vice versa. A few studies have also presented valid comparisons of the magnitude of costs and benefits in cross-modal, as compared to within-modal, presentations. These comparisons have provided estimates of amodal and within-modal interference under various conditions. While some of the above findings could potentially be attributed to the use of a preparatory task-set switching strategy by experimental participants, many findings cannot. As such, these results suggest the importance of interference at a central stage, after visual and auditory information have converged.

In the human mind, attention is a limited commodity. When attentional resources must be divided amongst tasks or stimuli, costs or benefits in response time or accuracy often result. Sometimes these costs or benefits result from within-modality attentional processing; sometimes they result from central (amodal) attentional processing; and sometimes they result from a combination of within-modality and central processing. Thus, for each new set of dual-task findings, it is important to distinguish among these alternatives. However, in the 1980s and 1990s, excitement over parallel processing, and disenchantment with single-channel information processing theories, has led to the common practice of attributing processing interference to limitations in within-modality resources. Given that so many of the attentional experiments performed in the 1980s and 1990s have used visual stimuli, many theories of visual attention have emerged, while there have been relatively fewer investigations of cross-modal attention. However, cross-modal investigations of attention have very recently gained popularity. These investigations have demonstrated interesting patterns of modality-specific and cross-modal attentional limitations and links.

This chapter focuses on examining cross-modal effects in dual-task temporal attention experiments. The evidence for specific sources of interference (such as within-modality processing) and general (central) sources of interference will be examined in dual-task paradigms using continuous stimuli tasks (such as auditory shadowing), and more discrete stimuli and tasks (such as those found in the Psychological Refractory Period). Many temporal attention paradigms will be discussed somewhat briefly, but the issue of modality-specific interference versus central (amodal) interference in the attentional blink paradigm will be discussed in detail. Across paradigms, a general pattern will emerge where both specific and general sources of interference can be seen to underlie the results of dual-task temporal attention paradigms.

Despite the focus on temporal attention, I begin with a brief discussion of some cross-modal findings within the spatial attention literature. This foray is considered worthwhile, given that several patterns in the temporal attention literature will correspond with patterns found in the spatial attention literature. These include: finding attentional effects in a variety of modalities, finding attentional effects with cross-modal stimuli, finding evidence for both within-modality and central sources of attentional interaction, and finding evidence that cross-modal matching can influence stimulus selection, at least under some conditions.

Cross-modal effects in spatial attention

In the real world, a signal in one sensory modality often acts as a cue as to the location or identity of a stimulus in another modality. For example, the honk of a car horn will cause your attention to be drawn to the car producing the noise, sometimes overtly (with eye-movements to that location) and sometimes covertly (without eye-movements). However, until recently, spatial cuing experiments investigating covert orienting had been performed almost exclusively with visual cues and targets using Posner's (1978) spatial cuing paradigm. Using this paradigm, participants have shown benefits in response time (RT) to targets presented at the cued location, and costs in RT to targets presented at the uncured location, relative to a neutral baseline (see Wright and Ward 1994 for a review). These cuing effects have been found both with uninformative exogenous cues that automatically direct attention to the desired location, and with central symbolic, informative endogenous cues that participants can use to maximize their target RT benefits and minimize their target RT costs (see Rafal *et al.* 1991 for how these may differ physiologically).

Spence and Driver (1994) reported covert orienting effects in the auditory modality, analogous to those found in the visual modality, when using both uninformative exogenous cues and informative endogenous cues. Participants were faster to report the target when it was presented to the cued side than when the target was presented on the side opposite the cue. These results are particularly compelling given that participants were cued to either the left or right direction, but the required response was an up–down target judgement, orthogonal to the cue direction so that the cue could not prime the target response. Using the same orthogonal cuing paradigm with exogenous cues, Spence and Driver (1997) demonstrated that auditory cues influenced RTs to auditory or visual targets, and visual cues influenced RTs to visual targets, but not to auditory targets (but see Ward 1994 for a different pattern of cross-modal results). Spence and Driver (1997) argued that visual covert orienting can be led by auditory covert orienting, but an asymmetry exists in that auditory covert orienting cannot be led by visual covert orienting. They argued that this may be due to links with overt orienting where such a pattern could have evolutionary advantages. Spence, Nicholls, Gillespie, and Driver (1998a) demonstrated that exogenous tactile cues led to faster RTs to the cued locations for tactile, auditory, or visual targets.

Spence and Driver (1996) also used the orthogonal cuing paradigm with informative endogenous cues, and demonstrated that visual attention shifted with auditory attention, even when it was not advantageous to make the shift in the visual modality. However, although links between visual and auditory attention were found, results suggested some degree of modality independence as well, since, at any given time, attention was not always focused on the same side for both modalities. Thus, although there are multiple cross-modal links in spatial attention, endogenous and exogenous orienting do not appear to operate within an exclusively amodal system.

The results of numerous experiments examining the cross-modal links in attention when visual, auditory, and tactile receptors are shifted in space relative to each other suggest that extensive cross-modal integration is necessary for humans to construct a representational space prior to directing attention (Driver and Spence 1998). Support for this assertion was also found when Driver (1996) asked participants to shadow one of two concurrent mono auditory messages, distinguished only by its synchrony with a face on a video monitor speaking the words. Shadowing performance was better when the face was presented on the opposite side to the message than when it was presented to the same side as the mono message. Presumably, the interference between the two auditory messages was reduced when the distant video pulled the corresponding auditory message to a more distant location. Given that illusory separation benefited auditory selection, Driver concluded that at least some of the cross-modal matching took place prior to auditory selection. Furthermore, Spence, Ranson, and Driver (2000) asked participants to shadow words presented behind them, while viewing either random chewing lips movements, or lip-movements that matched the to-be-shadowed auditory message. An irrelevant stream of auditory words was presented either on the same or the opposite side to the visual lips. Shadowing performance was reduced when the irrelevant auditory words were presented in the same hemifield as the lips, compared to when the irrelevant auditory words were presented in the opposite hemifield to the lips. Furthermore, this deficit in shadowing performance was greater when the lip movements matched the auditory message than when the lips showed chewing movements. This pattern of results provides additional evidence that at least some cross-modal matching takes place prior to auditory selection, and that the efficiency of selection is related to the spatial separation of the cross-modal stimuli.

Cross-modal inhibition of return (IOR) has been reported between all combinations of visual, auditory, and tactile cue-target pairs (Spence, Lloyd, and McGlone 1998b). Mondor, Breau, and Milliken (1998a) have also demonstrated auditory IOR based on both pitch information and location information. Rorden and Driver (1999) have shown that the direction of an upcoming saccade influenced the shift of auditory attention, just as it influenced the shift of visual attention. Orthogonal auditory discriminations were found to be faster to the side of an upcoming saccade than to the side opposite an upcoming saccade. Discriminations were also faster when fixating toward a sound as compared to fixating away from a sound. Furthermore, Perrott, Cisneros, McKinley, and D'Angelo (1996) provided evidence that localized free-field auditory cues facilitated target detection RTs in a visual search task when using an almost complete visual field.

Cross-modal spatial links have also been observed in cognitive neuropsychological experiments examining contralateral neglect or extinction. Hemispatial neglect and extinction are most commonly reported in the visual modality, but both tactile (di-Pellegrino *et al.* 1997) and auditory (De Renzi *et al.* 1989; Hugdahl *et al.* 1991; Soroker *et al.* 1997; Sterzi *et al.* 1996) neglect and extinction have also been found. Patients with visual hemispatial neglect often demonstrate auditory hemispatial neglect as well (Soroker *et al.* 1995a; Soroker *et al.* 1997), reinforcing the notion of a central biasing of spatial attention. Furthermore, performance with contralesional auditory targets increased when sounds were thought to come from a dummy speaker positioned in the ipsilesional hemifield (Soroker *et al.* 1995a), and when visual lips presented McGurk stimuli in the ipsilesional hemifield (Soroker *et al.* 1995b). These results again demonstrated the influence of cross-modal integration prior to stimulus selection. However, patients may also neglect or extinguish stimuli in only a single modality, and leave other modalities intact, providing evidence that the attentional bias can operate within and/or across modalities (Barbieri and De Renzi 1989).

Cross-modal links are also found when cross-modal stimulus presentations are used on each trial. For example, di Pellegrino *et al.* (1997) examined visual – tactile extinction in a single patient with right frontotemporal cortex damage resulting from a stroke. This patient exhibited severe left tactile extinction, in that he could report a touch on the contralesional (left) side of his body when it was presented alone, but not when it was presented with a simultaneous touch to the ipsilesional (right) side of his body. Interestingly, this patient also extinguished a contralesional tactile stimulus if it was presented with a simultaneous ipsilesional visual stimulus, thus demonstrating cross-modal extinction. Mattingley, Driver, Beschins, and Robertson (1997) also reported cross-modal extinction for three patients with right-hemisphere damage. Cross-modal extinction was found both when a visual target was presented in the ipsilesional field and a tactile target was presented in the contralesional field, and when a tactile target was presented in the ipsilesional field and a visual target was presented in the contralesional field.

The above discussion of cross-modal links in spatial attention is selective and brief. However, even such a brief presentation makes it clear that: (1) many spatial attention phenomena (such as spatial cuing, neglect, and extinction) that have commonly been associated with visual attention can also be found in modalities such as touch and audition; (2) these same spatial attention phenomena can be found with cross-modal stimulus presentations within a trial (for example when one stimulus is visual and another is auditory); (3) there appear to be both within-modality and central (amodal) sources of stimulus interference; and (4) under some circumstances, cross-modal matching can occur before stimulus selection. As we will see, these summary points from the discussion of cross-modal spatial attention will also apply to the discussion of cross-modal temporal attention. Indeed, there is some evidence to suggest that some of the modality asymmetries in the spatial attention literature also exist in the temporal attention literature.

Cross-modal effects in temporal attention

Several experiments spanning the last few decades have investigated the issue of central versus specific sources of dual-task interference. Some investigators have isolated specific sources of interference such as response-selection operations (McCann and Johnston 1992; Pashler 1984; Welford 1952); input modality (Treisman and Davies 1973); or response modality (McLeod 1977). For example, there is good evidence that limitations on response-selection operations underlie the Psychological Refractory Period (PRP) phenomenon (McCann and Johnston 1992; Pashler 1984). Also, Treisman and Davies (1973) asked participants to monitor two concurrent streams (both auditory, both visual, or one of each) for the occurrence of target words. Greater dual-task costs were observed when both streams were presented in the same modality than when one was visual and the other was auditory, demonstrating modality-specific processing limitations. However, even when streams were presented in different modalities, it was harder to monitor both streams than just one stream or the other, providing evidence for a central source of dual-task interference, in addition to modality-specific sources. McLeod (1977) found interference with a continuous visual input–manual output task when it was combined with a tone-discrimination task. However, the interference occurred only when the tone task required a manual response with the hand not involved in the visual task, not when tone responses were made vocally. This result provided evidence for specific response-modality interference. As mentioned by Bourke, Duncan, and Nimmo-Smith (1996), specific sources of dual-task inter-ference have also been found for common spatial encoding (Baddeley and Lieberman 1980), and common phonological encoding (Salame and Baddeley 1982), among others.

Some researchers argue that specific sources of dual-task interference are all that exist. Central sources of interference are thought to result from under-specified or overly simplistic single-channel theories, or are thought to represent a homunculus directing information processing within the human mind (cf. e.g. Allport 1980, 1989; Navon 1984). For example, Allport (1980, 1989) argued that there are no general, central limitations on attentional processing and that any dual-task interference results from specific sources of interference that may or may not be known to exist by researchers. Allport (1989) argued that attentional selectivity exists only in specialized subsystems and only to limit the information available to (potentially) control goal-directed actions (i.e. selection for possible action).

Claims such as Allport's have been supported by some experiments showing that even seem-ingly highly demanding tasks that avoid specific sources of interference could be performed con-currently with little or no dual-task costs (cf. e.g. Allport *et al.* 1972). In the first experiment of Allport *et al.* (1972) participants were given a prose passage to shadow as the primary task. The prose passage was paired with either auditory presentation of words, visual presentation of words, or visual presentation of pictures as the secondary task. Relative to undivided attention conditions, all three secondary tasks showed dual-task costs in recognition memory. Larger costs were observed for more similar tasks, such that recognition accuracy was lowest for auditory words, and highest for visual pictures. Shadowing performance showed the same pattern. Allport *et al.* con-cluded that the results were inconsistent with single-channel processing hypotheses. However, although the results clearly show the importance of modality- and stimulus-specific interference, they also support the existence of central interference, given that dual-task interference was observed even when stimuli and tasks shared no known sources of specific interference. In the more frequently cited experiment from Allport *et al.* (1972), participants (third-year undergraduate music students) shadowed a continuous speech message presented auditorily while at the same time sight-reading piano music and playing the corresponding tune. Despite the fact that both these tasks were assumed to require attentional processing, little or no dual-task costs were observed, leading Allport *et al.* to state that such results were incompatible with general-purpose central pro-cessing views of attentional resources.

If there are central sources of dual-task interference, how come Allport *et al.*'s (1972) partici-pants could perform both tasks concurrently without interference? An explanation may lie in the training of the participants themselves, prior to the experiment. Participants in the Allport *et al.* experiment were trained third-year undergraduate music students. It is established that experts can often chunk stimuli within their realm of expertise into meaningful larger units (cf. e.g. Chase and Simon 1973), and that extensive practice with a task can lead to an automa-tion of the task such that it does not require central processing resources (cf. e.g. Shiffrin and Schneider 1977). Thus, the considerable experience of Allport *et al.*'s participants at sight-reading and playing music may have resulted in the participants' being able to chunk the bars or pieces of music into larger units for processing, and/or allowed them to process the information with fewer central processing resources. Yet, Allport *et al.*'s participants did vary in the proficiency of their piano playing. The above explanation would predict that the most compe-tent piano players would show the fewest dual-task costs, and that the least competent piano players would show the most dual-task costs. Indeed, this pattern of dual-task costs was reported by Allport *et al.* Similarly, I would expect massive dual-task costs if beginning piano players were asked to perform the same concurrent tasks.

It is difficult to reconcile theories postulating only specific sources of interference with several studies showing interference with wildly different tasks that avoid all known sources of specific interference (e.g. Bourke *et al.* 1996; Jolicœur 1999b; Jolicœur and Dell' Acqua 1998, 1999; Posner *et al.* 1989; Reisberg 1983). For example, Reisberg showed that the performance of mental arithmetic, a 7-item digit load, and a counting task all slowed the ability of participants to reverse the identity of a reversible ambiguous figure. Posner *et al.* (1989) showed that shadowing digits interfered with covert shifts of attention to peripheral visual cues. Furthermore, Jolicœur (1999) demonstrated that a speeded pitch discrimination to an unmasked auditory tone interfered with subsequent masked letter discrimination for several hundred ms. Jolicœur and Dell' Acqua (1998, 1999) found the reverse effect, where discrimination of masked letters produced response slowing for speeded pitch judgements of tones. Furthermore, the slowing was larger as more letters were encoded. Bourke *et al.* (1996) examined evidence for a central source of interference using simple tasks with non-expert participants. They factorially combined four tasks (tone detection, random letter generation, a manual tactile task, and a visual recognition task) selected to avoid known specific sources of dual-task inference such as input modality (Treisman and Davies 1973), response selection (Pashler 1994), and response modality (McLeod 1977). Each of the four tasks was performed alone, performed as the primary task within the pair of tasks, or performed as the secondary task within the pair of tasks. The tasks differed in the amount of central processing resources required, and across the range of dual-task combinations, the pattern of interference observed was predicted by each task's overall resource demand. Bourke *et al.* concluded that the results provided strong support for a general factor underlying dual-task performance, in addition to the known sources of specific interference, but did not discuss the nature of this general factor.

The experiments reported in the section above demonstrate the inconsistent pattern of results that is found when continuous tasks (such as shadowing a passage or piano playing from music) are used in dual-task attention experiments. Sometimes the evidence favours the existence of specific sources of interference only, and at other times it favours the existence of both general (central) and specific sources of interference. Although the use of continuous tasks may mimic real-world activities, such tasks may also underestimate the dual-task costs involved. For example, call to mind the familiar feeling of retrieving from echoic memory what someone has said to you after just finishing an attention-demanding task. Continuous tasks may allow the participant to use a variety of strategies (for example, chunking, buffering or switching) to manoeuvre between tasks, as humans do adeptly every day, thereby minimizing dual-task interference. If the goal is to know how humans function with everyday dual-task situations where numerous strategies can be employed, then dual-tasks of a continuous nature would be preferred. However, if the goal is to examine the fundamental central processing limitations in the human mind, then dual-tasks with two discrete (distinct) targets (with or without additional distractors) presented at variable stimulus-onset asynchronies (SOAs) would be preferred, given their more simplified nature. Of course, the nature of the stimuli and tasks, and the relationship between T1 and T2 stimuli and tasks, must be taken into account. As McLeod (1978) pointed out, a uniform probe (T2) task does not suit every occasion, and should not be assumed to do so. Several paradigms have employed the logic of two discrete targets. These paradigms (negative priming, repetition blindness, the psychological refractory period, and the attentional blink) will be the focus of the remainder of this chapter.

Negative priming is found when participants are slower to report a target that was a to-be-ignored distractor on the previous trial (cf. e.g. Tipper 1985). Negative priming has been observed with a variety of visual stimuli such as coloured words (e.g. Neill 1977), non-coloured words (e.g. Lowe 1998), line drawings of objects (e.g. Tipper 1985), line drawings of non-objects (DeSchepper and Treisman 1996), locations of moving objects (Tipper *et al.* 1990), and across pictures and words (Tipper and Driver 1988). More recently, Banks, Roberts, and Ciranni (1995) demonstrated negative priming within the auditory modality, where both targets and distractors were auditory words. Also, Driver and Baylis (1993) have demonstrated cross-modal negative priming where ignored auditory digits slowed RTs to subsequent visual digit targets.

When two similar or identical targets are presented close together in time under conditions of rapid serial visual presentation (RSVP), the second target (T2) shows impaired recall or detection performance compared to trials where the two targets are unrelated, a phenomenon known as *repetition blindness* (RB) (Kanwisher 1987). As the name might suggest, RB was initially though to occur only in the visual modality (Kanwisher and Potter 1989). However, Miller and MacKay (1994) demonstrated that robust RB could be found with auditory stimuli, provided that the stimuli were presented in aprosodic word lists, as opposed to prosodic sentences, given that sentence structure and prosody help to overcome the effect. To this author's knowledge, there have been no published reports of experiments examining whether RB can also be found when the two critical targets are presented cross-modally.

When two targets (T1 and T2) are unmasked and require on-line speeded responses, the response times (RTs) to T2 are inflated at short target–target stimulus-onset-asynchronies (SOAs) as compared to long target–target SOAs. This effect is known as the *psychological refractory period* (PRP) phenomenon (cf. e.g. Pashler 1994; Smith 1967; Welford 1952). There is strong evidence that the PRP reflects a bottleneck on central response-selection operations (e.g. McCann and Johnston 1992; Pashler 1994; Welford 1952); but see Meyer and Kieras (1997) for an alternative opinion. According to this model, response-selection operations (mapping from stimulus identity to the required response) can only proceed for one stimulus at a time. If response-selection operations are busy with T1, then T2 must wait, and this waiting shows itself as increased RTs at short SOAs. Experiments employing locus of slack logic support this position, in that manipulations of perceptual or encoding factors have underadditive effects with SOA (e.g. McCann and Johnston 1992; Pashler 1984; Van Selst and Jolicœur 1994); yet response-selection and response-execution manipulations remain additive with SOA (e.g. McCann and Johnston 1992; Pashler 1984; Ruthruff *et al.* 1995). Owing to the relatively late (response-selection) locus of interference postulated for the PRP effect, it was no surprise that the PRP effect was found when T1 and T2 were presented cross-modally (Davis 1957). Indeed, studies of the PRP effect routinely present targets in two different modalities to eliminate any response-slowing that may result from perceptual interference between the two stimuli, so that interference occurring at response selection can be more properly examined (e.g. Karlin and Kestenbaum 1968; Pashler 1990; McCann and Johnston 1992). The quick and easy acceptance of cross-modal dual-task costs and the relatively late postulated locus of interference (response selection) in the PRP paradigm very probably resulted from the use of speeded responses. However, in dual-task paradigms where no speeded responses are required, acceptance has not come so easily, as we will see with the attentional blink paradigm.

The above discussion of cross-modal links in temporal attention is necessarily selective and brief. However, it can be seen that: (1) dual-task paradigms investigating temporal attentional processing have shown both specific and general (central) sources of dual-task interference; (2) dual-task costs can be observed when targets are presented in different stimulus modalities, even when all known specific sources of interference have been removed; and (3) some temporal attention phenomena that have commonly been associated with visual attention can also be found auditorily, or when using cross-modal stimuli. Therefore the key question to ask is not whether there are specific or central sources of dual-task interference, but rather what is the nature of these sources, and how do they combine to produce coherent conscious behaviour. A detailed examination of the attentional blink phenomenon will be used to help us address such questions.

Cross-modal attentional blinks

The background

When two masked targets (T1 and T2), both requiring attention for later unspeeded report, are presented within half a second of each other, accuracy of T2 report is poor—an Attentional Blink (AB). All the early AB experiments used only visual stimuli for T1 (the target in traditional AB terminology), T2 (the probe in traditional AB terminology), and filler characters (Broadbent and Broadbent 1987; Raymond *et al.* 1992; Weichselgartner and Sperling 1987). In the case of Raymond *et al.*, the use of visual stimuli was not necessarily because we thought the AB was uniquely visual in nature. Instead, it was a result of the fact that earlier studies had used rapid serial visual presentation (RSVP) in multi-task temporal attention experiments (Broadbent and Broadbent 1987; Weichselgartner and Sperling 1987), as well as in single-task temporal attention experiments (e.g. Intraub 1985; Lawrence 1971; McLean *et al.* 1983), and in repetition blindness experiments (e.g. Kanwisher 1987; Kanwisher and Potter 1990). As with other areas of attention, temporal attention research, performed mostly in the 1980s and 1990s, relied mostly on visual presentations of items. Relatively few experiments employed auditory or cross-modal presentations of stimuli. A notable exception is the Psychological Refractory Period (PRP) paradigm, where cross-modal targets have been commonly used since the 1950s to remove any possible perceptual interference between the two targets (e.g. Davis 1957; Karlin and Kestenbaum 1968; Pashler 1990).

Despite the fact that use of visual stimuli in early AB experiments was not necessarily theoretically driven, the visual nature of the stimuli and their processing became incorporated into various theories of the AB (e.g. Duncan *et al.* 1994; Raymond *et al.* 1992; Shapiro *et al.* 1994; Ward *et al.* 1996). For example, Raymond *et al.* (1992) postulated a two-stage detect-then-identify model where the AB was said to result from the shut-down of an early selection mechanism when the first target (T1) occupied the sensory pattern buffer during its identification. Further processing of visual high-level patterns was blocked until target identification had been completed. If the second target (T2) was presented before T1 identification was completed, then its processing was blocked. According to this theory, an AB would only be found when T1 and T2 contained visual pattern information. Duncan *et al.*'s (1994) and Ward *et al.*'s (1996) attentional dwell time theory also incorporated the visual nature of the stimuli. This theory suggested that the AB reflected the slow and laborious nature of visual attention, and that the AB could be taken as a measure of the time course of *visual* attention. It was suggested that conscious attentional processing could not be deployed to T2 until such processing has been completed for T1.

Shapiro *et al.* (1994) demonstrated that a robust AB could be found with even the simplest T1 detection task when the T1 contained visual pattern information (e.g., detecting a white dot pattern among black letters). However, no AB was found when T1 was a blank temporal gap containing no pattern information, even though the T1 tasks (detect blank gap and report whether gap was short or long in duration) produced much lower accuracy than the dot detection task (Shapiro *et al.* 1994). Also, when T2 was a coloured symbol presented amongst black letters, Shapiro, Arnell, and Drake (1991) and Shapiro, Raymond, and Taylor (1993) found an AB when participants were instructed to report the symbol's shape or size. However, no AB was found (and performance was at ceiling) when subjects were instructed to report the symbol's colour. To account for these and other findings, Shapiro *et al.* (1994) and Raymond, Shapiro, and Arnell (1995) put forward the similarity theory of the AB—a late-selection visual-interference model based on stimulus similarity. According to similarity theory, stream items that resemble T1 and T2 templates compete for report or transfer from visual short-term memory (VSTM). If T1 and T2 are similar to the filler items (especially the T1 +1 and T2 +1 filler items), then the filler items are more likely to gain access to VSTM. According to this model T2 may become lost or ignored owing to its relatively late presentation and the confusion in an overcrowded VSTM. Once T1 representations have passed through VSTM, then T2 is weighted more heavily and can be reported. This theory postulates that the AB is uniquely visual in nature given that the confusion that causes the AB is said to occur in a visual buffer (VSTM).

The first wave of cross-modal AB experiments

Given the proliferation of AB theories focusing on the visual nature of the AB, the obvious question to ask became, 'Could the AB be found when T1 and/or T2 were not presented visually?' In 1995, four sets of researchers independently presented data addressing this question. However, each set of researchers, using somewhat different methodologies, found a different pattern of results.

Four broad patterns of results were possible when the T1 modality (auditory or visual) and T2 modality (auditory or visual) were factorially combined to produce four conditions (VV, AA, VA, and AV).[1] If the AB is a uniquely visual phenomenon, as is suggested by several theories (e.g. Shapiro *et al.* 1994; Ward *et al.* 1996), then an AB should only be observed when both T1 and T2 are presented in the visual modality (VV condition), but should not be found in any other modality combination (AA, VA, or AV). If the AB is modality-specific in nature, but not confined to visual stimuli, then an AB should be observed when T1 and T2 are presented in the same modality (VV or AA), but not when T1 and T2 are presented in different modalities (VA or AV). If the AB results from central, amodal processing interference alone (i.e. with no within-modality processing interference), then an AB should be observed in all four modality combinations, and should be just as large when T1 and T2 are presented in different modalities as when T1 and T2 are presented in the same modality. If the AB results from a combination of central (amodal) processing interference and modality-specific interference, then an AB should be found in all four modality combinations, but the magnitude of the AB should be greater

---

[1] Abbreviations such as VV will be used to refer to the modality conditions. The first letter of the abbreviation corresponds to the first letter of the T1 modality (auditory or visual), and the second letter of the abbreviation corresponds to the first letter of the T2 modality (visual or auditory). Therefore VV will be used to refer to the visual T1–visual T2 condition, AA will be used to refer to the auditory T1–auditory T2 condition, VA will be used to refer to the visual T1–auditory T2 condition, and AV will be used to refer to the auditory T1–visual T2 condition.

when T1 and T2 are presented in the same modality than when T1 and T2 are presented in different modalities.

Potter, Chun and Muckenhoupt (1995) and Potter, Chun, Banks, and Muckenhoupt (1998) presented either streams of visual digits in RSVP or streams of auditory compressed-speech digits in rapid auditory presentation (RAP), and asked participants to report the two letter targets (T1 and T2) presented amongst the digit distractors. Auditory and visual streams were matched for T1 identification accuracy, resulting in a presentation rate of 8.33 items/s for visual streams and 7.41 items/s for auditory streams. They found robust AB in the visual modality, but no AB in the auditory modality. In the auditory modality T2 accuracy did not vary as a function of T1–T2 SOA, but was lower overall than accuracy for a single target control condition where T2 was matched for stream position. This finding was replicated when both visual and auditory streams were presented at a rate of 7.41 items/s. In a separate experiment, Potter *et al.* (1995, 1998) examined cross-modal AB[2] by presenting stimuli in the first half of the stream either visually or auditorily, and then switching the modality of stream presentation after the T1 + 1 stimulus. For example, in the VA condition visual characters would be presented in RSVP up to, and including, the T1 + 1 digit, and then, beginning with the T1 + 2 stimulus, auditory characters would be presented in RAP. Modality order was blocked, and participants were asked to report the letter targets presented amongst the digit distractors. Again, a presentation rate of 8.33 items/s was used for visual streams and one of 7.41 items/s was used for auditory streams. Potter *et al.* found no AB in either cross-modal condition (VA or AV). T2 accuracy did not vary as a function of T1–T2 SOA, and was quite high, at about 90 per cent correct for both modality conditions. Potter *et al.* concluded that visual stimuli are uniquely susceptible to the AB, owing to the relatively short duration of iconic memory, and the relatively small capacity of VSTM.

Shulman and Hsieh (1995) used similar procedures to those of Potter *et al.* except that the T1 task was to report the identity of a novel target letter, distinguished from filler letters by a high-pitched voice in the auditory modality, or a high-luminance character in the visual modality. The T2 task was to determine whether a pre-specified letter occurred in the post-target stream. Also, at 7.94 items/s for both visual and auditory streams, the auditory presentation rate was slightly faster than that of Potter *et al.* (1995). Shulman and Hsieh reported finding an AB in both within-modality conditions (VV and AA), and in the AV cross-modal condition, but no AB in the VA cross-modal condition. While significant, the AB found in the AA condition was reduced in magnitude (about 10 per cent difference between experimental and control at its maximum) relative to the magnitude of the AB in the VV and AV conditions (at least 20 per cent difference between experimental and control at its maximum). These authors concluded that there were both within-modality and amodal sources of interference on stimulus encoding.

Duncan, Martens, and Ward (1997) produced a pattern of AB results different from both Potter *et al.* (1995) and Shulman and Hsieh (1995). Duncan *et al.* presented two streams of stimuli for each T1–T2 modality combination (VV, AA, VA, and AV) with a single target in

---

[2] Expressions such as 'auditory-AB', 'visual-AB', and 'cross-modal AB' are simply shorthand labels that indicate the modality of the T1 and T2 stimuli. For example, 'auditory AB' stands for 'an AB observed using an auditory T1 and T2', and 'cross-modal AB' stands for 'an AB observed when T1 and T2 are presented in different modalities'. These shorter labels are not meant to indicate the locus of the processing limitation, but only the modality of target presentation.

each stream. Therefore, in the VV condition both streams were visual, in the AA condition both streams were auditory, and in each of the cross-modal conditions one stream was visual and the other was auditory. In the AA condition, the two auditory streams of compressed speech were distinguished from each other by pitch. In the VV condition, the two visual RSVP streams were distinguished from each other by spatial location. Each stream had a presentation rate of 4 items/s, and pairs of streams were maximally offset, so that the second stream ran 125 ms after the first stream. Targets were words ('cod' or 'cot' and 'nab' or 'nap'), included among repeated presentations of the phonemic syllable 'guh' in auditory streams, and repeated presentations of 'xxx' in visual streams. Reliable AB was observed in both within-modality conditions, but no AB was observed in either cross-modal condition. Results for both cross-modal conditions showed lower T2 accuracy overall for dual-target conditions than for single-target control conditions, but this deficit was equally pronounced at all T1–T2 SOAs. Duncan *et al.* (1997) concluded that the restrictions to sensory attention and awareness do exist outside the visual modality, but are modality-specific in nature.

In contrast to all of the above findings, Arnell and Jolicœur (1995, 1999) found an AB in all four auditory–visual modality combinations. In our experiment, all participants experienced concurrent visual RSVP letter streams and auditory compressed-speech RAP streams, where visual and auditory letter onsets were simultaneous, and each stream was presented at a rate of 10.7 items/s. Only the instructions to attend to T1 and T2 in specific modalities differentiated the four modality conditions. Letters in the visual and auditory streams were independent, with the exception that the visual and auditory T1 targets occurred simultaneously. T1 digit identity, T2 presence/absence, and T2 position were all independent across the visual and auditory streams. The independence across streams did not allow participants to use the stimuli in one modality to assist them with their response to stimuli in the other modality. T1 modality (visual or auditory) was fully crossed with T2 modality (visual or auditory), and modality combinations were administered between participants. The T1 task was to identify a digit (1, 2, 3, or 4) from the stream of letters. The T2 task was to report whether an 'X' was present or absent in the post-target letter stream. Participants were informed of the modality or modalities in which T1 and T2 were to be presented. In the control blocks participants were instructed to ignore the digit (T1) and just to report the presence/absence of the X (T2). In the experimental blocks participants were instructed to report the identity of the digit (T1) and to also report the presence/absence of the X (T2). All responses were made in an unspeeded manner.

Figure 8.1 displays T2 accuracy in each modality combination as a function of control and experimental instructions and T1–T2 SOA. A reliable AB was found in each modality combination, but the magnitude of the AB was statistically and meaningfully larger in the visual T2 conditions than in the auditory T2 conditions. However, the most interesting aspect of the results was that the AB magnitude in the within-modality conditions was roughly equal to the AB magnitude in the cross-modal conditions (see panel E of Fig. 8.1). Finding robust ABs in the cross-modal conditions provided good evidence that central (amodal) processing limitations underlie the AB. Finding that the AB was as large when T1 and T2 were presented in different modalities as it was when T1 and T2 were presented in the same modality suggested that there was no within-modality component to the AB. Arnell and Jolicœur (1999) concluded that the AB is not a uniquely visual phenomenon, and that central limitations on stimulus consolidation are responsible for the AB.

Arnell and Jolicœur (1999) performed a follow-up experiment to ensure that the cross-modal ABs reported above did not result from participants' attending to the to-be-ignored T1 presented in the same modality as T2. Given that the visual and auditory T1s were presented simultaneously, it was possible that participants attended to the irrelevant T1 presented in the

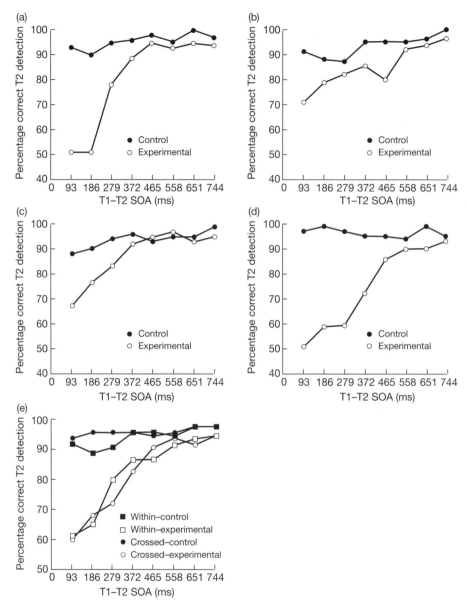

Fig. 8.1 From Arnell and Jolicœur (1999). Mean percentage of correct T2 responses to T2 present trials as a function of T1–T2 SOA. Open symbols represent performance in the experimental condition (identify T1 number and report T2 presence/absence) and closed symbols represent performance in the control condition (report T2 presence/absence only). Panel (a) contains means obtained in the visual T1–visual T2 condition. Panel (b) contains means obtained in the auditory T1–auditory T2 condition. Panel (c) contains means obtained in the visual T1–auditory T2 condition. Panel (d) contains means obtained in the auditory T1–visual T2 condition. To facilitate the comparison of AB magnitudes within and across modalities, Panel (e) shows the T2 means collapsed into within and cross-modality conditions where VV and AA means are averaged for the within-modality condition, and VA and AV means are averaged for the crossed-modality condition.

same modality as T2, and that the observed cross-modal AB was in fact within-modality AB. For example, in the AV cross-modal condition, all or part of the visual T2 accuracy deficit may have resulted from participants' attending to the visual T1 instead of the temporally coincident auditory T1. When the irrelevant, to-be-ignored T1 digit was removed and replaced with a random letter, reliable ABs were still observed in both cross-modal conditions. These blinks were reduced only slightly in magnitude compared to the cross-modal blinks reported above, and the magnitude of the cross-modal AB was still statistically equivalent to the magnitude of the within-modality AB.

The next wave of cross-modal AB experiments

As a result of these four discrepant sets of findings, researchers began investigations to reconcile the divergent pattern of cross-modal AB results. As we will see, these experiments have revealed several factors that appear to be responsible for the pattern of cross-modal AB results, and have helped to elucidate further the nature of the specific and central sources of interference involved.

In order to maintain the tenability of a VSTM confusion theory, Shapiro and Terry (1996) suggested that, regardless of the modality of presentation (visual or auditory), the information in each stimulus is converted to a visual representation that then undergoes processing limitations in a visual buffer such as VSTM. It is not clear why visual representations should necessarily be elicited for auditory stimuli. Also, it is not known whether the visual representation created from the auditorily presented stimulus would be as strong as one created from a visually presented stimulus. In addition, it is not clear why the initial auditory representation itself would not be enough to save a T2 stimulus from undergoing AB. However, given that all the initial cross-modal and auditory AB experiments used alphanumeric stimuli with a readily accessible visual representation, such a theory was possible. To test this claim, Arnell and Jolicœur (1999) performed an AV cross-modal experiment where all the auditory stimuli, including T1 and T2, were pure tones. Pure tones were used because they have no corresponding visual representation. If cross-modal blinks result from VSTM confusion after converting an auditory stimulus to a visual representation, then no AB should be found for pure tones, given that participants would be unable to convert them to a visual representation and have them reside in VSTM. In the experiment, T1 was one of two high-pitched tones presented in a stream of tones with slightly lower pitch. T2 was the visual letter 'X'. Visual and auditory streams ran concurrently at a rate of 10 items/s for each stream. On control blocks participants were instructed to ignore the tones, and just report whether the X was present or absent in the visual stream. On experimental blocks participants were instructed to report whether the T1 tone was high or low in pitch, and then report whether the X was present or absent in the visual stream. The results showed that T2 accuracy was good at all SOAs in the control condition, but in the experimental condition T2 accuracy was reliably poor at short T1–T2 SOAs (from 100 to 500 ms). The presence of a cross-modal AB using pure tones provides good evidence that the AB is not uniquely visual in nature, and strongly suggests that auditory and cross-modal ABs do not result from visual interference after all stimuli are converted to visual representations, as was suggested by Shapiro and Terry (1996).

Arnell and Jolicœur (1999) noticed an intriguing pattern in the auditory and cross-modal AB results presented in 1995. Although there were many differences between the stimuli and procedures used in the Arnell and Jolicœur (1995, 1999) experiments, those of Potter *et al.* (1995, 1998), and those of Shulman and Hsieh (1995) that could account for the apparently discrepant findings, the strength of the effects was inversely related to the rate of presentation. Potter *et al.* (1995, 1998) had the slowest rate of presentation (7.41 items/s), and found no evidence for an AB in their AA condition, and no cross-modal AB. Shulman and Hsieh (1995) had a slightly faster rate (7.94 items/s), and found a small AB in the AA condition, but found no cross-modal AB in the VA condition. Arnell and Jolicœur (1995, 1999) used the fastest rate (10.72 items/s) and demonstrated an AB in all conditions, with no reduction in the amount of AB across modalities compared to the amount of AB within modality. Potter *et al.* argued that the lack of an auditory AB in their experiment did not result from the relatively slow presentation rate, because a robust visual AB was found when using the same presentation rate of 7.41 items/s in RSVP streams. However, visual and auditory AB need not have the same boundary conditions or timing requirements, even if they originate from a common central processing limitation. The temporal resolution of the auditory system is higher than the temporal resolution of the visual system (Eddins and Green 1995). This suggests that auditory streams may require faster rates of presentation than visual streams to produce equivalent effects. The difference in required presentation rates would probably operate through T1 and/or T2 masking, in that the faster the presentation rate the less time T1 and T2 have to be processed. Giesbrecht and Di Lollo (1998) and Jolicœur (in press) have demonstrated that pattern masking of T2 is necessary to observe an AB effect. Also, the strength of T1 masking has been shown to modulate the magnitude of the AB (Chun and Potter 1995; Raymond *et al.* 1992, 1995; Seiffert and Di Lollo 1997).[3] If visual and auditory streams are presented at the same rate, then visual streams may be receiving stronger T1 masking, thereby increasing the difficulty of the T1 task, and stronger T2 masking, thereby effectively reducing the interval during which the probe is available for processing. This speculation could also explain the consistent pattern of finding larger AB effects with visual T2s than with auditory T2s. There is good evidence that T2 masking has more influence over the magnitude of the AB effect than does T1 masking. Although the magnitude of the AB effect is often markedly reduced when T1 is not effectively masked (cf. e.g. Raymond *et al.* 1992; Seiffert and Di Lollo 1997) a relatively large AB can sometimes still be found (cf. e.g. Chun and Potter 1995; Seiffert and Di Lollo 1997). However, when T2 is not effectively masked, no AB is ever observed (Giesbrecht and Di Lollo 1998; Jolicœur, in press). If effective masking of T2 is more important to the production of an AB than effective masking of T1, then we might expect the modality of T2 to be more influential in setting the presentation rate threshold. In this manner, conditions with visual T2s would be able to show AB effects at shorter SOAs than conditions with auditory T2s, and for a given presentation duration the AB observed with visual T2s would be larger than that observed for auditory T2s.

To test empirically the notion that the presentation rate necessary to observe an AB may be different for the visual and auditory modalities, Arnell and Jolicœur (1999) presented visual or auditory streams at rates of 9.52, 8.33, 7.41, and 6.67 items/s (105, 120, 135, and 150 ms per item respectively). No cross-modal conditions were examined. Modality (auditory or visual) and rate of presentation were manipulated between subjects. Unlike the other experiments of Arnell and Jolicœur, participants in the visual condition experienced visual streams, but no auditory streams were played. Similarly, participants in the auditory condition experienced auditory streams, but no visual streams were presented. Control and experimental instructions

---

[3] However, McLaughlin, Klein and Shore (2001) present evidence that other masking manipulations do not affect AB magnitude.

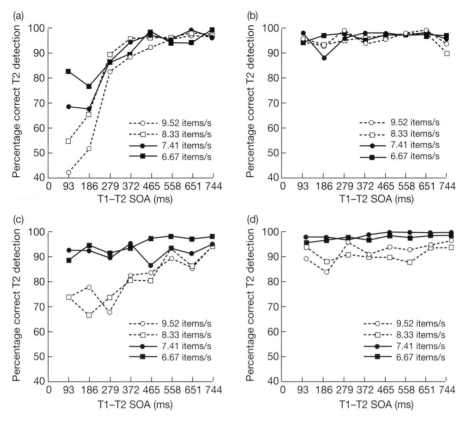

Fig. 8.2 From Arnell and Jolicœur (1999). Mean percentage of correct T2 responses to T2 present trials as a function of presentation rate and T1–T2 SOA. Panel (a) contains means obtained in the visual modality with experimental instructions (identify visual T1 number and report presence/absence of visual T2). Panel (b) contains means obtained in the visual modality with control instructions (report presence/absence of visual T2 only). Panel (c) contains means obtained in the auditory modality with experimental instructions (identify auditory T1 number and report presence/absence of auditory T2). Panel (d) contains means obtained in the auditory modality with control instructions (report presence/absence of auditory T2 only).

were the same as in previous experiments; just report presence/absence of the letter 'X' (T2) for control blocks, and report the identity of the digit (T1) and the presence/absence of the letter 'X' for experimental blocks.

As is shown in Fig. 8.2, presentation rate did affect the AB magnitude observed in both the visual and auditory modalities. In the visual modality the magnitude of the AB decreased monotonically as the presentation rate increased, but a reliable AB was found at all four presentation rates. However, in the auditory modality a reliable AB was found for the two fastest presentation rates (9.52 and 8.33 items/s), but no AB was found for the two slowest presentation rates (7.41 and 6.67 items/s). Note that in this experiment no visual stimuli were presented to participants performing under the auditory condition. Therefore this finding also provides good evidence that an auditory AB can be found without the presence of visual stimuli. Again, the auditory AB is smaller in magnitude than the visual AB. The pattern of findings is exactly what we would expect if the inconsistent pattern of results across laboratories resulted, at least in part, from the various presentation rates employed. Finding a significant visual AB, yet no significant auditory AB, at a rate of 7.41 items/s replicates Potter *et al.*'s (1995, 1998) pattern of findings at the same rate. Finding a reliable auditory AB at the two fastest presentation rates is what one would expect given that Arnell and Jolicœur (1999) reliably found auditory ABs at a rate of 10.72 items/s, and Shulman and Hsieh (1995) found auditory ABs with a presentation rate of 7.94 items/s. Given that no auditory AB was observed at a rate of 7.41 items/s, Shulman and Hsieh's rate may have been just fast enough to produce an auditory AB that was smaller than those of Arnell and Jolicœur, yet statistically reliable. Given these results, Arnell and Jolicœur concluded that visual patterning was neither necessary nor sufficient to produce an AB, and that auditory ABs could be reliably observed provided that T1 and T2 were adequately and quickly masked, for example by using a fast presentation rate for auditory streams.

Rate of presentation, however, cannot explain the auditory AB found by Duncan *et al.* (1997). In their experiment, each visual or auditory stimulus had a presentation rate of 4 items/s—a rate that is slower than Arnell and Jolicœur (1995, 1999), Potter *et al.* (1995, 1998), and Shulman and Hsieh (1995). However, in the auditory condition two auditory items were presented concurrently, just offset in time. It is possible that, in addition to effects of masking, the total rate of information transmission is also important. By presenting two items every 250 ms, Duncan *et al.* (1997) may have approximated the information rates that we would have produced by presenting one item every 125 ms. At this rate we would still expect to observe an auditory AB, given the results of Shulman and Hsieh (1995).

Recall that Duncan *et al.* (1997) found no cross-modal AB. It is also possible that the relatively slow auditory presentation rate used by these researchers was not fast enough to produce either cross-modal AB or auditory AB effects. The auditory AB found by Duncan *et al.* would then need to be considered artefactual. This supposition is possible given that T2 accuracy showed a deficit only at the shortest SOA in the AA condition, and this was the only cell using one or more auditory stimuli where the two targets were presented in the same modality, and overlapped in time, thus causing a unique potential for low-level perceptual masking or confusion effects unrelated to the AB. Indeed, in this cell both auditory targets were combined into a single stimulus for computer presentation. This supposition is supported by the relatively low T1 accuracy at the shortest SOA in the AA condition. Recently, Arnell, Transgrud, Hammes, and Larson (2000) were able to replicate the AA results of Duncan *et al.* (1997) using identical procedures and stimuli. However, they found that the AB pattern remained intact even when all non-target stimuli were removed, suggesting that the AA effect did not represent an AB.

Potter *et al.* (1998) noticed another pattern amongst the conflicting auditory and cross-modal AB results. Participants in the Potter *et al.* experiments performed the same task for T1 and T2, and the experiment was designed in such a way that participants did not have to reconfigure their task set after T1 and before T2. In contrast, in all other experiments (Arnell and Jolicœur 1999; Duncan *et al.* 1997; Shulman and Hsieh 1995), participants did have to reconfigure their task set to varying degrees after the presentation of T1. In the Potter *et al.* experiments, participants were asked to report the two letter targets presented amongst the digit distractors. Participants performed the same task for both T1 and T2; therefore they did not have to switch tasks after identifying T1. Furthermore, the identity of T1 gave them no additional information about the identity of T2. In contrast, in the experiments performed by Arnell and Jolicœur (1995, 1999) and Shulman and Hsieh (1995), participants performed one task for T1, and a different task for T2. For example, in most of the Arnell and Jolicœur (1999) experiments participants were asked to report the identity of the lone digit as the T1 task, and to detect the presence/or absence of an 'X' for the T2 task. Given that participants were told that the digit would always be presented before the 'X', participants probably adopted a strategy where they looked or listened for the digit, and then, after seeing or hearing the digit, switched tasks, and looked or listened for the 'X'. Also, because modality combination was blocked or manipulated between participants, on cross-modal trials participants knew the order of the modalities, and not only reconfigured their attention to the T2 task, but also reconfigured their attention to another (T2) modality. In the Duncan *et al.* (1997) experiments participants were asked to report the two words that were presented amongst the filler non-word distractors. Furthermore, the AV and VA cross-modal conditions were randomly mixed within blocks of trials. In this manner the experiment involved no task switching. However, only the cross-modal conditions (AV and VA) were presented within a block of trials, and one set of target words was used for the auditory targets ('cot' or 'cod') and another for the visual targets ('nap' or 'nab'). Suppose you were a participant in the cross-modal experiment. At the start of each trial you needed to prepare for the possibility that T1 might be either visual or auditory; therefore you needed to monitor both modalities, and keep all four words in mind as templates. However, once T1 had been presented, you could then confidently reconfigure your attention to the modality opposite to that of T1, and keep only the two modality-relevant words in mind as templates. Such reconfiguration is also advantageous for the participant in the AA condition. In the AA condition participants received a high-pitched and a low-pitched stream, with a single target (either T1 or T2) occurring in each stream on each trial. Furthermore, the high-pitched stream always contained the words 'cot' or 'cod', and the low-pitched stream always contained the words 'nap' or 'nab'. At the start of each trial, participants had to monitor both streams, and keep all four target words in mind as templates. However, once T1 had been presented then the participants could confidently reconfigure their attention to the auditory stream opposite to that of T1, and keep only the two stream-specific words in mind as templates. Thus, in this manner there was switching even in the Duncan *et al.* (1997) experiments where the T1 and T2 tasks were the same.

As Potter *et al.* (1998) explained, a change in perceptual or preparatory set from T1 to T2 could mimic an AB effect. Improvement in T2 performance with increasing SOAs could be due to recovery from dual-task interference produced by T1, and this is the interpretation that has been assumed in most dual-task studies. However, under these conditions such results could also be due to a switching of task-set (cf. e.g. Allport *et al.* 1994; Allport and Hsieh, this volume, Chapter 3; Rogers and Monsell 1995). For example, Allport *et al.* (1994) have shown than even when the participants expect a task switch, as in the above AB experiments, performance on a second target presented just after the switch is impaired. If states of attentional priority or response preparation are configured in one manner before the presentation of T1, and then reconfigured in anther manner after the presentation of T1, then improvements in T2 performance across SOA could reflect this process of task-set reconfiguration, rather than dual-task interference from the processing of T1. It is worth noting that this issue is not confined to cross-modality AB experiments. Indeed, the majority of visual AB experiments have also used designs and tasks that promote task-set reconfiguration after T1. Experiments have clearly shown that visual AB can be found when no switch in preparatory set is employed (e.g. Chun and Potter 1995). However, when investigating the AB in any modality it is preferable to use designs that remove the need for changes in preparatory set within a trial, so that any resulting T2 deficit can be unambiguously attributed to dual-task interference.

Potter *et al.* (1998) argued that the results of Arnell and Jolicœur (1999) (and Duncan *et al.* (1997) and Shulman and Hsieh (1995)) reflected two different types of attentional deficits.

Potter *et al.* posited that the T2 accuracy deficit that was observed at short SOAs in the VV condition was a true AB caused by dual-task interference. However, they argued that the T2 accuracy deficit observed at short SOAs in the cross-modal and auditory conditions was the result of task-set switching, not dual task-interference, and that the T2 deficits were artefactual and only mimicked the AB. They also used the shape of the T2 accuracy function across SOAs to argue for the true, versus artefactual, ABs. In the cross-modal and auditory conditions of Arnell and Jolicœur, T2 accuracy was lowest at the shortest T1–T2 SOA, and did not show the characteristic U-shaped function associated with the AB. In the VV condition T2 accuracy was lowest and equal at the two shortest SOAs. Potter *et al.* argued the AB effects should be U-shaped, and that the monotonic patterns arose because the cross-modal and auditory T2 accuracy deficits were the results of task set-switching alone. The VV pattern was said to result from a combination of task set switching and dual-task interference. Potter *et al.* reasoned that if switches in task set were responsible for the cross-modal and auditory T2 accuracy deficits observed by Arnell and Jolicœur (1999), then they should be able to replicate Arnell and Jolicœur's findings when using a task set switching design, but should find no cross-modal or auditory AB in a similar design when task set switching was removed.

Potter *et al.* (1998) performed a replication of Arnell and Jolicœur's initial experiment that investigated the AB in four modality combinations (VV, AA, VA, and AV). They presented concurrent visual and auditory streams, with a presentation rate of 8.33 items/s for each stream. Modality combination was manipulated between participants, and participants were instructed to report the lone digit amongst letters (T1 task) and detect the presence/absence of a subsequent 'X' (T2 task). Potter *et al.* replicated Arnell and Jolicœur's results, finding that T2 accuracy was reduced at short SOAs in all four modality combinations. In the VV condition T2 performance was U-shaped, but in the other three modality combinations T2 performance was lowest at the shortest SOA. Potter *et al.* (1998) then modified the above experiment to remove task set switching. The experiment was kept the same except that participants now searched for two letter targets amongst digit distractors. A different pattern of results was found. T2 accuracy was reduced at short SOAs for the VV condition, and significantly (but minutely) in the VA condition. T2 accuracy was flat across SOA in the AA and AV conditions. Potter *et al.* concluded that there are two distinct types of interference that can occur with two temporally spaced targets. One is a strictly visual dual-task interference that produces a U-shaped AB function. The other is an amodal interference, independent of the AB, which results from reconfiguring task preparatory set after T1.

However, there were more differences than just task set switching in the two Potter *et al.* experiments. For one, the first experiment had participants report a letter T1 amongst digit distractors, and the second had participants report digit targets from amongst letter distractors. Potter *et al.* (1998) addressed this concern by performing the AA condition with no task set switching where participants listened for letter targets amongst digit distractors. No AB was found, supporting the contention that it did not matter whether letters or digits acted as targets or distractors. However, in the replication of Arnell and Jolicœur (1999), the distractors and T2 were letters, and T1 was a digit. Thus, T2 was from the same alphanumeric class as the distractors, and T1 was from a different class. However, in the experiments that removed task set switching, both T1 and T2 were letters, and the distractors were digits, therefore both targets were from a different alphanumeric class than the distractors. While T1 was in a different alphanumeric class from the distractors in all of the Potter *et al.* (1998) experiments, the only experiment where T2 was in the same alphanumeric class as the distractors was the experiment that replicated the results of Arnell and Jolicœur (1999). This critical difference could also produce the pattern of findings reported in Potter *et al.* by influencing T2 masking. Giesbrecht and Di Lollo (1998) have demonstrated that pattern masking of T2 is necessary to observe an AB effect. Furthermore, Chun and Potter (1995) demonstrated that the AB effect was larger when the T2 letter appeared among digit distractors than when it appeared amongst symbol distractors, showing that T2–distractor conceptual discriminability modulates the magnitude of the AB. Isaak, Shapiro and Martin (1999) have also shown that when the physical similarity is kept constant, the conceptual relationship between T2 and the distractors influences the magnitude of the AB, in such a way that a larger AB is found when more distractors are conceptually similar to T2.

It is possible that the small T2 accuracy deficits seen in the task set switching experiments of Potter *et al.* when T2 was an 'X' among letters could easily be reduced to null effects when T2 was a letter among digits that stood out more from the background of distractors, and received less conceptual masking than the 'X' received from the same class of alphanumeric stimuli. It is impossible to distinguish whether task set switching explanations or the T2-distractor relationship can explain Potter *et al.*'s pattern of findings, given that both co-vary in these experiments.

Cross-modal experiments without task set switching—hybrid AB–PRP experiments

Regardless of whether task set switching or the T2-distractor relationship can explain the pattern of findings in the Potter *et al.* (1998) experiments, AB designs that avoid within-trial changes in task preparatory set can provide a cleaner estimate of AB magnitude. One such set of experiments was performed by Arnell and Duncan (in press). These experiments were not typical AB experiments; speeded responses were required to unmasked auditory targets, and unspeeded responses were required to masked visual targets, making it a hybrid AB–PRP experiment. Such experiments have been discussed in more breadth and detail by Jolicœur (this volume, Chapter 5); but the Arnell and Duncan experiments will be discussed here, given that these experiments contained no task-set switching, and were specifically designed to estimate the relative influences of within-task or modality limitations and central, amodal processing limitations. Two successive targets (T1 and T2) were presented on each trial at a variety of SOAs, randomly in all possible combinations: both targets unspeeded/visual or both speeded/ auditory on within-task/modality trials; unspeeded/visual then speeded/auditory, or speeded/ auditory then unspeeded/visual on crossed-task/modality trials. Modality combination varied unpredictably and randomly from trial to trial, and the identity or modality of T1 provided no information as to the identity or modality of T2. At the start of each trial, participants had to monitor both modalities for all potential targets. Because T1 identity and modality was independent of the identity and modality of T2, even after T1 had been presented, participants had to monitor both modalities for all potential targets, thereby removing any preparatory changes in task-set, as discussed by Potter *et al.* (1998). In both experiments visual targets were masked digits (1–9), requiring an unspeeded response indicating their identity at the end of the trial. In the first experiment, auditory stimuli were pure tones presented in low, medium or high pitch, requiring participants to make a speeded button press indicating the tone's pitch immediately after presentation of the tone. In the second experiment, the auditory stimuli were unmasked compressed-speech words ('cot', 'mug', 'yes') spoken in a low-, medium-, or high-pitched voice. This task required participants to make a speeded button press indicating the word's identity, the word's pitch, or both immediately after presentation of the word.

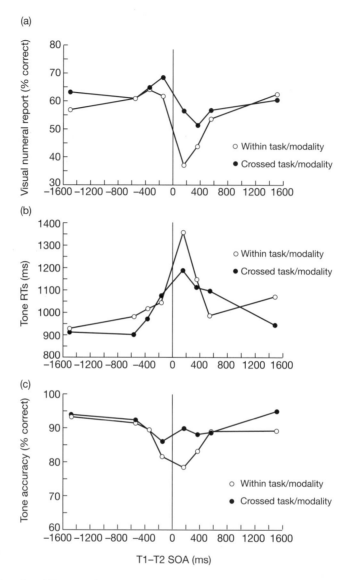

Fig. 8.3 From Arnell and Duncan (in press). Panel (a) displays mean accuracy rates for masked visual numerals in the first experiment as a function of T1–T2 SOA, and task/modality within vs. crossed. Panel (b) displays unmasked auditory tone RTs from the first experiment as a function of T1–T2 SOA, and task/modality within vs. crossed. Panel (c) displays unmasked auditory tone accuracy rates from the first experiment as a function of SOA, and task/modality within vs. crossed. Positive SOAs reflect T2 performance at each SOA after T1, and negative SOAs reflect T1 performance at each SOA backward from T2.

In both experiments, large and robust response slowing was observed for auditory T2s at short SOAs. This result was found both when auditory T2s were preceded by auditory T1s, and when auditory T2s were preceded by visual T1s not requiring a speeded response (see Figs. 8.3B and 8.4B). However, in both experiments, greater auditory T2 response slowing was observed in the

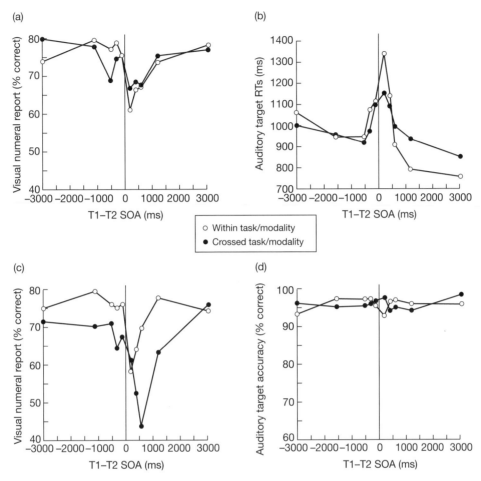

Fig. 8.4 From Arnell and Duncan (in press). Panel (a) displays masked visual numeral accuracy rates averaged from the 'single dimension' trials (report pitch only or identity only) in the second experiment as a function of T1–T2 SOA, and task/modality within vs. crossed. Panel (c) displays masked visual numeral accuracy rates from the 'both dimension' (report pitch and identity) trials in the second experiment as a function of T1–T2 SOA, and task/modality within vs. crossed. Panel (b) displays unmasked auditory RTs averaged from the 'single dimension' trials in the second experiment as a function of T1–T2 SOA, and task/modality within vs. crossed. Panel (d) displays unmasked auditory accuracy rates averaged from the 'single dimension' trials in the second experiment as a function of T1–T2 SOA, and task/modality within vs. crossed. Positive SOAs reflect T2 performance at each SOA after T1, and negative SOAs reflect T1 performance at each SOA backward from T2.

within-task/modality condition (both T1 and T2 speeded and auditory) than in the crossed-task/modality condition (T1 unspeeded and visual, T2 speeded and auditory). In the first experiment, dual-task accuracy costs (i.e. an AB) were observed for visual T2s at short SOAs in the within-task/modality condition (both T1 and T2 unspeeded and visual), but not in the crossed-task/modality condition (T1 speeded and auditory, T2 unspeeded and visual; see Fig. 8.3A). In contrast, in the second experiment, both within-task/modality and crossed-task/modality conditions produced large and robust visual T2 accuracy costs at short SOAs. Furthermore, accuracy costs were at least as large in the crossed-task condition as in the within-task condition (see Fig. 8.4A), especially when participants made speeded responses to two separate dimensions of an auditory T1 (see Fig. 8.4C).

The results provide clear evidence that both shared central resources and within-task and/or within-modality resources contribute to dual-task costs. Finding dual-task interference at short SOAs for crossed-task/modality conditions where reconfiguration of task set should not occur, provides strong evidence for central processing limitations. Finding that dual-task interference was greater in the within-task/modality conditions than in the crossed-task/modality conditions provides strong evidence for additional within-task-and/or-modality processing limitations. Future experiments will need to address whether these limitations are modality-specific, task-specific, or both. The fact that dual-task interference was found when one task was speeded but unmasked, and the other task was unspeeded but masked, provides evidence that both response selection and stimulus consolidation require the same central processing resources, at least in part. Furthermore, the visual T2 accuracy deficits could be made larger and longer arbitrarily by increasing the response-selection demands of the speeded auditory T1 task. This finding provides good evidence that response selection requires central capacity resources that are also used for unspeeded stimulus identification. This will be discussed further below.

Cross-modal experiments without task set switching—AB experiments

The Arnell and Duncan (in press) experiment provides strong evidence for central resource limitations underlying dual-task interference experiments. However, because of the use of speeded responses to auditory targets, it does not speak to whether an AB *per se* can be found outside the visual modality, when task set switching has been removed, in more typical AB experiments where there are no speeded responses.

Goddard, Isaak, and Slawinski (1997) reported an auditory AB using pure tones in an experiment that involved no task set switching. Target and distractor stimuli were all pure tones of widely varying pitch, presented at a rate of 11 tones/s. On experimental trials, two tones (T1 and T2), separated by variable SOAs, were noticeably louder than the filler tones that were all of equal and lower amplitude. Participants were asked to make unspeeded responses identifying the pitch (high, medium, or low) of the T1 tone and the T2 tone. On control trials, only the louder T2 tone was presented, and participants were asked to make unspeeded responses identifying the pitch (high, medium, or low) of the T2 tone. Notice that there is no task set switching in this experiment, given that participants perform the same task in the same modality for both T1 and T2. Also, the pitches of T1 and T2 were independent, meaning that the possible response alternatives do not change from T1 to T2 after T1 is known.

Results showed a large and robust auditory AB. Control trials showed very high T2 accuracy rates at all T1–T2 SOAs, but experimental trials showed very poor T1 accuracy at the three shortest SOAs (under 20 per cent at the shortest SOA), with T2 accuracy recovering to near-control levels by the longest SOA. In the experimental condition, T2 accuracy was lowest at the shortest SOA, replicating the AB pattern found in Arnell and Jolicœur's auditory condition. Potter *et al.* (1998) claimed that T2 performance was poorest at the shortest SOA in the Arnell and Jolicœur (1999) experiments because the poor T2 accuracy was the result of task set switching, not true AB, and that true AB should show a U-shaped pattern. However, the experiment of Goddard *et al.* (1997) used auditory stimuli in an experiment that contained no task set switching, yet also showed the lowest T2 performance at the shortest T1–T2 SOA. Such results do not support the claim of Potter *et al.* that AB without task switching should necessarily show a U-shaped function, and that the presence or absence of the U-shaped function can be taken as evidence for the presence or absence of task set switching in a given experiment.

Mondor (1998) also provided strong evidence for an auditory AB in an experiment where no task set switching should have operated. Mondor used tones presented in RAP streams at a rate of 10 tones/s. Distractors were pure tones varying in pitch, and a pure tone of noticeably higher pitch acted as T1. T2 was a complex tone composed of five log-related frequencies. T1 was present on half of all trials, as was T2. T1 and T2 presence/absence was fully crossed, producing T1-only trials, T2-only trials, trials where both T1 and T2 were present, and trials where neither T1 nor T2 were present. At the end of each trial, participants made unspeeded responses indicating whether T1 was present or absent, and whether T2 was present or absent in the auditory stream. Since T1 was present on only half of all trials, task set switching was reduced in this experiment. At the start of each trial participants had to monitor the auditory stream for both T1 and T2. The high rate of T2 accuracy observed at all T2 positions when T1 was absent showed that participants were indeed alert for both T1 and T2. However, some amount of task set switching was possible, if, after hearing T1, participants reconfigured their task set to listen just for the T2 tone.

A reliable auditory AB was found in this study. T2 accuracy was equally high (approximately 95 per cent correct) at all T1–T2 SOAs when T1 was absent, but when T1 was present T2 accuracy was significantly reduced for approximately half a second following T1. Replicating Arnell and Jolicœur (1999) and Goddard *et al.* (1997), T2 accuracy was lowest (approximately 77 per cent) in the dual-task condition at the shortest T1–T2 SOA. Mondor (1998) performed a replication of the above experiment, with the exception that the T1+1 tone was removed from the auditory stream. Removal of the T1+1 item in visual streams has resulted in the absence of (Raymond *et al.* 1992) or attenuation of (Chun and Potter 1995; Seiffert and Di Lollo 1997) the AB. However, in contrast with the RSVP results, Mondor found no attenuation of the AB when the T1+1 tone was removed from the auditory stream. On T1-absent trials, T2 accuracy was high (approximately 92 per cent) at all T1–T2 SOAs, but on T1-present trials, T2 accuracy was reduced (approximately 70 per cent at minimum) for approximately 500 ms after T1 presentation. However, one difference in the AB pattern did emerge in the two experiments: a U-shaped function was observed for T2 accuracy in the second experiment. Mondor concluded that the visual AB theories of Chun and Potter (1995), Raymond *et al.* (1992) and Shapiro and Raymond (1994) were inconsistent with the presence of auditory AB, and with the finding that the auditory AB was not attenuated when the T1+1 item was removed from the auditory stream. Instead, Mondor suggested that the AB may result from a bottleneck on response-selection operations, as in the PRP paradigm.

The results of Goddard *et al.* (1997) and Mondor (1998) demonstrate that robust auditory AB can be observed in experiments that allow little or no task set switching from participants. Arnell and Larson (submitted) performed an experiment to examine whether an AB could be found in all four T1 and T2 modality combinations (VV, AA, VA, and AV), under conditions where task set switching should not have operated. Visual and auditory streams were independent, yet ran concurrently at a rate of 12.0 items/s for each modality stream. In the Arnell and Larson experiment, modality combination was not manipulated between participants or blocked within participants as it was in all other published cross-modal AB experiments (with the exception of Arnell and Duncan, submitted). Instead, each participant performed all four modality combinations, and the modalities of T1 and T2 were independent and unpredictable from trial to trial. Furthermore, the identities of T1 and T2 were completely independent, such

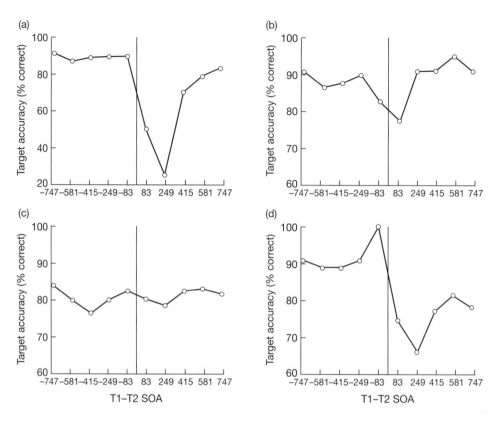

Fig. 8.5 From Arnell and Larson (submitted). Mean percentage of correct T1 and T2 responses as a function of T1–T2 SOA. Panel (a) contains means from visual targets when T1 and T2 were presented in the same modality (i.e. in the VV condition). Panel (b) contains means from auditory targets when T1 and T2 were presented in the same modality (i.e. in the AA condition). Panel (c) contains means from auditory targets when T1 and T2 were presented in different modalities (auditory T1 data from AV trials, and auditory T2 data from VA trials). Panel (d) contains means from visual targets when T1 and T2 were presented in different modalities (visual T1 data from VA trials, and visual T2 data from AV trials). Positive SOAs reflect T2 performance at each SOA after T1, and negative SOAs reflect T1 performance at each SOA backward from T2. Note the difference in scale in panel A, as compared to panels B, C, and D.

that the identity of T1 did not in any way change the possible identity of T2. At the start of each trial, participants received no information as to the modality of T1 or T2, so they needed to monitor both modalities for all potential target types. Because the modalities of T1 and T2 were independent, even after T1 had been presented, the participant still had no knowledge of T2 modality or identity, and needed to continue monitoring both modalities for all potential target types. Thus, preparatory changes in task-set were removed in Arnell and Larson's design, given that participants had no knowledge of T1 or T2 modality or target identity at any point in the trial prior to each target presentation. Visual targets (T1 and T2) were randomly selected to be any of the 26 letters except 'K' and 'L', and were presented in blue amongst black letter distractors. Auditory targets were the randomly selected letters 'K' or 'L', and were presented amongst auditory letter distractors that did not contain Ks or Ls. Auditory targets were presented in the right ear, and a 50 ms pure tone started at the same time was played in the left ear. Participants were told that the visual targets would always be blue, and that the auditory targets would always be either a 'K' or an 'L' and would be coincident with the tone. No tone was played for auditory distractor letters, which were presented binaurally. At the end of each trial, participants were prompted to provide unspeeded responses indicating the identity of T1 and T2 in either order, guessing if unsure.

T1 and T2 accuracy for each modality combination are presented in Fig. 8.5 as a function of T1–T2 SOA. T1 performance (especially at the longest SOAs) was used to establish perform-ance baselines under conditions of no dual-task processing interference. T2 accuracy was reduced at short T1–T2 SOAs in the VV, AA, and visual-crossed modality combinations, indi-cating the presence of an AB for these modality combinations. T2 performance was approx-imately equal at all SOAs in the auditory—crossed condition, indicating no AB for this condition. The presence of a VV AB without preparatory task set switching replicates Chun and Potter (1995) and Arnell and Duncan (in press). The presence of an AA AB without preparatory task set switching replicates Goddard et al. (1997) and Mondor (1998), and adds to the litera-ture demonstrating that reliable, robust ABs can be found in the auditory modality, even under conditions where task set switching does not operate. The presence of an AB for visual T2s pre-ceded by auditory T1s provides evidence that cross-modality ABs can be found even when preparatory changes of task set cannot be said to underlie the T2 accuracy reductions at short SOAs. Finding reliable AA and AV ABs without task set switching argues strongly against the-ories that interpret the AB as a uniquely visual phenomenon (Duncan et al. 1994; Potter et al. 1998; Raymond et al. 1992, 1995; Shapiro et al. 1994; Shapiro and Raymond 1994; Shapiro and Terry 1996; Ward et al. 1996), and directly contradicts Potter et al.'s (1998) conclusion that auditory and cross-modal ABs cannot be found when changes in preparatory task-set are removed from experiments.

Furthermore, the U-shaped pattern, characteristic of the AB, is found in the AV condition of Arnell and Larson's results. Here T2 accuracy was lower at the 249 ms SOA than at the 83 ms SOA. Suppose that Potter et al. (1998) were correct in stating that task-set switching in the experiments of Arnell and Jolicœur (1999) contributed to the absence of a U-shaped pattern, and that the U-shaped function demonstrates true AB without the influence of set switching. Then the results of the AV condition in Arnell and Larson provide compelling evidence that true AB can be found across T1–T2 modalities, and that task set switching did not operate in the Arnell and Larson experiment. Indeed, the presence of the U-shaped function in the AV condi-tion makes it difficult to explain the results in terms of any sort of preparatory or stimulus-driven switch costs, given that such switch costs are maximal at the shortest SOAs. As such, Arnell and Larson's AV results strongly suggest that central, amodal processing limitations con-tribute to the AB.

However, although there is evidence in the Arnell and Larson experiment for the importance of central processing limitations, there is also evidence of the importance of within-modality or perceptual limitations. The VV AB is obviously larger than the ABs found in the other modality conditions. In the Arnell and Jolicœur (1999) experiments with task set switching, modality conditions with visual T2s produced larger ABs than modality conditions with auditory T2s. However, within each T2 modality the magnitude of the AB was equivalent for within-modality and crossed-modality conditions. This is not the case for the Arnell and Larson findings. Analyses showed that, for both visual and auditory T2 modalities, the AB magnitude was larger when T1 and T2 were presented in the same modality than when T1 and T2 were presented in different modalities. Finding an AB at short SOAs in the visual-crossed condition provides strong evidence for central processing limitations. Finding that dual-task interference was greater in the within-modality conditions than in the crossed-modality conditions provides some evidence for additional within-modality processing limitations. It is possible that preparatory task and/or modality switching in the Arnell and Jolicœur (1999) experiments contributed to the magnitude of the cross-modal ABs, thereby causing them to approximate to the magnitude of the within-modality ABs. It is also possible that Arnell and Jolicœur were correct in concluding that only central, amodal processing limitations contribute to the AB, but that lower-level perceptual interference contributed to the magnitude of the within-modality effects in the Arnell and Larson experiment. Future experiments are required to decide between these alternatives. However, the existence of both within-modality and central, amodal processing limitations underlying the AB would support the findings of Arnell and Duncan (in press). It would also help to explain why it is often easier to find robust within-modality dual-task costs than robust cross-modality dual-task costs in typical AB paradigms (Duncan *et al.* 1997; Shulman and Hsieh 1995), in hybrid AB–PRP paradigms where one of the responses is speeded (e.g. Pashler 1989; Ruthruff and Pashler, submitted), and more generally in dual-task interference experiments (e.g. Treisman and Davies 1973; Rollins and Hendricks 1980).

Empirical conclusions for cross-modal AB

The question of whether AB can be found outside the visual modality has been answered, and the answer is 'yes'. It is clear that visual, auditory, and cross-modal ABs can be found under conditions where preparatory task-set switching can be assumed to operate, and under conditions where preparatory task-set switching can be assumed not to operate. However, as a former colleague used to say, 'Auditory attention is not just visual attention through the ears.' This is illustrated by the results showing that auditory and cross-modal ABs do not appear to have the same boundary conditions as the visual AB. Furthermore, stimulus manipulations can sometimes have quite different effects on visual and auditory AB (e.g. Mondor 1998 showed no reduction in auditory AB magnitude when the T1+1 item was removed from the stream). Thus, we can return to the question of what stimuli and tasks will produce an AB in different modalities, and across modalities, and what these can tell us about the nature of interference in temporal attention paradigms.

To note that the conditions that produce an AB are not always the same for different modalities does not suggest that central processing limitations do not underlie the AB. Selection of targets in rapid streams of information will be influenced by: the temporal resolution of each of the modalities; the stimulus dimensions that can be effectively selected in each modality; and the stimulus dimensions that are effectively masked in each modality. For example, as was discussed above, the temporal resolution of the auditory system is greater than that of the visual system (Eddins and Green 1995), suggesting that auditory targets may be less effectively masked than visual targets when presented in rapid auditory and visual streams presented at the same rate. This may make it more difficult to find an AB in conditions containing one or more auditory targets, and may require faster auditory presentation rates. Furthermore, the strength of target masking contributes to the magnitude of the AB (Chun and Potter 1995; Giesbrecht and Di Lollo 1998; Seiffert and Di Lollo 1997), yet the rapid serial presentation can mask some characteristics but not others, and these differ across modalities. For example, in an initial pilot attempt to examine auditory and cross-modal AB, Pierre Jolicœur and I defined the auditory T1 by pitch, asked participants to identify the letter spoken in the higher pitch, and presented T1 amongst distractors of equal and lower pitch. This manipulation was chosen because increasing the pitch of T1 in the auditory modality seemed superficially analogous to increasing the luminance of T1 in the visual modality. However, no AB was found in the AV condition, and participants reported that they often identified the T1 pitch at the end of the stream to facilitate T2 performance. It appeared that high-pitched T1 was not masked by the homogeneous lower-pitched filler stimuli, and that participants could retrieve the T1 identity from echoic memory at the end of the stream. This finding corresponds with those of Mondor, Zatorre, and Terrio (1998*b*), who have provided evidence that in audition both pitch and location function much like spatial location does in vision (but not like colour does in vision), in that attention easily creates separate streams based on these attributes. In the same manner that one would expect to find a dramatically attenuated AB if a visual T1 was unmasked and presented in an obviously different visual location (cf. e.g. Raymond *et al.* 1995), one would also expect little or no AB in our auditory AB task. This example highlights the fact that we must recognize that the characteristics of each modality will influence the selection and masking of targets. Such differences may produce different AB patterns, even if the AB is influenced by central processing limitations.

Research thus far strongly suggests that presentation rate is one of the factors that is different for visual and auditory modalities, in that auditory streams require faster minimum presentation rates to produce an AB (approximately 8 items/s or faster), as compared to visual streams, where a robust AB is still found at rates of 6.7 items/s or lower (Arnell and Jolicœur 1999). It is very likely that the exact rates required to produce an auditory or visual AB depend on many factors, such as the difficulty of the task, the nature of the stimuli, and even participant characteristics. However, it is also likely that auditory streams will still require a faster minimum presentation rate to produce an AB than will visual streams, regardless of the specifics involved.

The presence or absence of preparatory task set switching has also been demonstrated to be an important factor in the pattern of cross-modal and auditory AB findings (Potter *et al.* 1998). While reliable auditory and cross-modal ABs have been observed under conditions where preparatory task-set switching should not operate, the presence or absence of task set switching does appear to influence the magnitude of the AB. Potter *et al.* (1998) are correct in asserting that the possibility of task-set switching should be removed to allow for a cleaner estimate of cross-modal and auditory AB magnitude. To avoid inflated estimates of AB magnitude, researchers should attempt to remove preparatory task-set switching from experiments, not only when investigating auditory or cross-modal AB, but also in purely visual AB experiments. Of course, many other factors will be influential as well, especially as investigators begin to look for AB effects in modalities other than vision and audition (for example Shapiro *et al.* 1998 in the tactile modality).

Theoretical implications for the AB

Revisiting earlier ideas

Results demonstrating robust, reliable ABs in the auditory modality and across modalities invalidate claims that the AB is a uniquely visual phenomenon (e.g., Potter *et al.* 1998) and theories based on the uniquely visual nature of the AB (Duncan *et al.* 1994; Potter *et al.* 1998; Raymond *et al.* 1992, 1995; Shapiro *et al.* 1994; Shapiro and Raymond 1994; Shapiro and Terry 1996; Ward *et al.* 1996). For example, Shapiro and Raymond (1994) contend that the AB results from confusion in VSTM, where the visual representation of the probe becomes lost when VSTM is overcrowded with targets, stimuli temporally adjacent to targets, and stimuli closely resembling target templates. If the AB does result from VSTM confusion, then no AB should be found when one or more targets is presented outside the visual modality. Thus, the several findings of auditory and cross-modal AB (Arnell and Jolicœur 1999; Arnell and Larson, submitted; Goddard *et al.* 1997; Mondor 1998; Shulman and Hsieh 1995) falsify the VSTM confusion theory. One cannot claim that auditory stimuli are simply recoded into visual representations and then confused in VSTM, as Shapiro and Terry (1996) have suggested, given that several studies have used pure tones that have no corresponding visual representation (Arnell and Jolicœur 1999; Goddard *et al.* 1997; Mondor 1998). One could attempt to extend visual theories of the AB by positing that the source of the AB interference is not uniquely visual in nature, but is modality-specific in nature, and that the AB may result from modality-specific restrictions to concurrent attention. Indeed, Duncan *et al.* (1997) extended the visual dwell-time theory of Duncan *et al.* (1994) and Ward *et al.* (1996) in exactly this manner. While such extensions could explain the auditory AB results, they are not consistent with the findings of cross-modal AB where T1 and T2 are presented in different modalities (e.g. Arnell and Jolicœur 1999; Arnell and Larson, submitted; Shulman and Hsieh 1995). In short, uniquely visual or modality-specific theories of the AB are inconsistent with results showing reliable auditory and cross-modal ABs, especially those found under conditions where changes in preparatory task-set cannot operate.

If the AB is not uniquely visual in nature, why then did early experiments with visual, but non-patterned, T1s or T2s not produce any AB? Recall that Shapiro *et al.* (1994) found no AB when T1 was a temporal gap with no visual stimulation, and participants were asked either to detect the presence or absence of the T1 gap, or to determine whether the T1 gap was long or short in duration. This was despite the fact that such tasks were quite difficult. Also, Shapiro *et al.* (1991) and Shapiro *et al.* (1993) found no AB for a coloured symbol when participants were asked to report the symbol's colour, but did find an AB when participants were asked to report the symbol's shape or size. These null findings may have resulted from the fact that the particular characteristics being judged (temporal duration, or colour) were not effectively masked by the RSVP stream. The nature of RSVP streams is such that items are effectively masked only when other items in the stream vary on the to-be-reported dimension and/or effectively replace the dimensional attribute in iconic memory. In the Shapiro *et al.* experiments only the T1 or T2 item possessed the specific to-be-reported attribute, with all other stream items being homogeneous and different from at least one of the targets; thus the characteristics of temporal duration or colour may have been effectively unmasked. If a target is not effectively masked, it is likely that it is not subject to on-line processing and can therefore be dealt with off-line after processing resources are no longer limited. This suggestion is supported by the results of Ross and Jolicœur (1999), who demonstrated an AB for a T2 colour-discrimination when they asked participants to make fine colour discriminations and used distractor colours similar to T2 that varied randomly item to item within a stream.

Visiting newer ideas

While the AB does appear to result at least in part from central, amodal processing limitations, results suggest that there may also be a modality-specific contribution to the AB (Arnell and Duncan, submitted; Arnell and Larson, submitted; Isaak, Shapiro, and Martin, 1999; Shulman and Hsieh 1995). Cross-modal AB results provide good evidence for central processing limitations. Finding larger ABs when T1 and T2 are presented in the same modality than when T1 and T2 are presented in different modalities also provides evidence for within-modality sources of interference. However, at this point it is not clear whether within-modality limitations result from lower-level perceptual interference, task-specific interference, or interference within modality-specific buffers such as VSTM or its auditory equivalent, ASTM.

If both central and modality-specific interference contribute to the AB, and different tasks result in different central or modality-specific resource consumption, then this could help to explain some of the inconsistencies in the auditory and cross-modal AB results reported above. For crossed-modality conditions the magnitude of the AB would depend only on central processing limitations. For within-modality conditions, the magnitude of the AB would still be influenced by central processing limitations, but the AB would also depend on modality-specific processing limitations. The existence of central (amodal) processing limitations would explain cross-modal AB findings (Arnell and Jolicœur 1999; Arnell and Larson, submitted; Shulman and Hsieh 1995). The existence of modality-specific processing limitations, and the assumption that resource-consumption predicts the magnitude of the AB together could explain earlier cases of null AB results for crossed-modality conditions, but robust AB for within-modality conditions (Potter *et al.* 1998; the VA condition in Arnell and Larson, submitted; the VA condition in Shulman and Hsieh 1995). The difference in AB magnitude between within and crossed-modality conditions would be attributable to limitations on within-task resources. The absence of cross-modal AB observed in Potter *et al.* (1998) and the VA conditions of Arnell and Larson (submitted) and Shulman and Hsieh (1995) could be explained if central processing resources were not activated for a sufficient time, or to a sufficient degree. This logic has also been used by Arnell and Duncan (in press) to explain the pattern of inconsistencies found in the hybrid AB–PRP literature.

To explain why AV blinks are found more often than VA blinks it is necessary again to point out that visual T2s appear to be more susceptible to the AB than auditory T2s, perhaps owing to the better temporal resolution of the auditory modality. Potter *et al.* (1998) may also be correct in asserting that the relatively small capacity and relatively short duration of iconic memory make visual T2s prone to the AB, and that the relatively larger capacity and relatively longer duration of echoic memory work against finding an AB with auditory T2s. Therefore, even if conditions are such that cross-modal AB can be found, visual T2s are more susceptible to the processing interference. However, it is also possible that the VA AB found by Arnell and Jolicœur (1999) resulted entirely from task-set switching, and that visual T1 processing cannot interfere with the processing of auditory T2s. In this manner the pattern of results would mimic those of Spence and Driver (1997), where exogenous visual cues could not influence RTs to auditory targets, despite the finding that all other combinations of cue–target modality influenced target RTs.

The nature of central processing interference

Future experiments will be required to decide whether the within-modality processing limitations that contribute to both within-modality and cross-modal AB result from: lower-level perceptual interference, task-specific interference, or interference within modality-specific buffers, such as VSTM or ASTM. But can we say anything about the nature of the central (amodal) processing limitations? First of all, the existence of central processing limitations that underlie the AB is consistent with electrophysiological findings concerning the post-perceptual locus of the AB (cf. e.g. Luck *et al.* 1996; Vogel *et al.* 1998). As has been discussed by Luck and Vogel (this volume, Chapter 7) the AB shows a very distinctive signature when event-related potentials (ERPs) to T2 are measured. This ERP signature shows an intact N400 for T2s presented during the AB interval, but a diminished P300 for T2s presented during the AB interval (Luck *et al.* 1996; Vogel *et al.* 1998). Luck *et al.* (1996) state that the ERP signature indicates that T2 has been processed to the level of semantics (as indicated by the intact N400), but has not been encoded into working memory for conscious awareness (as indicated by the absence of the P300). Thus, ERP experiments have greatly constrained the locus of the processing limitation, placing it after perceptual processing, and prior to response-selection operations.

Behavioural findings have also provided evidence that a semantic representation for T2 is activated even when T2 cannot be reported. Shapiro, Caldwell, and Sorensen (1997a) found that the AB was attenuated when one's own name was the T2 stimulus. Shapiro, Driver, Ward, and Sorensen (1997b) found that when three targets were presented in RSVP streams, repetition priming was found for a related T3 when T2 was not reported. The evidence that T2 is processed semantically, yet is not encoded into working memory for later report, is consistent with the relatively late (after visual and auditory information converge) central interference proposed here. ERP experiments, analogous to those performed by Vogel *et al.* (1998), are currently under way in my laboratory to examine whether the ERP signature is the same when T1 and/or T2 are presented outside the visual modality, as when T1 and T2 are both presented visually.

Positing central processing limitations is also consistent with behavioural findings from hybrid AB–PRP tasks (e.g. Arnell and Duncan, (in press); De Jong and Sweet 1994; Jolicœur 1998, 1999a,b, Jolicœur and Dell Acqua 1998, 1999). Providing a speeded on-line response to an unmasked T1 has been shown to produce robust accuracy deficits to a cross-modal masked T2 at short SOAs (Arnell and Duncan, in press; De Jong and Sweet 1994; Jolicœur, 1999a; Jolicœur and Dell Acqua, 1999). Furthermore, the duration of T1 response-selection operations has been shown to modulate the accuracy deficit. Larger and longer dual-task accuracy costs have been observed at short SOAs following relatively longer T1 RTs (Jolicœur 1999b). Also, unspeeded encoding of the identity of a masked T1 has been shown to produce robust response-slowing to a cross-modal T2 at short SOAs (Arnell and Duncan, in press; Jolicœur and Dell' Acqua 1998, 1999). Results from these experiments provide strong evidence that response-selection operations, and operations to consolidate a stimulus in short-term memory, require the same central processing resources, at least in part (e.g., Arnell and Duncan, in press; Jolicœur 1998, 1999a,b; Jolicœur & Dell Acqua 1998, 1999; but see also Pashler 1989).

If consolidation in working memory and response-selection operations require the same central resources, then one might expect that the P300 would be reduced in magnitude during the AB and hybrid AB–PRP paradigms that produce accuracy deficits to masked T2s, and intact, but delayed, during the PRP and hybrid AB–PRP paradigms that produce response slowing to unmasked T2s. In this manner T2 accuracy costs would reflect the inability to encode T2 into working memory at short SOAs when T2 was masked, and T2 response slowing would reflect the delay in encoding T2 into working memory at short SOAs when T2 was unmasked. The P300 is reduced in magnitude during the AB interval (Luck *et al.* 1996). However, Luck has recently provided evidence that the P300 is not delayed in PRP tasks, despite substantial RT slowing (Luck 1998). Instead, the P300 appears to be attenuated (but not eliminated) at short SOAs. Also, Osman and Moore (1993) have shown that the lateralized readiness potential (LRP) component, which is a measure of response-selection processes, is delayed at short SOAs in the PRP paradigm.

Thus far, it appears that the AB reflects deficits in stimulus consolidation, and that the PRP reflects a delay in response selection. However, it seems that the same central processing resources are required for both consolidation and response-selection operations. If response-selection or stimulus-consolidation operations are under way for T1, then response-selection or stimulus-consolidation operations for T2 must wait, or suffer processing deficits. In the AB paradigm, while STM consolidation is under way for T1, T2 consolidation is impaired (as evidenced by the absent P300). Because unspeeded, off-line responses are required in the AB paradigm, there are no response-time measures or delays. In the PRP paradigm, while response-selection operations are under way for T1, T2 response-selection operations are delayed (as evidenced by the delayed LRP), and T2 STM-consolidation operations are attenuated but not delayed (as evidenced by the reduced, but not delayed, P300).

This relatively late locus of dual-task interference is inconsistent with most AB theories that postulate that the interference is post-perceptual, yet occurs in modality-specific stores before visual and auditory information converge (e.g. Duncan *et al.* 1994; Shapiro and Raymond 1994; Shapiro *et al.* 1994; Ward *et al.* 1996). However, a relatively late locus of interference is consistent with the dual-task interference (DTI) theory of Jolicœur (Arnell and Jolicœur 1999; Jolicœur 1998, 1999a; Jolicœur and Dell Acqua 1998, 1999). Jolicœur proposes that the early stages of sensory encoding (SE) and perceptual encoding (PE) are relatively free of capacity limitations, as have others (Duncan 1980; Potter 1976, 1993; Chun and Potter 1995; Pashler 1994). However, Jolicœur proposes that encoding information in STM requires short-term consolidation (STC), which relies on central processing resources. Therefore, while STC is taking place for one target, processing of a second target requiring STC or response-selection operations will be postponed. If T2 is masked (as it is in the AB paradigm), then the perceptual representation of T2 will not be able to wait long before it is overwritten by the T2 mask. If central processing resources are not available before T2 is overwritten, then T2 will be lost and will not be consolidated in STM. However, if (as in the PRP paradigm) T2 is unmasked, then the perceptual representation of T2 will receive bottom-up support from the continued presence of T2, or from persistence in iconic memory (Coltheart 1980). The postponement will result in response-selection delays, but the persistence will allow T2 to remain available for central processing operations.

According to this logic either masking or speeded responding to each target is sufficient to produce dual-task costs, but neither is necessary, given that each will force on-line processing (Arnell and Duncan 1997). However, if both masking and speeded responding are absent for either target, in such a way that one or both targets may be buffered in a perceptual store, and attended to after processing resources are freed, then no dual-task costs should be observed (Arnell and Duncan 1997; Giesbrecht and Di Lollo 1998).

This suggests that the role of the central interference may be to limit the contents of conscious awareness for organized control of behaviour. In one form or another Duncan *et al.* (1994), Jolicœur (1999a), Arnell and Jolicœur (1999), Jolicœur and Dell'Acqua (1998, 1999), and Chun and Potter (1995) have all proposed that the AB resulted from the sustained attentional processing that was required to consolidate a stimulus in awareness so that it might influence behaviour. Response-selection operations also clearly require awareness in selecting the correct response. Stimulus analysis of all target or potential target attributes (such as location, semantics, phonology, etc.) could occur in parallel in distributed systems. In contrast, creating object files (Kahneman *et al.* 1992) or tokens (Kanwisher 1987), where attributes are integrated into a conscious episodic percept, would require central resources that protect the contents of conscious awareness from overload. Indeed, Raymond and Sorensen (1994) reported that no AB was found when T1 was seen as the same object as previous distractors and did not require a new object file. However, an AB was found when T1 was seen as a new object, and a new object file was opened. Furthermore, behavioural (Shapiro *et al.* 1997*a,b*) and ERP (Luck *et al.* 1996; Vogel *et al.* 1998) results indicated semantic knowledge of T2s that could not be reported, showing that T2 stimulus attributes were available even without conscious knowledge of the target.

Cross-modality interactions in dual-task paradigms

For the AB literature, and indeed the temporal and spatial dual-task literature more generally, a pattern has emerged where neither central interference alone, nor specific interference alone (e.g. within-modality interference) can explain the range of dual-task findings. Instead, the pattern strongly suggests both central and specific levels of interference. The existence of specific levels of interference would explain why dual-task interference is often larger within tasks or modalities than across tasks or modalities (cf. e.g., Arnell and Duncan, in press; Arnell and Larson, submitted; Duncan *et al.* 1997; Pashler 1989; Treisman and Davies 1973). The existence of central interference would explain why dual-task interference is often observed even when two targets are presented in different modalities (e.g., Arnell and Jolicœur 1999; Arnell and Larson, submitted; Shulman and Hsieh 1995; Treisman and Davies 1973) and when tasks and stimuli share no known sources of specific interference (Arnell and Duncan, in press; Bourke *et al.* 1996; De Jong and Sweet 1994; Jolicœur 1999b; Jolicœur and Dell'Acqua 1998, 1999). The existence of both specific and central sources of interference would also explain why dual-task costs are sometimes found within task and/or modalities, but dual-task costs are sometimes not found across tasks and/or modalities (cf. e.g. Duncan *et al.* 1997; Pashler 1989; Potter *et al.* 1998). Dual-task costs within task/modality may result from a combination of central interference and specific task or modality interference, whereas dual-task costs across tasks and modalities may result only from central interference. If one or more of the dual tasks does not engage central resources for a sufficient time or to a sufficient degree, then no dual-task costs are observed across tasks and modalities. If both dual tasks engage central resources sufficiently, then dual-task costs are observed across tasks and modalities. Indeed, Jolicœur (1999b) and Arnell and Duncan (in press) have demonstrated that the amount of dual-task costs observed across tasks and modalities varies with the response-selection demands of the first task. While the exact nature of the central source of interference is not yet clear, behavioural and ERP evidence with the AB and hybrid AB–PRP paradigms has provided evidence that response-selection operations and stimulus consolidation in working memory are impaired and/or bottlenecked by this central interference (Arnell and Duncan, in press; Jolicœur 1999; Jolicœur and Dell'Acqua, 1998, 1999; Luck *et al.* 1996; Osman and Moore 1993; Vogel *et al.* 1998).

Acknowledging that interference can be both general and specific in nature means that we must be careful in theorizing sources of interference for newly observed dual-task costs. Just as it was not safe to assume that the AB resulted from visual interference, simply because visual stimuli were presented, it is unwise to assume that other dual-task costs result from specific sources of interference alone, unless central sources of interference can be ruled out. Similarly, when using tasks or stimuli that share specific sources of interference, we should not assume that new dual-task costs reflect only central sources of interference, unless specific sources of interference can be ruled out. Experimental designs such as those of Treisman and Davies

(1973), Arnell and Larson (submitted), and Arnell and Duncan (in press) are ideally suited for isolating and estimating specific and central sources of interference, and I believe that such designs should be employed regularly in future dual-task research. In this manner, we may begin to understand better the nature of central and specific sources of dual-task interference in temporal attention paradigms.

References

Allport, D. A. (1980). Attention and performance. In G. Claxton (ed.), *Cognitive psychology: new directions*, pp. 112–53. London: Routledge and Kegan Paul.

Allport, D. A. (1989). Visual attention. In M. I. Posner (ed.), *Foundations of cognitive science*, pp. 631–82. Cambridge, MA: MIT Press.

Allport, D. A., Antonis, B., and Reynolds, P. (1972). On the division of attention: a disproof of the single channel hypothesis. *Quarterly Journal of Experimental Psychology*, **24**, 225–35.

Allport, D. A., Styles, E. A., and Hsieh, S. (1994). Shifting intentional set: exploring the dynamic control of tasks. In C. Umilta and M. Moscovitch (eds), *Attention and Performance XV*, pp. 421–52. Hillsdale, NJ: Erlbaum.

Arnell, K. M., and Duncan, J. (1997, November). Speeded responses or making can produce an attentional blink. Poster presented at the 38th Annual Meeting of the Psychonomic Society, Philadelphia, PA.

Arnell, K. M., and Duncan, J. (in press). Specific and general sources of dual-task cost in stimulus identification and response selection. *Cognitive Psychology*.

Arnell, K. M., and Jolicœur, P. (1995). Allocating attention across time and stimulus modality: evidence from the attentional blink phenomenon. Poster presented at the 36th Annual Meeting of the Psychonomic Society, Los Angeles, CA, November.

Arnell, K. M., and Jolicœur, P. (1999). An attentional blink across stimulus modalities: evidence for central processing limitations. *Journal of Experimental Psychology: Human Perception and Performance*, **25**(3), 1–19.

Arnell, K. M. and Larson, J. M. (submitted). A cross-model attentional blink without task switching. Manuscript submitted for publication.

Arnell, K. M., Trangsrud, H. B., Hammes, J., and Larson, J. M. (2000, Nov.), Perceptual vs. attentional sources of dual-task costs. Poster presented at the 41st annual meeting of the Psychonomic Society, New Orleans, LA.

Baddeley, A. D., and Lieberman, K. (1980). Spatial working memory. In R. S. Nickerson (ed.), *Attention and Performance VIII*, pp. 521–39. Hillsdale, NJ: Erlbaum.

Banks, W. P., Roberts, D., and Ciranni, M. (1995). Negative priming in auditory attention. *Journal of Experimental Psychology: Human Perception and Performance*, **21**, 1354–61.

Barbieri, C., and De Renzi, E. (1989). Patterns of neglect dissociation. *Behavioural Neurology*, **2**, 13–24.

Bourke, P. A., Duncan, J., and Nimmo-Smith, I. (1996). A general factor involved in dual-task performance decrement. *Quarterly Journal of Experimental Psychology*, **49A**, 525–45.

Broadbent, D. E., and Broadbent, M. H. P. (1987). From detection to identification: response to multiple targets in rapid serial visual presentation. *Perception and Psychophysics*, **42**, 105–13.

Chase, W. G., and Simon, H. A. (1973). Perception in chess. *Cognitive Psychology*, **4**, 55–81.

Chun, M. M., and Potter, M. C. (1995). A two-stage model for multiple target detection in rapid serial visual presentation. *Journal of Experimental Psychology: Human Perception and Performance*, **21**, 109–27.

Coltheart, M. (1980). Iconic memory and visible persistence. *Perception and Psychophysics*, **27**, 183–228.

Davis, R. (1957). The human operator as a single-channel information system. *Quarterly Journal of Experimental Psychology*, **9**, 119–29.

De Jong, R., and Sweet, J. B. (1994). Preparatory strategies in overlapping-task performance. *Perception and Psychophysics*, **55**, 142–51.

De Renzi, E., Gentilini, M., and Barbieri, C. (1989). Auditory neglect. *Journal of Neurology, Neurosurgery, and Psychiatry*, **52**, 613–17.

DeSchepper, B., and Treisman, A. (1996). Visual memory for novel shapes: implicit coding without attention. *Journal of Experimental Psychology: Learning, Memory, and Cognition*, **22**, 27–47.

di-Pellegrino, G., Ladavas, E., and Farne, A. (1997). Seeing where your hands are. *Nature*, **388**, 730.

Driver, J. (1996). Enhancement of selective listening by illusory mislocation of speech sounds due to lip-reading. *Nature*, **381**, 66–8.

Driver, J., and Baylis, G. C. (1993). Cross-modal negative priming and interference in selective attention. *Bulletin of the Psychonomic Society*, **31**, 45–8.

Driver, J., and Spence, C. (1998). Cross-modal links in spatial attention. *Philosophical Transactions of the Royal Society B*, **353**, 1319–31.

Duncan, J. (1980). The demonstration of capacity limitation. *Cognitive Psychology*, **12**, 75–96.

Duncan, J., Ward, R., and Shapiro, K. L. (1994). Direct measurement of attentional dwell time in human vision. *Nature*, **369**, 313–15.

Duncan, J., Martens, S., and Ward, R. (1997). Restricted attentional capacity within but not between sensory modalities. *Nature*, **387**, 808–10.

Eddins, D. A., and Green, D. M. (1995). Temporal integration and temporal resolution. In B. C. J. Moore (ed.), *Hearing*, pp. 207–42. San Diego, CA: Academic Press.

Giesbrecht, B. L., and Di Lollo, V. (1998). Beyond the attentional blink: visual masking by item substitution. *Journal of Experimental Psychology: Human Perception and Performance*, **24**, 1454–66.

Goddard, K. M., Isaak, M. I., and Slawinski, E. B. (1997). Attention blinks in pure tone rapid auditory presentation. Poster presented at the 133rd meeting of the Acoustical Society of America, State College, PA, USA, June.

Hugdahl, K., Wester, K., and Asbjørnsen, A. (1991). Auditory neglect after right frontal lobe and right pulvinar thalamic lesions. *Brain and Language*, **41**, 465–73.

Intraub, H. (1985). Visual dissociation: an illusory conjunction of pictures and forms. *Journal of Experimental Psychology: Human Perception and Performance*, **11**, 431–42.

Isaak, M. I., Shapiro, K. L., and Martin, J. (1999). The attentional blink reflects retrieval competition among multiple RSVP items: tests of the interference model. *Journal of Experimental Psychology: Human Perception and Performance*, **25**, 1774–92.

Jolicœur, P. (1998). Modulation of the attentional blink by on-line response selection: evidence from speeded and unspeeded Task 1 decisions. *Memory and Cognition*, **26**, 1014–32.

Jolicœur, P. (1999b). Restricted attentional capacity between sensory modalities. *Psychonomic Bulletin and Review*, **6**, 87–92.

Jolicœur, P. (1999a). Dual-task interference and visual encoding. *Journal of Experimental Psychology: Human Perception and Performance*, **25**, 596–616.

Jolicœur, P., and Dell' Acqua, R. (1998). The demonstration of short-term consolidation. *Cognitive Psychology*, **36**, 138–202.

Jolicœur, P., and Dell' Acqua, R. (1999). Attentional and structural constraints on memory encoding. *Psychological Research,* **62**, 154–164.

Kahneman, D., Treisman, A., and Gibbs, B. J. (1992). The reviewing of object files: object specific integration of information. *Cognitive Psychology*, **24**, 175–219.

Kanwisher, N. G. (1987). Repetition blindness: type recognition without token individuation. *Cognition*, **27**, 117–43.

Kanwisher, N. G., and Potter, M. C. (1989). Repetition blindness: the effects of stimulus modality and spatial displacement. *Memory and Cognition*, **17**, 117–24.

Kanwisher, N. and Potter, M. (1990). Repetition blindness: levels of processing. *Journal of Experimental Psychology: Human Perception and Performance*, **16**, 30–47.

Karlin, L., and Kestenbaum, R. (1968). Effect of number of response alternatives on the psychological refractory period. *Quarterly Journal of Experimental Psychology*, **20**, 167–78.

Lawrence, D. H. (1971). Two studies of visual search for word targets with controlled rates of presentation. *Perception and Psychophysics*, **10**, 85–9.

Lowe, D. (1998). Long-term positive and negative identity priming: evidence for episodic retrieval. *Memory and Cognition*, **26**, 435–43.

Luck, S. J. (1998). Sources of dual-task interference: evidence from human electrophysiology. *Psychological Science*, **9**, 223–7.

Luck, S. J., Vogel, E. K., and Shapiro, K. L. (1996). Word meanings can be accessed but not reported during the attentional blink. *Nature*, **383**, 616–18.

McCann, R. S., and Johnston, J. C. (1992). Locus of the single-channel bottleneck in dual-task interference. *Journal of Experimental Psychology: Human Perception and Performance*, **18**, 471–84.

McLaughlin, W Klein, and Shore, (2001). The attentional blink is immune to masking induced data limits. *Quarterly Journal of Experimental Psychology*, **54A**, 169–96.

McLean, J. P., Broadbent, D. E., and Broadbent, M. H. P. (1983). Combining attributes in rapid serial visual presentation tasks. *Quarterly Journal of Experimental Psychology*, **35A**, 171–86.

McLeod, P. (1977). A dual task response modality effect: support for multiprocessor models of attention. *Quarterly Journal of Experimental Psychology*, **29**, 651–67.

McLeod, P. (1978). Does probe RT measure central processing demand? *Quarterly Journal of Experimental Psychology*, **30**, 83–9.

Mattingley, J. B., Driver, J., Beschins, N., and Robertson, I. H. (1997). Attentional competition between modalities: extinction between touch and vision after right hemisphere damage. *Neuropsychologia*, **35**, 867–80.

Meyer, D. E., and Kieras, D. E. (1997). A computational theory of executive cognitive processes and multiple-task performance: 1. Basic mechanisms. *Psychological Review*, **104**, 3–65.

Miller, M. D., and MacKay, D. G. (1994). Repetition deafness: repeated words in computer-compressed speech are difficult to encode and recall. *Psychological Science*, **5**, 47–51.

Mondor, T. A. (1998). A transient processing deficit following selection of an auditory target. *Psychonomic Bulletin and Review*, **5**, 305–11.

Mondor, T. A., Breau, L. M., and Milliken, B. (1998*a*). Inhibitory processes in auditory selective attention: evidence of location-based and frequency-based inhibition of return. *Perception and Psychophysics*, **60**, 296–302.

Mondor, T. A., Zatorre, R. J., and Terrio, N. A. (1998*b*). Constraints on the selection of auditory information. *Journal of Experimental Psychology: Human Perception and Performance*, **24**, 66–79.

Navon, D. (1984). Resources—a theoretical soup store? *Psychological Review*, **91**, 216–34.

Neill, W. T. (1977). Inhibitory and facilitatory processes in attention. *Journal of Experimental Psychology: Human Perception and Performance*, **3**, 444–50.

Osman, A., and Moore, C. M. (1993). The locus of dual-task interference: psychological refractory effects on movement-related brain potentials. *Journal of Experimental Psychology: Human Perception and Performance*, **19**, 1292–1312.

Pashler, H. (1984). Processing stages in overlapping tasks: evidence for a central bottleneck. *Journal of Experimental Psychology: Human Perception and Performance*, **10**, 358–77.

Pashler, H. (1989). Dissociations and dependencies between speed and accuracy: evidence for a two-component theory of divided attention in simple tasks. *Cognitive Psychology*, **21**, 469–514.

Pashler, H. (1990). Do response modality effects support multiprocessor models of divided attention? *Journal of Experimental Psychology: Human Perception and Performance*, **16**, 826–42.

Pashler, H. (1994). Dual-task interference in simple tasks: data and theory. *Psychological Bulletin*, **2**, 220–44.

Perrott, D. R., Cisneros, J., McKinley, R. L., and D'Angelo, W. R. (1996). Aurally aided visual search under virtual and free-field listening conditions. *Human Factors*, **38**, 702–15.

Posner, M. I. (1978). *Chronometric explorations of mind*. Hillsdale, NJ: Erlbaum.

Posner, M. I., Sandson, J., Dhawan, M., and Shulman, G. L. (1989). Is word recognition automatic? A cognitive–anatomical approach. *Journal of Cognitive Neuroscience*, **1**, 50–60.

Potter, M. C. (1976). Short-term conceptual memory for pictures. *Journal of Experimental Psychology: Human Learning and Memory*, **2**, 509–22.

Potter, M. C. (1993). Very short-term conceptual memory. *Memory and Cognition*, **21**, 156–61.

Potter, M. C., Chun, M. M., and Muckenhoupt, M. (1995). Auditory attention does not blink. Paper presented at the 36th Annual Meeting of the Psychonomic Society, Los Angeles, California, USA, November.

Potter, M. C., Chun, M. M., Banks, B. S., and Muckenhoupt, M. (1998). Two attentional deficits in serial target search: the visual attentional blink and an amodal task-switch operation. *Journal of Experimental Psychology: Learning, Memory, and Cognition*, **24**, 979–92. 979–92.

Rafal, R., Henik, A., and Smith, J. (1991). Extrageniculate contributions to reflex visual orienting in normal humans: a temporal hemifield advantage. *Journal of Cognitive Neuroscience*, **3**, 322–8.

Raymond, J. E., and Sorensen, R. E. (1994). Motion, object selection, and the attentional blink. Poster presented at the 35th Annual Meeting of the Psychonomic Society, St Louis, MO, November.

Raymond, J. E., Shapiro, K. L., and Arnell, K. M. (1992). Temporary suppression of visual processing in an RSVP task: an attentional blink? *Journal of Experimental Psychology: Human Perception and Performance*, **18**, 849–60.

Raymond, J. E., Shapiro, K. L., and Arnell, K. M. (1995). Similarity determines the attentional blink. *Journal of Experimental Psychology: Human Perception and Performance*, **21**, 653–62.

Reisberg, D. (1983). General mental resources and perceptual judgements. *Journal of Experimental Psychology: Human Perception and Performance*, **9**, 966–79.

Rogers, R. D., and Monsell, S. (1995). Costs of a predictable switch between simple cognitive tasks. *Journal of Experimental Psychology: General*, **124**, 207–31.

Rollins Jr., H. A., and Hendricks, R. (1980). Processing of words presented simultaneously to eye and ear. *Journal of Experimental Psychology: Human Perception and Performance*, **6**, 99–109.

Rorden, C., and Driver, J. (1999). Does auditory attention shift in the direction of an upcoming saccade? *Neuropsychologia*, **37**, 357–77.

Ross, N. E., and Jolicœur, P. (1999). Attentional blink for color. *Journal of Experimental Psychology: Human Perception and Performance*, **25**, 1483–1494.

Ruthruff, E., and Pashler, H. The role of central operations in the attentional blink. Manuscript submitted for publication.

Ruthruff, E., Miller, J., and Lachmann, T. (1995). Does mental rotation require central mechanisms? *Journal of Experimental Psychology: Human Perception and Performance*, **21**, 552–70.

Salame, P., and Baddeley, A. D. (1982). Disruption of short-term memory by unattended speech: implications for the structure of working memory. *Journal of Verbal Learning and Verbal Behavior*, **21**, 150–64.

Seiffert, A., and Di Lollo, V. (1997). Low-level masking in the attentional blink. *Journal of Experimental Psychology: Human Perception and Performance*, **23**, 1061–73.

Shapiro, K. L., and Raymond, J. E. (1994). Temporal allocation of visual attention: inhibition or interference? In D. Dagenbach and T. H. Carr (eds.), *Inhibitory mechanisms in attention, memory, and language*, pp. 151–87. San Diego: Academic Press.

Shapiro, K. L., and Terry, K. (1996). The attentional blink: the eyes have it (but so does the brain). In R. Wright (ed.), *Visual attention*. Oxford: Oxford University Press.

Shapiro, K. L., Arnell, K. M., and Drake, S. H. (1991). Stimulus complexity mediates target detection in visual attention search. Poster presented at 1991 ARVO meeting: Sarasota, FL. [Abstract] *Investigative Ophthalmology and Visual Science*, **32**(4), 1040.

Shapiro, K. L., Raymond, J. E., and Taylor, T. (1993). The attentional blink suppresses size and shape but not colour information. Presented at 1993 ARVO meeting: Sarasota, FL. [Abstract] *Investigative Ophthalmology and Visual Science*, **34**(4), 1232.

Shapiro, K. L., Raymond, J. E., and Arnell, K. M. (1994). Attention to visual pattern information produces the attentional blink in Rapid Serial Visual Presentation. *Journal of Experimental Psychology: Human Perception and Performance*, **20**, 357–71.

Shapiro, K. L., Caldwell, J., and Sorensen, R. E. (1997a). Personal names and the attentional blink: a visual 'cocktail party' effect. *Journal of Experimental Psychology: Human Perception and Performance*, **23**, 504–14.

Shapiro, K. L., Driver, J., Ward, R., and Sorensen, R. E. (1997b). Priming from the attentional blink: a failure to extract visual tokens but not visual types. *Psychological Science*, **8**, 95–100.

Shapiro, K. L., Hillstrom, A., and Spence, C. (1998). Is there an attentional blink in taction? Paper presented at the 39th Annual Meeting of the Psychonomic Society, Philadelphia, PA, November.

Shiffrin, R. M., and Schneider, W. (1977). Controlled and automatic information processing: II. Perception, learning, automatic attending and a general theory. *Psychological Review*, **84**, 127–90.

Shulman, H., and Hsieh, V. (1995). The attention blink in mixed modality streams. Paper presented at the 36th Annual Meeting of the Psychonomic Society, Los Angeles, November.

Smith, M. C. (1967). Theories of the psychological refractory period. *Psychological Bulletin*, **67**, 202–13.

Soroker, N., Calamaro, N., and Myslobodsky, M. S. (1995*a*). Ventriloquist effect reinstates responsiveness to auditory stimuli in the 'ignored' space in patients with hemispatial neglect. *Journal of Clinical and Experimental Neuropsychology*, **17**, 243–55.

Soroker, N., Calamaro, N., and Myslobodsky, M. S. (1995*b*). McGurk illusion to bilateral administration of sensory stimuli in patients with hemispatial neglect. *Neuropsychologia*, **33**, 461–70.

Soroker, N., Calamaro, N., Glicksohn, J., and Myslobodsky, M. S. (1997). Auditory inattention in right-hemisphere-damaged patients with and without visual neglect. *Neuropsychologia*, **35**, 249–56.

Spence, C., and Driver, J. (1994). Covert spatial orienting in audition: exogenous and endogenous mechanisms facilitate sound localization. *Journal of Experimental Psychology: Human Perception and Performance*, **20**, 555–74.

Spence, C., and Driver, J. (1996). Audiovisual links in endogenous covert spatial attention. *Journal of Experimental Psychology: Human Perception and Performance*, **22**, 1005–30.

Spence, C., and Driver, J. (1997). Audiovisual links in exogenous covert spatial orienting. *Perception and Psychophysics*, **59**, 1–22.

Spence, C., Nicholls, M. E. R., Gillespie, N., and Driver, J. (1998*a*). Cross-modal links in exogenous covert spatial orienting between touch, audition and vision. *Perception and Psychophysics*, **60**, 544–57.

Spence, C., Lloyd, D. M., and McGlone, F. P. (1998*b*). Crossmodal inhibition of return (IOR) between touch, vision, and audition. Presented at the 29th Annual Meeting of the Psychonomic Society, Dallas, TX, November.

Spence, C., Ranson, J., and Driver, J. (2000). Crossmodal selective attention: on the difficulty of ignoring sounds at the locus of visual attention. *Perception and Psychophysics*, **62**, 410–24.

Sterzi, R., Piacentini, S., Polimeni, M., Liverani, F., and Bisiach, E. (1996). Perceptual and premotor components of unilateral auditory neglect. *Journal of the International Neuropsychological Society*, **2**, 419–25.

Tipper, S. P. (1985). The negative priming effect: inhibitory effects of ignored primes. *Quarterly Journal of Experimental Psychology*, **37**A, 571–90.

Tipper, S. P., and Driver, J. (1988). Negative priming between pictures and words: evidence for semantic analysis of ignored stimuli. *Memory and Cognition*, **16**, 64–70.

Tipper, S. P., Brehaut, J. C., and Driver, J. (1990). Selection of moving and static objects for the control of spatially based attention. *Journal of Experimental Psychology: Human Perception and Performance*, **16**, 492–504.

Treisman, A. M., and Davies, A. (1973). Divided attention to ear and eye. In S. Kornblum (ed.), *Attention and performance IV*, pp. 101–17. London: Academic Press.

Van Selst, M., and Jolicœur, P. (1994). Can mental rotation occur before the dual-task bottleneck? *Journal of Experimental Psychology: Human Perception and Performance*, **20**, 905–21.

Vogel, E. K., Luck, S. J., and Shapiro, K. L. (1998). Electrophysiological evidence for a postperceptual locus of suppression during the attentional blink. *Journal of Experimental Psychology: Human Perception and Performance*, **24**, 1656–74.

Ward, L. M. (1994). Supramodal and modality-specific mechanisms for stimulus-driven shifts of auditory and visual attention. *Canadian Journal of Experimental Psychology*, **48**, 242–59.

Ward, R., Duncan, J., and Shapiro, K. L. (1996). The slow time-course of visual attention. *Cognitive Psychology*, **30**, 79–109.

Weichselgartner, E., and Sperling, G. (1987). Dynamics of automatic and controlled visual attention. *Science*, **238**, 778–80.

Welford, A. T. (1952). The 'psychological refractory period' and the timing of high-speed performance: a review and theory. *British Journal of Psychology*, **43**, 2–19.

Wright, R. D. and Ward, L. M. (1994). Shifts in visual attention: an historical and methodological overview. *Canadian Journal of Experimental Psychology*, **48**, 151–66.

9

Getting beyond the serial/parallel debate in visual search: a hybrid approach

Cathleen M. Moore and Jeremy M. Wolfe

Abstract

The question of whether visual search involves at least one item-by-item serial processing stage or whether instead it is an entirely parallel process has been debated for decades. Recently, estimates of 'attentional dwell-time', which is the time required to reallocate attention from one item to another, have been brought to bear on this question. Here. we review this and other classes of evidence that favor serial or parallel models of visual search, and conclude that hybrid models that are neither strictly serial nor strictly parallel are better candidates for describing human visual search. We end the chapter with a sketch of one such model, and some of its possibilities.

If you look for a 2 among the 5s in Fig. 9.1, it will take you longer to find the 2 in 1B than in 1A and even longer to confirm that there is no 2 present in 1C. How should we understand the dependence of response time (RT) in this sort of task on the number of items (set size)? The answer to this question has been phrased in essentially dichotomous terms for a generation. It could be that each item is processed one after the other in series (cf. e.g. Treisman and Gelade 1980; Wolfe *et al.* 1989; Wolfe 1994). It could be that information is accumulated from multiple items at the same time in parallel (Grossberg *et al.* 1994; Humphreys and Muller 1993; Palmer and McLean 1995). Which is it? Is attention deployed to one item at a time in visual search or is it distributed across many or all items?

These sound like dramatically different alternatives. Why then has it proved so difficult for proponents of serial models or proponents of parallel models to gain the upper hand in this debate? One explanation has been that, for all their apparent differences, serial and parallel models can be made to predict similar experimental outcomes (e.g. Townsend 1971, 1976, 1990). We will argue for a somewhat different position. There is a class of plausible models

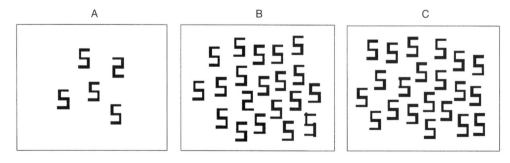

Fig. 9.1 An example of a fairly difficult visual search task. Finding the 2 among the 5s takes longer in B than in A, and confirming that there is no 2 in C takes longer still. Can response times in tasks like this reveal anything about the underlying processes that occur during visual search?

within which the distinction between serial and parallel deployment of attention is like the distinction between wave and particle theories of light. Measured one way, light behaves like a wave. Measured another way, it behaves like a particle. So far as we can tell, the reality is that light is both wave and particle. In a similar manner, we will argue that processing in visual search is, in a non-trivial sense, both serial and parallel. The analogy to light is intended as a loose one, in that light, so far as we can tell, is never particle and wave at the same time, and the models of search to which we refer are simultaneously serial and parallel.

This is not simply a claim that vision contains both serial and parallel components. It is relatively trivial to note that the earliest steps in visual processing are performed in parallel (for example, the registration of the image on the retina) and that many of the possible output steps (for example, eye movements, response execution) are necessarily serial. We are focused here on the role of selective attention, which is that process by which items are selected for the purpose of classifying them as target or non-target during visual search, which we will argue can be both serial and parallel in nature at the same time.

Before describing this alternative, however, we discuss some of the research that has given rise to this apparent dichotomy-in-need-of-resolution. This is not intended as an exhaustive survey of the literature, but rather as an illustration of some of the main points of unresolved contact between the two camps. After describing some of the theoretical reasons for proposing a serial process in visual search, we briefly review the empirical basis for inferring serial mechanisms from RT data in visual search experiments. Next, we discuss some of the empirical evidence that argues against serial processing in search, favouring instead entirely parallel models. Finally, we describe a way of looking at the data that seeks to explain how visual search can appear to be serial and parallel at the same time.

Why propose serial selection in visual search?

All else being equal, parallel processing would seem to be preferable to serial processing. Why do one thing at a time if you can do many? Recent ideological energy behind serial models of visual search comes from the conviction that, beyond a certain point, it becomes computationally impossible to perform vision in parallel (cf. e.g. Tsotsos 1990). Earlier, and more general, impetus for a serial stage to processing comes from the work of Broadbent (1958) and others, who began conceptualizing human information-processing in terms of formal information theory. Under that scheme, the human system was thought of as a limited-capacity information channel, which required that processing be reduced from parallel to serial (or nearly serial) at some point within the system.

Within the domain of visual processing, there are at least two specific aspects of search that have been hypothesized as requiring serial processing. First, Treisman and others have argued that to 'bind' the features of an object correctly together, the selection of individual objects is necessary (e.g. Treisman and Gelade 1980). Second, object recognition during search has a necessarily serial flavour to it (cf. e.g. Duncan 1980; Wolfe *et al.* 1989).

Figure 9.2 provides a standard illustration of the apparent need for serial selection when searching for a conjunction target. While it is easy enough to detect the black square in panels A and B, it is more difficult to do so in panel C. This, according to Treisman, is because the black square is defined with regard to its distractors on the basis of a single feature in each of the two upper displays, but only by the conjunction of features in the lower display. Therefore,

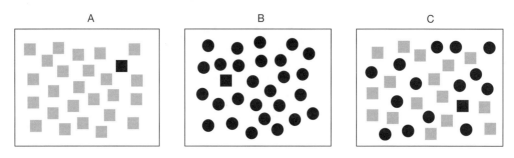

Fig. 9.2 An example of a standard conjunction search. In all panels the target is a black square. It is fairly easy to find the black square in panels A and B, where the target differs from the distractors on the basis of a single feature (colour in panel A, and shape in panel B). It is more difficult to find the black square in panel C, however, where the target differs from the distractors only on the basis of a conjunction of features (black *and* square). Treisman and Gelade (1980) argued that while feature information can be extracted simultaneously for all items in the display (i.e., in parallel), conjoining features requires t h e allocation of attention to the individual objects (i.e., in series).

according to the theory, attention must be separately allocated to each individual feature to detect the black square as a black square.

Figure 9.3 illustrates a variation on the basic conjunction theme that suggests that the proper conjoining of features is preceded by a loose, unstructured, grouping of features on the basis of rough object representations. If you look for black horizontal lines, you will find it easier to find one in the right display than in the left display (Wolfe and Bennett 1997). This finding is consistent with the idea that preattentive processes (i.e., processes prior to the conjunction of features) parse the world up into candidate objects, which are essentially loose collections (think 'lists') of basic features. On the right, each object has only one orientation and only one colour (ignoring the central grey patch that is present in all items). So the preattentive object representations might consist of something like $Object_1 = \{black, vertical\}$ or $Object_1 = \{white, horizontal\}$. In contrast on the left, each object is both black and white and both vertical and

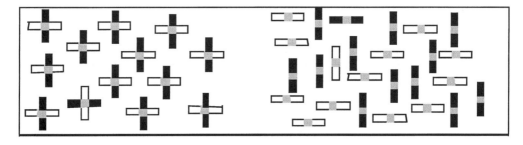

Fig. 9.3 An example of a variation on the standard conjunction search. Finding the black horizontal rectangle in the display on the right is easier than finding it in the display on the left (Wolfe and Bennett 1997). This suggests that while the proper conjoining of features with each other may require attention, preattentive processes parse the world up into loose structured collections of basic features, which may serve ascandidates for serial selection.

horizontal. So all the preattentive object representations would look something like Object$_1$ = {black, white, vertical, horizontal}. As such, all the objects on the left are preattentively identical, whereas all of the objects on the right are not.

The binding argument for seriality accounts for effects like those in Figs. 9.2 and 9.3 in so far as it holds that, until attention selects an item, it is impossible to determine what colour is bound to what orientation or shape. If attention is not allocated to individual items, binding failures or confusions can result in the illusory conjunction of features into items that are not in the display, but whose features are present across a number of different items in the display (Ivry and Prinzmetal 1990; Prinzmetal 1995; Treisman and Schmidt 1982). An example of this would be the illusory perception of a grey circle in Fig. 9.2C.

At a theoretical level, object recognition, like the conjunction of features, seems like a plausible candidate for serial processing. To recognize an object, it is necessary to form some connection between the visual representation of that object and its representation in memory. As will be discussed below, it may be that only one such link can be maintained at any instant in time (Wolfe *et al.* 2001; see also Carrier and Pashler 1996). To the extent that visual search requires object recognition, this would be another theoretical justification for a mandatory serial stage in the processing of items in visual search.

Theory aside, what are the empirical foundations of serial models? In the next section, we consider some of the evidence that has been taken as supportive of the idea that visual search involves a serial component.

Empirical foundations of serial models

One root of the dichotomy between parallel and serial processing during visual search can be found in the early work of Neisser (1967) and Treisman (Treisman and Gelade 1980), some of the theoretical consequences of which were described in the preceding section (see also: Egeth *et al.* 1972; Hoffman 1979; and van der Heijden 1975). Following the ideas of Broadbent (1958), Neisser introduced the distinction between a parallel 'preattentive' stage of processing and other ensuing bottleneck processes that can handle only one stimulus, or a very few stimuli, at a time. Attention, as we are using the term, refers to the process that selects items to pass through those bottlenecks.

Within the context of visual search, Treisman developed these ideas into Feature Integration Theory (FIT). In its early form (Treisman and Gelade 1980), FIT held that a limited set of visual search tasks could be performed in parallel. These were tasks in which the target was defined by a single basic stimulus feature, such as colour, size, or orientation (e.g., Figs. 9.2A and 9.2B). Other search tasks, notably those requiring the binding of simple features into more complex stimuli, required the serial deployment of attention from item to item until the target was found (e.g., Fig. 9.2C). FIT has since been modified to include the possibility that feature information can be used to guide the deployment of attention (Wolfe *et al.* 1989; Wolfe 1994; Treisman and Sato 1990). Thus for example, search for a red number among black and red letters might be serial; but it would be serial through the set of red letters. Attention would be guided toward the red items, and not 'wasted' on black items (Egeth *et al.* 1984; Moore and Egeth 1998; Zohary and Hochstein 1989). While models of this sort admit to an early parallel stage of processing, they also have a commitment to a later, item-by-item, serial deployment of attention. It is through this item-by-item engagement of processing that stimuli are classified as target or non-target during the search process.

How can such models be tested? The original Feature Integration Theory proposed that some searches were parallel and others were serial, so one could seek evidence concerning the existence of two different modes of search. Toward this end, the slope of the function relating RT to the number of items in the display (*set size*) was offered as a metric for categorizing specific search tasks as serial or parallel (cf. e.g. Treisman and Gelade 1980; Sternberg 1975). The logic is that if search is parallel (and unlimited in processing capacity), then adding items to the display should not affect the amount of time it takes to find the target. Therefore, the slopes relating RT to set size should be near zero. In contrast, if search is serial, then each item that is checked should add a more-or-less constant amount of time to the total time required to find the target. So in this case, RT should increase linearly with set size. A serial self-terminating model—one in which search ceases when the target is found—makes the additional, more specific, prediction that target-absent slopes will be twice as large as target-present slopes. This follows because if the target is present, then, on average, it will be found after checking half the items in the display. In contrast, in order to report that the target is absent, all the items in the display must be checked.

Consistent with the basic tenets of FIT, both patterns of results have been found (see Wolfe 1994; 1998 for reviews). That is, some searches have produced near zero RT by set size functions, and others have produced linearly increasing functions with something close to a 2:1 slope ratio between target-absent and target-present trials. These results have been taken as evidence that visual search sometimes includes an item-by-item serial component.

The search-slope logic is appealing at many levels, and has certainly been invoked in many studies. There are, however, serious problems with it as an arbiter of the serial/parallel debate (see for example Palmer 1994; Townsend 1971, 1990; Wolfe 1998). First of all, while a serial self-terminating model predicts that RT will increase as a linear function of set size, and that there will be a 2:1 slope ratio between target-absent and target-present trials, this logic does not work in reverse. That is, finding this pattern of results does not mean that search was necessarily serial and self-terminating. The pattern may have been caused by some other type of process. Townsend and others, for example, have shown that models that include no serial component, but do include a limited-capacity parallel component, can produce the same pattern of results (e.g. Townsend 1971, 1990; Ward and McClelland 1989). Moreover, on the basis of accuracy versions of this task, Palmer and others have shown that even models including only unlimited-capacity parallel processes may be able to produce the signature pattern of results (Eckstein 1998; Palmer and McClean 1995). One might appeal to parsimony, and argue that the assumptions that are necessary for the serial self-terminating model are fewer and simpler than those required for the alternative parallel models. However, even if true, this argument would not eliminate the logical problems. Moreover, despite Occam's edict, nature does not necessarily submit to the same priorities as those to which theoreticians aspire; more parsimonious does not necessarily imply more accurate.

In addition to the logical problems with inferring search mechanism from slope data, the data themselves raise niggling doubts about this enterprise. If search slopes, by themselves, were diagnostic of search mechanism, and searches could be divided into 'parallel' searches and 'serial' searches, then the distribution of slopes across the set of all search tasks might be expected to be bimodal. That is, there might be some principled slope value that could be established such that searches producing slopes above that level could be reliably labelled as 'serial', while those with slopes below this magic value could be labelled as 'parallel'. In fact, with something like this in mind, ten ms/item has become a conventional value for this magic slope. The data, however, reveal no such bimodality. Wolfe (1998) analysed the slopes of hundreds of

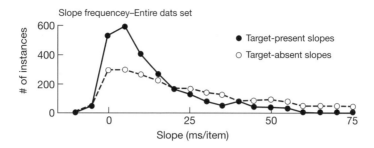

Fig. 9.4 Frequency distributions of subject slopes from several different standard visual search tasks. Each subject contributed a target-absent slope and a target-present slope. There is no evidence of bimodality in these distributions. (Taken from Wolfe 1998.)

visual-search experiments representing about 1,000,000 individual search trials. The data set included several different types of standard search tasks. Yet the distributions of slopes did not even hint at bimodality (see Fig. 9.4).

Note that the failure to find a bimodal distribution does not, by itself, falsify the hypothesis that there are 'parallel' and 'serial' searches. While a bimodal distribution would favour the conclusion that there were two different modes of search, the failure to find bimodal distributions does not support the conclusion that there is only one mode of search, though it is consistent with that conclusion. This follows because the distributions shown in Fig. 9.4 could represent a mixture of two underlying 'serial' and 'parallel' distributions that obscures the underlying distributions. While there are analytic tools for extracting component distributions from mixed distributions (Meyer *et al.* 1986; Yantis *et al.* 1992), unless the underlying distributions are substantially far apart, it is fairly difficult to do so in practice (Yantis *et al.* 1992). In any case, it is clear that the empirically unimodal distributions seen in Fig. 9.4 indicate that the mechanism of search cannot be trivially inferred (for example, by setting a criterion of 10 ms/item) from the slope of the RT × set size function.

In addition to the question of whether there are two qualitatively different modes of search, the data also raise doubts about the more specific conclusion that search is sometimes serial and self-terminating. First, in Wolfe's (1998) overview analysis of the many visual-search experiments, he also looked at the distribution of target-absent to target-present slope ratios. He found that, over a wide range of target-present slopes, the ratio was reliably greater than the 2.0 that the serial self-terminating models predict. Moreover, searches yielding increasing search slopes have produced target-present and target-absent slopes that are nearly identical (cf. e.g. Atkinson *et al.* 1969; Townsend and Roos 1973). Pashler (1987) has shown that while the 2:1 slope ratio between target-present trials can be observed across large ranges of display size, it can be shown to be closer to 1:1 when the range is small (i.e., from about 1 to 8 items). On the basis of these data, Pashler suggested the possibility of something like a serial-by-clumps model, wherein clumps of items are selected and processed in parallel, but the clumps are selected in series (see also Grossberg *et al.* 1994). Finally, Ward and McClelland (1989) noted that the variances that are typically observed across the target-present and target-absent conditions of visual-search experiments are often inconsistent with a simple serial self-terminating model.

Again, however, keeping with the theme that slope data from visual search tasks make for a poor arbiter of the serial versus parallel question, the failure to find 2:1 slope ratios is less conclusive than one might think or hope. RTs on target-absent trials are RTs of unsuccessful searches for a target. If these were all simple exhaustive searches through the display, modelling would be uncomplicated. There is no reason to think, however, that these are all simple exhaustive searches. First, subjects make errors. In some cases, these errors seem to reflect an early abandonment of unpromising searches, which complicates matters (Chun and Wolfe 1996). Much worse, the simple 2:1 story relies on the assumption that rejected distractors are marked so that they are not revisited by attention during the search. This is a vexed topic (Klein 1988; Klein and MacInnes, in press; Klein and Taylor 1994; Wolfe and Pokorny 1990); but the very real possibility exists that any such marking is either imperfect (Arani *et al.* 1984) or nonexistent (Horowitz and Wolfe 1998). With all these complications, serial models do not make unambiguous predictions about slope ratios.

In summary, while there may be theoretical reasons to suppose that there is a serial stage during visual search and while RT search data can be modelled as a parallel stage feeding a serial stage, the RT data do not unambiguously support the existence of both stages. Other related evidence has been mustered in defense of a serial stage of processing (e.g. Kwak *et al.* 1991), but these have not been adequate to provide a resolution to the debate.

Empirical foundations for parallel models

Parallel models of visual search, in contrast to serial models, hold that attention can be distributed across multiple objects and that information about the identity of those objects can be developed simultaneously. There are many variants of parallel models (e.g. Grossberg *et al.* 1994; Bundesen 1990; Kinchla 1974; Ward and McClelland 1989). While the details of these models vary, they all reject the notion of a mandatory serial selection of individual items at any level of processing. A generic parallel model of the difficult search illustrated in Fig. 9.1 would have stimulus identity information accumulating for all objects at the same time. When one item crossed some threshold for identification as a 2, a target-present response would be generated. When all (or almost all) items had crossed a threshold for identification as 5's a target-absent response would be generated. This then, in brief oversimplification, is the crux of the debate. Remaining with the 2 vs. 5 example, a parallel model would hold that it is possible to accumulate information concerning identification for multiple characters at one time, whereas a strict serial model would insist that identification of a 2 or a 5 requires attentional selection of that character and only that character for at least a moment of time. What evidence is there to support a parallel model over a serial model?

The unimodel distribution of search slopes that is shown in Fig. 9.4 raises the possibility of a single-factor model of search efficiency. Perhaps a limited-capacity parallel process mediates all visual searches. Different combinations of targets and distractors impose different demands on this process. Low-demand conditions produce efficient search with shallow slopes. High-demand conditions produce inefficient search with steep slopes, while intermediate conditions produce intermediate results (e.g. Duncan and Humphreys 1989; Humphreys and Müller 1993). A model of this sort would eliminate the need for the item-by-item serial deployment of attention in visual search. Consistent with such a model, several lines of investigation, which we describe in detail below, raise the possibility that stimuli can at least sometimes be identified in parallel. More than being consistent with a parallel limited-capacity model, definitive evidence on this point would constitute at least indirect evidence against serial models. This follows because if a process as late as stimulus identification can occur in parallel, then it seems unlikely that serial search should ever be required, except in cases that require individual eye fixations of each stimulus.

Using what we will refer to as the *simultaneous/sequential method*, Shiffrin and Gardner (1972; see also Eriksen and Spencer 1969) provided evidence suggesting that stimulus identification might occur in parallel. In Experiment III, for example, they used a visual search task in which the target was a T or an F, and the distractors were T/F hybrid characters. The task was to report whether the target was a T or an F. Display size was held constant at four items. What was varied in this study was the timing of information presentation within a trial. In the *simultaneous* condition, all the stimuli were presented at once. In the *sequential* condition, half the stimuli were presented at one time and then, after a 500 ms delay, the other half of the stimuli were presented. The total amount of time that a given piece of information was present was the same across the two conditions. The logic was that if the processing of one item interfered with the simultaneous processing of other items (for example, if processing had to be serial), and if attention could be switched within the time period between frames in the sequential condition (500 ms in this case), then there should be an advantage for the sequential condition over the simultaneous condition. Instead, Shiffrin and Gardner's data were consistent with the idea that multiple stimuli can be processed simultaneously without interfering with each other, or that attention could not be moved in time; they found that target identification was no more accurate in the sequential condition than in the simultaneous condition.

Pashler and Badgio (1987) extended the work of Shiffrin and Gardner (1972) by using a task that was intended to ensure that all of the stimuli in the display had to be identified. In this task—known as the *highest-digit task*—displays of digits were presented, and subjects reported the identity of the highest digit in the display. Unless the highest digit in the display was also the highest digit in the whole set, all the digits in the display had to be identified in order to answer correctly. (Those trials on which the highest digit in the display was the highest in the set were analysed separately.) With regard to timing, Pashler and Badgio's experiments were more like Experiments I and II of Shiffrin and Gardner's study than Experiment III, in that stimulus presentations were separated in the sequential condition by only 50 ms (the stimulus duration). The results, however, were very similar. Specifically, across several different experiments, Pashler and Badgio found little or no advantage for the sequential condition over the simultaneous condition. These results suggest one of two things: processing all the way up through stimulus identification occurred in parallel, unaffected by the processing of other stimuli, or attention could not be shifted from one frame to the next in the time available in the sequential condition.

For proponents of parallel models, the results from experiments using the simultaneous/sequential method suggest that processing all the way up through stimulus identification can occur without an item-by-item serial engagement of individual stimuli, and therefore, that there is no need to propose a serial mechanism in visual search. As is so often the case in these serial/parallel debates, however, matters are not clear-cut. First, the studies that found little or no advantage for the sequential condition over the simultaneous condition used set sizes of only four items. When set size was increased to sizes that were closer to those used in standard visual-search tasks, a sequential advantage was observed (Fisher 1984). Second, sequential advantages were also observed when the target-distractor discrimination was more complex (Duncan 1987; Kleiss and Lane 1986; Shiu and Pashler, described in Pashler 1998). Finally, subjects were not performing all that well in either the sequential or the simultaneous versions of this task. For instance, in various versions of the Pashler and Badgio (1987) study, accuracy was in the range of 50–67 per cent. This suggests that subjects could acquire information that was adequate to identify between 2 or 3 of the 4 items. In the sequential condition, the failure to process more items might reflect a failure to redeploy attention successfully from location to location at the required speed (see Shiffrin and Gardner 1972 for a similar argument). Even if attention can deploy quite rapidly (say, every 50 ms), there is evidence that it cannot be forced to specific locations at those high rates by exogenous prompts (e.g. Reeves and Sperling 1986; Weichselgartner and Sperling 1987; Wolfe and Alvarez 1999). We will return to this problem of the speed of attentional deployment later.

Another source of evidence cited as support for the idea that stimuli can be identified in parallel during visual search is in the spirit of the original search-slope logic. The idea is to introduce various stimulus manipulations and observe their effects on the pattern of RT. If a manipulation increases the time required for a serial step in processing, then each item in the display will take longer to process and the slope of the RT × set size function will increase. If a manipulation reduces the speed of a parallel stage, then the overall RT should increase, but it should not necessarily do so more for displays with many items than for displays with fewer items. Thus, the effects of the manipulation would be over-additive in the serial case and additive in the parallel case. Note, however, that a limited-capacity parallel process might (again) mimic a serial process if the basic cost in RT was greater for larger set sizes than smaller set sizes, not because of an item-by-item increment of cost, but because of a greater cost with greater numbers of items.

Applying this logic, Pashler and Badgio (1985) introduced manipulations of stimulus quality that were aimed at slowing down early perceptual processes and/or stimulus-identification processes. Examples of these manipulations include high-contrast versus low-contrast and added-noise versus no added-noise. In several different versions of the experiment, Pashler and Badgio obtained additive effects of stimulus quality and set size on RT, suggesting that the perceptual processes that were affected by the manipulations occurred in parallel.

Pashler and Badgio's (1985) results rule out any model that holds that the perceptual processing of one item must wait until the identification of any other item that has been engaged is completed. Notice, however, that this is a stronger serial model than is usually assumed. Specifically, models of visual search that include a serial component, like Feature Integration Theory and Guided Search, also include an early parallel perceptual stage of processing. Pashler and Badgio's data are consistent with a model that has an early parallel stage that is affected by the stimulus quality manipulations, and that feeds into a later serial stage. If stimuli are 'cleaned up' in the parallel stage then the cost of clean-up could be incurred once for the whole display, rather than for each item in turn as they are engaged by a later serial process. If one considers Pashler and Badgio's results in light of Sternberg's (1969) additive-factors logic, they suggest that the processes affected by the stimulus-quality manipulations and those affected by the set-size manipulation occur at two different stages of processing: one may be parallel and one may be item-by-item serial.

A variation of this technique applied by Egeth and Dagenbach (1991) gets around some of these difficulties of interpretation. They used a constant set size of two items, and manipulated the stimulus quality of the two items independently (Townsend and Nozawa 1988 described a similar method). Thus, there were three different types of display, one in which both items were of high quality, one in which both were of low quality, and one in which one item was of high quality and the other was of low quality. They reasoned that if the two items were identified serially, then the effects of the two quality manipulations (one for each item) should be additive. That is, relative to having one low-quality stimulus and one high-quality stimulus, having two high-quality stimuli should be twice as good, and having two low-quality stimuli should be twice as bad. In contrast, if the two items are identified in parallel, then effects of the two quality manipulations should be underadditive. Once the system has to deal with one low-quality stimulus, it will not incur any more cost for having to deal with two.

In two different applications of this technique—search for an X among Os (or vice versa) and search for a T among Ls (or vice versa)—Egeth and Dagenbach (1991) obtained under-additive effects of the two quality manipulations. If they had obtained only underadditive effects under all applications of their technique, then their diagnostic would be subject to the same interpretative worry as that of Pashler and Badgio (1985). That is, a model in which serial stimulus identification is preceded by a parallel 'clean-up' process could account for the results. Interestingly, however, they did not obtain only underadditive effects. Rather, when search was for a rotated T among rotated Ls (or vice versa), they obtained additive effects. A parallel-clean-up model cannot account for these results, because the clean-up procedure would have to be engaged whenever there was any low-quality stimulus, one or two. Instead, the results suggest that the target was identified through the engagement of a serial process. Again, however, as is perhaps always the case, a parallel limited-capacity model can also accommodate the results. It is possible that increasing numbers of low-quality stimuli results in increasingly high costs on processing of low-quality stimuli (for example, it takes more time to 'clean up' more stimuli). Such a model could produce additive effects like those observed for the rotated Ts and Ls search task.

Mordkoff, Yantis, and Egeth (1990, Experiment 3) provide perhaps the strongest evidence in favour of parallel identification of targets in a search task. They used a redundant-targets strat-egy. Their target was defined by a conjunction of features. Thus, under most models that include a serial component, it should have engaged serial processing. The twist to their study was that, on some of the trials, there were two targets in the display. As is usually the case with redundant targets, target-present responses were faster when there were two targets in the display than when there was only one (see also Pashler 1987; van der Heijden 1975). The redundant-target advantage alone, however, does not help to distinguish between parallel and serial models of search, because both predict it. For serial models, the probability that a given item that engages the serial process is a target is greater when there are two targets in the display than when there is only one. Therefore, a target will be found early more often when there are two than when there is only one, and assuming a self-terminating model, mean RT will be lower. For parallel models, depending on the specifics of the model, there are a number of reasons why a redundant-target trial might enjoy an advantage over single-target trials. One of those possibilities is that the evidence that accrues for the two targets in parallel somehow combines, allowing the 'target-present' threshold to be reached faster than if only one target were present. Such a model is known as a *coactivation* model, because multiple stimuli co-activate the target representation.

The insight that Mordkoff *et al.* (1990) had was to apply Miller's (1982) Race Model Inequality to their search situation. They noted that, even though serial and parallel models make similar predictions concerning mean response time, they make different predictions con-cerning the distributions of RTs. Specifically, under a serial model, the fastest RTs for the redundant-target trials can never be faster than the fastest RTs for the single-target trials. The lower mean RTs for redundant target trials would come from the fact that the fastest times occur more often on redundant-target trials than on the single-target trials; the fastest times, however, will not differ across the two conditions. In contrast, under a parallel model, in which the two targets both contribute to activation growth, the fastest response times for the redundant-target trials may well be faster than those in the single-target trials. Any evidence of this would rule out a strict serial model, and favour a coactivation version of a parallel model. In fact, Mordkoff *et al.* did find that the fastest RTs on the redundant-target trials were faster than those on the single-target trials, suggesting that the advantage came from coactivation of the target representation during a parallel identification process, rather than from serial selection of items.

A potential limitation to Mordkoff *et al.*'s (1990) study is that, like the earlier simultaneous/ sequential experiments, it used only small set sizes (<=6). Nonetheless, to the extent that it is evidence of a cooperative relationship between two or more items, it falsifies strict serial models that hold that identification of one item is entirely independent of identification of others.

The problem of attentional dwell-time

A different source of evidence that has been cited as favouring parallel models over serial models concerns the *attentional dwell-time*. The attentional dwell-time is the amount of time that is required for attention to be redeployed once it has already been deployed to some item or location (Duncan *et al.* 1994; Moray 1969). Posner and others performed studies in which they attempted to measure the time that is required to remove attention from one item and deploy it to another. These estimates tended to be of the order of hundreds of milliseconds (e.g., Posner 1980; Posner and Cohen 1984; Posner and Presti 1987). Using rapid serial visual presentation (RSVP) Reeves and Sperling (1986) found that it took several hundred milliseconds to switch attention from one stream of stimuli to another. Similarly, many studies have shown that the identification of one target in an RSVP stream interferes with the identification of another target for several hundred milliseconds (e.g. Broadbent and Broadbent 1987; Raymond *et al.* 1992; Shapiro *et al.* 1994; Chun and Potter 1995), an effect referred to as the *attentional blink* (see many other chapters of the present volume).

In contrast to these long estimates of the attentional dwell time, serial models of attentional deployment in visual search seem to require fairly rapid processing of items. In the absence of required eye movements, searches described as 'serial' have slopes in the range of 20–30 ms/ item for target-present trials and about twice that for target-absent trials (e.g. Treisman and Gelade 1980; Wolfe 1998). If, for simplicity, search is assumed to be serial and self-terminating, then subjects must process half the items on an average target-present trial, and all the items on a target-absent trial. That means that the rate of processing is given by twice the target-present slope, which should be roughly equal to the target-absent slope.

Estimating the rate of processing from slopes, however, is tricky. On the one hand, fifty ms/item might be too fast an estimate of the rate of processing. Subjects do make errors, and we may assume that some of these errors reflect a trade-off of accuracy for speed. Also, estimating processing rate this way assumes that search is unguided, and that on target-absent trials, it really is exhaustive. These assumptions are likely to be incorrect (Chun and Wolfe 1996). On the other hand, 50 ms/item might be too slow an estimate of the rate of processing. This estimate assumes that each rejected distractor is marked in a way that prevents the redeployment of attention to rejected distractors. While 'inhibition of return' (IOR) has been proposed as such a mechanism (Posner *et al.* 1985; Klein 1988; Tipper *et al.* 1996; see also Watson and Humphreys 1997; Theeuwes *et al.* 1998), there are questions concerning this interpretation of IOR (Klein and Taylor 1994; Wolfe and Pokorny 1990). Moreover, several studies suggest that whatever labelling may occur during search, it must be incomplete (Arani *et al.* 1984; Courtney and Guan 1998). Most extremely, results reported by Horowitz and Wolfe (1998) suggest that there may be no labelling at all. If this is true, then serial models would predict that items would be sampled at random from the display without regard to prior selection, yielding estimates of about 25 ms/item.

While some of the specific claims surrounding the various estimates of processing rate in search are controversial, all that matters for present purposes is that, even taking a broad view, serial models require an ability to process an item every 25–60 milliseconds. Studies of the 'attentional dwell time' would seem to falsify such models by estimating that attention is tied up with each stimulus for much longer than the models seem to allow.

Duncan, Ward, and Shapiro (1994; see also Ward *et al.* 1996) were the first to question directly the feasibility of such a fast-switching attentional mechanism in visual search, by citing long estimates of the attentional dwell time (see also Chapter 10 in the present volume). They estimated the attentional dwell-time using a dual-task method in which two targets (T1 and T2) were presented in two different locations on each trial. A mask followed each target, and, at the end of a trial, subjects identified both targets as best as they could. The time between the two targets (*stimulus onset asynchrony, SOA*) varied from 0 milliseconds to about 1 second. (Note the similarity to the simultaneous/sequential method described earlier.) The key finding was that, at the shorter SOAs, the identification accuracy for the second of the two targets (*T2*) was impaired relative to a single-task (report only T2) control condition. T2 accuracy was especially low at the shorter SOAs. Assuming that the dual-task interference arose because processing of the first target (*T1*) interfered with the processing of the second target, Duncan *et al.* interpreted the duration of this interference as a measure of the attentional dwell-time. In several different experiments, substantial T2 impairment was observed as late as 300 ms after the presentation of T1, and some interference was often still observed as much as 500 ms after the presentation of T1. This implies that the attentional dwell-time is approximately 300–500 ms long, which is clearly longer than the 25–60 ms estimate that serial models of visual search require.

How should we understand these long estimates of the attentional dwell-time, and how should we reconcile them with the slopes of the visual search experiments? One approach would be to accept the long dwell-time and declare that search slopes of less than 300–500 ms/ item can only be obtained by processing multiple items in parallel during the dwelling of attention, and that therefore there can be no item-by-item serial component to search. Some caution is required, however, before leaping to this conclusion. To begin, it is possible that the especially long attentional dwell-time observed by Duncan *et al.* (1994; Ward *et al.* 1996) was specific to the conditions under which it was measured, conditions that are different in many ways from those of standard spatial visual-search. Moore, Egeth, Berglan, and Luck (1996), for example, showed that, using the same procedure as Duncan *et al.*, estimates of the attentional dwell-time could be reduced from about 450 ms to about 200 ms simply by manipulating whether or not the first target was masked. In addition, the RSVP studies described above suggest that rejecting targets, which is what is happening during visual search, can occur at a rate as fast as 100 ms/item. If it could not, then finding even T1 in an attentional-blink task, in which stimuli are often presented at this rate (e.g. in Chun and Potter 1995) would be impossible.

In summary, it seems impossible to generalize from the extremely long estimates of dwell-time in Duncan *et al.*'s (1994; Ward *et al.* 1996) experiments, or the RSVP experiments (e.g., Raymond *et al.* 1992; Reeves and Sperling 1986; Shapiro *et al.* 1994) to standard visual search. The conditions are very different, and there are strong indications that, as the conditions are brought closer together, the estimates of the attentional dwell-time are also brought closer together. Still, it must be noted that in no case have estimates of dwell-time gone as low asthe 25–60 ms processing times that are suggested by search slopes; this remains an inconsistency.

Beyond serial and parallel

The inconsistency with which we were left in the preceding section disappears if we make one crucial observation about search slopes. Slopes are not dwell-times. Slopes are rates. That is, a slope of 50 ms/item is not a claim that an item can be identified in 50 ms. It is a claim that 20 such items can be identified every second. It might take several hundred ms to process each item fully. As long as the beginning of the processing of item N does not have to wait for the end of processing of item N-1, however, dwell-time—or total processing time—can be substantially longer than the processing rate.

An assembly line is a useful metaphor for illustrating this point (see Fig. 9.5). Think of the process of seeing and identifying a stimulus as being analogous to the processes of making a car on an assembly line. The line may be capable of delivering a car every ten minutes, but it does not follow from this that it takes only ten minutes to make a car. Rather, parts are fed into the system at one end. They are bound together in a process that takes an extended period of time, and cars are released at some rate (e.g., one car every ten minutes) at the other end of the system. Is this a serial process or a parallel process? Cars enter and emerge from the system in a serial manner. Moreover, many component processes could be strictly serial. For instance, it might be possible to paint only one car at a time. So, in these senses, the assembly line is a serial processor. However, if we ask how many cars are being built at the same time, it becomes clear that this is also a parallel processor.

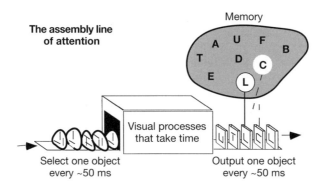

Fig. 9.5 A cartoon illustrating the assembly-line analogy for the class of search models that are both serial and parallel. There are processes that take some finite amount of time, and for which items are engaged serially. However, the engagement of one item need not wait on the completion of those processes for a previously engaged object. This model is serial; items are engaged individually. This model is parallel; items are processed simultaneously. Such a model could yield a rate of processing (i.e., the number of items per unit time that are engaged and output by the system) of approximately 20 per second, or one every 50 ms, while yielding a dwell-time (i.e., total processing time) that is considerably longer (e.g., 500 ms or more).

One way of conceptualizing this type of model (though not the only way) is to recognize it as item-by-item serial, but functionally parallel. By this we mean that items engage processing individually, but are analysed, interpreted, and transformed, at least partially, at the same time. Models of this class lie between strict serial models and strict parallel models. Returning to our metaphor, a strict serial model would hold that only one car could be on the assembly line at a time. Under this model, the search slope rate would in fact match the dwell-time. So any reliable mismatch between those two measures would falsify it. In a thoroughly parallel model, each point on the assembly line would be capable of handing more than one item at a time, and all cars could start any given process simultaneously. So, any reliable evidence that any given step in the process is serial, in the 'one item at a time' sense, would falsify this model.

Note that the assembly-line model is not a two-stage model in the manner of Neisser (1967). It is not a proposal for distinct serial and parallel stages of processing. Rather, it is an architecture that is at once both serial and parallel. It does not preclude the possibility of other serial and parallel stages of processing lying before or after the assembly line. For instance, parallel preattentive processing of basic visual features could still occur prior to its beginning.

The assembly-line notion is not without precedent. Miller (1988, 1990, 1993) has discussed related issues regarding the question of whether information flow is continuous or discrete. While he draws a distinction between continuous versus discrete information flow, on the one hand, and sequential versus overlapping processing stages, on the other, many of the issues are similar. The main message is that it is a continuum rather than a dichotomy. Moreover, it is not even a simple continuum, because different aspects of the information flow can be, at the same time, both more continuous and more discrete. We are similarly arguing that visual processing can be, at the same time, both serial and parallel.

More directly relevant to the present discussion is that models resembling the assembly-line idea have been proposed for visual search in particular, yet have somehow failed to prevent the development of an apparent dichotomy between serial and parallel models. For example, Harris, Smith, and Bates (1979) proposed a model in which items are loaded in series at a very fast rate (10 ms/item) into a set of asynchronous parallel processors that take a minimum of 40 ms to 'encode' a stimulus. In their model, actual encoding time for an individual item depends on the number of items currently being processed in parallel.

Returning to the world of metaphor, Harris *et al.*'s (1979) model proposes a multi-lane assembly line, and serves to illustrate the fact that there are a dizzying number of theoretical options available to the modeller within this class of model. For instance, if items are loaded in series, must they remain in the same sequence until they are off-loaded at the other end, or is it possible for an easy item to be selected second, but recognized first (an implication of multi-lane architectures)? Can the same item be on the assembly line in more than one place at a time? This makes no sense in a real assembly line (unless one imagines the different parts of the unassembled car in different places at one time). In the case of the assembly line of attention, though, the 'cars' are stimuli in the visual image. As another example, if items are loaded every 50 ms or so and then processed for 200–500 ms, could an item be selected by attention at time 0 and then selected again at time 150 ms while the first 'copy' is still on the assembly line? Something like this possibility is implied by Horowitz and Wolfe's (1998) claim that selected items are not marked or inhibited during the course of a visual search. This idea is also similar to the re-entrant object-substitution model that Di Lollo and Enns and colleagues have applied to the attentional blink and other visual phenomena (Di Lollo *et al.*, 2000; Enns and Di Lollo 1997; Giesbrecht and Di Lollo 1998) Finally, if an object can be represented by two copies on the assembly line, would it be recognized more quickly? If so, what if a *type* of object were represented twice, as in the redundant target experiments of Mordkoff *et al.* (1990)?

The resolution of questions like these is beyond the scope of this chapter. For present purposes, the important point is that an architecture of this sort would produce data that look either 'serial' or 'parallel' depending on how the experimental question is asked. In the next section we consider some of the data we reviewed above in the light of an assembly-line-like model.

What can an assembly-line type model do for the serial/parallel debate?

At this stage, we can offer some thoughts as to how models of the assembly-line class could deal with various phenomena. For example, we have already seen how such a model could accommodate the apparent conflict between rates of processing that are estimated from search slopes and the attentional dwell-time. Specifically, it is possible within such a model that items in a search display engage the system individually, at a rate of approximately 50 ms/item or so, and yet require several hundred ms for full processing.

Assembly-line type models might also be able to handle the data from the simultaneous/ sequential method. Recall that the basic finding here was that small numbers of stimuli could be identified just as well when they were presented to subjects sequentially as when they were presented simultaneously (cf. e.g. Pashler and Badgio 1987; Shiffrin and Gardner 1972). When larger numbers of stimuli were presented, however, sequentially presented stimuli were processed better than simultaneously presented stimuli (Fisher 1984). Also, when stimulus identification was made more complicated, sequentially presented stimuli were processed better than simultaneously presented stimuli (Duncan 1987; Kleiss and Lane 1986; Shiu and Pashler, described in Pashler 1998). All these findings can be accommodated within an assembly-line type of model. Assume that the serial engagement of stimuli occurs until all the slots on the line are filled. As stimuli are processed and released from the system, slots free up for new stimuli to be engaged. If the number of to-be-processed stimuli is smaller than the number of slots available on the line, then no limits to processing will be revealed. When more stimuli are presented, however, the line may fill up, providing an opportunity for processing limitations to be observed. These limits might occur for several different reasons. The first reason is straightforward; with more stimuli, more slots will be filled up, leaving fewer or perhaps even no slots available. Unengaged to-be-processed stimuli would then have to wait until previously engaged stimuli are released from the line, freeing up slots. Second, as more slots fill up, items may flow through the system at a slower rate. This would mean that previously engaged items would be released from the system at a slower rate, which in turn would mean that slots would be freed up for still-to-be-engaged stimuli at a slower rate. Exactly this sort of dependence of what they call 'encoding time' on the number of other stimuli engaged is incorporated in Harris *et al.*'s (1979) model. Finally, another way of slowing up the system, and therefore reducing the rate at which slots become available for still-to-be-engaged stimuli, would be to increase the complexity of the processing for a given stimulus. This would then lead to the observation of processing limits when the identification task is more complex (cf. e.g. Kleiss and Lane 1986). While this scenario clearly involves a number of assumptions, and requires further specification, it does illustrate how an assembly-line type of model, which is neither serial nor parallel, could, in principle, accommodate the overall pattern of results from the simultaneous/sequential literature.

Another result that an assembly-line type of model, like that of Harris *et al.* (1979), could accommodate is the basic redundant-targets advantage reported by Pashler (1987) and van der

Heijden (1975), as well as the more specific characteristics of redundant-target effects reported by Mordkoff *et al.* (1990). Specifically, the two copies of the target would be engaged individually, but they would both be present within the system at the same time, albeit at different phases of processing, thereby providing the opportunity for the processing of one to affect the processing of the other. It is this simultaneous processing that allows for both the general advantage of two targets over one (Pashler 1987; van der Heijden 1975) and for the fastest of the two-target trials being faster than the fastest of any of the one-target trials (Mordkoff *et al.* 1990).

Exactly where in processing the advantage for redundant targets would come from, would depend on the specifics of the model and could, in principle, be identified through either analytic or simulation techniques, which are beyond the scope of the present chapter. Still, appealing once again to analogy, we can offer one possible scenario. Imagine that there are two types of car that are built by the assembly line: an American version with the steering wheel on the left, and a British version with the steering wheel on the right. Every time a particular frame comes through the system, the steering-wheel assembler 'recognizes' the frame as being of a given type and sets itself up to, for example, put the wheel on the right. If the next car in line happens also to be of the British sort, then the set-up process need not be repeated, as it is already set to go. Now imagine that there is some characteristic of these cars for which there are many versions (for example, paint colour). The process responsible for that characteristic would benefit when individual cars (items in the display) are of the same version (redundant stimuli). Notice that according to this (admittedly loose) explanation, one would expect 'redundant distractor' advantages as well as redundant-target advantages, which do in fact occur (cf. e.g. Duncan and Humphreys 1989).

Object recognition as a possible item-by-item process within an assembly-line type of model

The preceding section has shown how assembly-line models might account for some of the evidence that points toward parallel processing in search. In this final section we consider *object recognition* as a specific example of a mandatory serial process within search.

Whether it is serial or parallel, object recognition is a mandatory step in most visual search tasks. In Fig. 9.5, recognition is cartooned as a link between the item on the assembly line and the cloud labelled 'memory'. Recognition must involve such a link. Consider the other possibilities: (1) A visual stimulus can be as fully processed as possible and yet, if it is not in contact with its representation in memory, it cannot be said to be recognized in the present moment. (2) By the same token, if a node in memory is active in the absence of the appropriate visual stimulus, the content of that node may be thought about or remembered, but it is not seen and recognized.

Finally, (3), it cannot be that mere simultaneous activity of visual system and memory node constitutes recognition. After all, it is possible to see a chair and think about an elephant without spuriously 'recognizing' the chair as an elephant. Some link between the visual and mnemonic representations would seem to be required.

That being the case, how many such links can be maintained at one time? For the sake of simplicity, consider a scene containing a tree, a horse, and a rock. If multiple links can be maintained, then tree, horse, and rock could be recognized simultaneously. With the links thus established, suppose that we asked a subject if a horse was present in the display. The query 'horse' would activate the horse node in memory. The horse node would already be linked to the perceptual representation of the horse in the display, and so the subject could make a positive response, unaffected by the presence of other objects in the scene. In contrast, if only one link can be maintained at a given time, then upon first presentation of the display and the query 'horse?' a standard visual search would be required to determine if a horse was present. In particular, the horse node in memory would need to be compared to the representation of each item in the display in turn, until the horse was 'recognized' (i.e. was linked with the perceptual representation of the horse in the display). Suppose we added to the display a windmill, a rhinoceros, and a cow. It follows that increasing the number of objects would cause RTs to increase at the usual rate of 20–30 ms/item on target-present trials. Note, however, that this 20–30 ms is not dwell-time. There is much more to processing stimuli in a visual search task than the formation of links between memorial and perceptual representations.

How could we test such an account? The critical experiment involves repeated search through the same, unchanging scene. After multiple searches for horse, tree, bird, cow, etc., all the objects must have been recognized at one time or another. If multiple recognition links between vision and memory can be maintained, then search slopes should drop from 20–30 ms/item to near zero as search is repeated. If, on the other hand, links must be established individually, then search slopes should reach an asymptotic minimum. In a host of repeated search experiments (Wolfe et al., 2000), this is exactly what happened; that is, slopes simply did not drop below the standard 20–30 ms/item value. Specifically, if a search was inefficient on the first appearance of the stimuli, then it was inefficient on the fifth, fifteenth, and fiftieth search through the same stimuli. This is not to say that observers did not learn anything about the display. After a few repetitions, they could perform the task in the absence of the stimulus, just as you, as a reader, could answer if questioned about the presence of a horse in the hypothetical scene that you have not seen. Despite this memorization of the scene, the efficiency of target detection (i.e., the search slope) did not improve. If it were possible to maintain multiple recognition links, then search should become efficient with repeated search through the same stimuli. As it did not, it follows that multiple links are not possible, and that the act of recognition has an irreducibly serial aspect to it.

As a final point, notice that this specific type of assembly-line model can account for phenomena like the attentional blink. Consider what happens to items when they emerge off this assembly line. Some items, like the target in a visual search task, are taken up from the end of the line and made the subject of other processes (for example, response processes). Other items (for example the distractors) fall into a Jamesian abyss and are lost (James 1983 [1890] Chapter 16). In the stable world of natural vision, a lost stimulus can be recovered by another search, if needed (O'Regan 1992). In the RSVP stream of an attentional blink experiment, however, if some act of processing T1 takes 300 ms, then 300 ms-worth of stimuli will go irredeemably into the abyss. A similar account can be offered for the failure to recognize changes in scenes (e.g. Pashler 1988; Rensink et al. 1997; Simons and Levin 1997) and the failure to recall aspects of a display that were unattended at the time of presentation (Mack and Rock 1998; Moore and Egeth 1997; Wolfe, 1999).

Conclusions

In summary, there is good evidence for serial processes of attentional selection in visual search tasks. There is also good evidence for parallel processing of multiple items in visual search tasks. In recent years, this has been seen as a contradiction, a conflict in need of a victory by one side or another. We believe, however, that Newell's (1973, p. 283) worry that 'you can't play 20 questions with nature and win' applies here, and in line with earlier thinking (e.g. Harris *et al.* 1979; Miller 1988, 1990), we propose that the mechanisms of selective attention in visual search are both serial and parallel at the same time. Different experiments will reveal a more serial or a more parallel pattern of results. We hold that future research should focus on constraining the still large set of possible architectures of this sort, rather than attempting to resolve a false dichotomy.

References

Arani, T., Karwan, M. H., and Drury, C. G. (1984). A variable-memory model of visual search. *Human Factors*, **26**, 631–9.

Atkinson, R. C., Holmgren, J. E., and Juola, J. F. (1969). Processing time as influenced by the number of elements in a visual display. *Perception and Psychophysics*, **6**, 321–6.

Broadbent, D. E. (1958). *Perception and communication*. New York: Pergamon Press.

Broadbent, D. E., and Broadbent, M. H. P. (1987). From detection to identification: response to multiple targets in rapid serial visual presentation. *Perception and Psychophysics*, **42**, 105–13.

Bundesen, C. (1990). A theory of visual attention. *Psychological Review*, **97**, 105–13.

Carrier, L. M., and Pashler, H. (1996). Attentional limits on memory retrieval. *Journal of Experimental Psychology: Learning, Memory and Cognition*, **21**, 1339–48.

Chun, M. M., and Potter, M. C. (1995). A two-stage model for multiple target detection in rapid serial visual presentation. *Journal of Experimental Psychology: Human Perception and Performance*, **21**, 109–27.

Chun, M. M., and Wolfe, J. M. (1996). Just say no: How are visual searches terminated when there is no target present? *Cognitive Psychology*, **30**, 39–78.

Courtney, A., and Guan, L. (1998). Assessing search performance with a simulation model. *Human Factors and Ergonomics in Manufacturing*, **8**, 251–63.

Di Lollo, V., Enns, J. T., and Rensink (2000). Competition for consciousness among visual events: The psychophysics of reentrant visual processes. *Journal of Experimental Psychology: General*, **129**, 481–507.

Duncan, J. (1980). The locus of interference in the perception of simultaneous stimuli. *Psychological Review*, **87**, 272–300.

Duncan, J. (1987). Attention and reading: wholes and parts in shape recognition–A tutorial review. In M. Coltheart (ed.), *Attention and performance XII: The psychology of reading*, pp. 39–61. Hillsdale, NJ: Erlbaum.

Duncan, J., and Humphreys, G. W. (1989). Visual search and stimulus similarity. *Psychological Review*, **96**, 433–58.

Duncan, J., Ward, R., and Shapiro, K. (1994). Direct measurement of attentional dwell time in human vision. *Nature*, **369**, 313–15.

Eckstein, M. P. (1998). The lower visual search efficiency for conjunctions is due to noise and not serial attentional processing. *Psychological Science*, **9**, 111–18.

Egeth, H., and Dagenbach, D. (1991). Parallel versus serial processing in visual search: further evidence from subadditive effects. *Journal of Experimental Psychology: Human Perception and Performance*, **17**, 551–60.

Egeth, H. E., Jonides, J., and Wall, S. (1972). Parallel processing of multielement displays. *Cognitive Psychology*, **3**, 674–98.

Egeth, H. E., Virzi, R. A., and Garbart, H. (1984). Searching for conjunctively-defined targets. *Journal of Experimental Psychology: Human Perception and Performance*, **10**, 32–9.

Enns, J. T., and Di Lollo (1997). Object substitution: a new form of masking in unattended visual locations. *Psychological Science*, **8**, 135–9.

Eriksen, C. W., and Spencer, T. (1969). Rate of information processing in visual perception: some results and methodological considerations. *Journal of Experimental Psychology*, **80**, 489–92.

Fisher, D. L. (1984). Central capacity limits in consistent mapping, visual search tasks: four channels or more? *Cognitive Psychology*, **16**, 449–84.

Giesbrecht, G., and Di Lollo, V. (1998). Beyond the attentional blink: visual masking by object substitution. *Journal of Experimental Psychology: Human Perception and Performance*, **24**, 1454–66.

Grossberg, S., Mingolla, E., and Ross, W. D. (1994). A neural theory of attentive visual search: interactions of boundary surface, spatial, and object representations. *Psychological Review*, **101**, 470–89.

Harris, J. R., Shaw, M. L., and Bates, M. (1979). Visual search in multicharacter arrays with and without gaps. *Perception and Psychophysics*, **26**, 69–84.

Hoffman, J. E. (1979). A two-stage model of visual search. *Perception and Psychophysics*, **25**, 319–27.

Horowitz, T. S., and Wolfe, J. M. (1998). Visual search has no memory. *Nature*, **394**, 575–7.

Humphreys, G. W., and Müller, H. J. (1993). Search via Recursive Rejection (ERR): a connectionist model of visual search. *Cognitive Psychology*, **25**, 43–110.

Ivry, R. B., and Prinzmetal, W. (1990). Effect of feature similarity on illusory conjunctions. *Perception and Psychophysics*, **49**, 105–16.

James, W. (1983 [1890]). *The principles of psychology*. Cambridge, MA: Harvard University Press.

Kinchla, R. A. (1974). Detecting targets in multi-element arrays: a confusability model. *Perception and Psychophysics*, **15**, 149–58.

Klein, R. (1988). Inhibitory tagging system facilitates visual search. *Nature*, **334**, 430–1.

Klein, R. M., and MacInnes, W. J. (1999). Inhibition of return is a foraging facilitator in visual search. *Psychological Science*, **10**, 346–52.

Klein, R. M., and Taylor, T. L. (1994). Categories of cognitive inhibition with reference to attention. In D. Dagenbach and T. H. Carr (eds), *Inhibitory processes in attention, memory and language*. New York: Academic Press.

Kleiss, J. A., and Lane, D. M. (1986). Locus of persistence of capacity limitations in visual information processing. *Journal of Experimental Psychology: Human Perception and Performance*, **12**, 200–10.

Kwak, H., Dagenbach, D., and Egeth, H. (1991). Further evidence for a time-independent shift of the focus of attention. *Perception and Psychophysics*, **49**, 473–80.

Mack, A., and Rock, I. (1998). *Inattentional blindness*. Cambridge, MA: MIT Press.

Meyer, D. E., Yantis, S., Osman, A. M., and Smith, J. K. (1986). Temporal properties of human information processing: tests of discrete versus continuous models. *Cognitive Psychology*, **17**, 445–518.

Miller, J. (1982). Divided attention: evidence for coactivation with redundant signals. *Cognitive Psychology*, **14**, 247–79.

Miller, J. O. (1988). Discrete and continuous models of human information processing: theoretical distinctions and empirical results. *Acta Psychologica*, **67**, 1–67.

Miller, J. O. (1990). Discreteness and continuity in models of human information processing. *Acta Psychologica*, **74**, 297–318.

Miller, J. (1993). A queue-series model for reaction time, with discrete-stage and continuous-flow models as special cases. *Psychological Review*, **100**, 702–15.

Moore, C. M., and Egeth, H. (1997). Perception without attention: evidence of grouping under conditions of inattention. *Journal of Experimental Psychology: Human Perception and Performance*, **23**, 339–52.

Moore, C. M., and Egeth, H. (1998). How does feature-based attention affect visual processing? *Journal of Experimental Psychology: Human Perception and Performance*, **24**, 1296–1310.

Moore, C. M., Egeth, H., Berglan, L., and Luck, S. J. (1996). Are attentional dwell times inconsistent with serial visual search? *Psychonomic Bulletin and Review*, **3**, 360–5.

Moray, N. (1969). *Attention: selective processing in vision and hearing*. London: Hutchinson.

Mordkoff, J. T., Yantis, S., and Egeth, H. E. (1990). Detecting conjunctions of color and form in parallel. *Perception and Psychophysics*, **48**, 157–68.

Neisser, U. (1967). *Cognitive psychology*. New York: Appleton-Century-Crofts.

Newell, A. (1973). You can't play 20 questions with nature and win: projective comments on the papers of this symposium. In W. G. Chase (ed.), *Visual information processing*. New York: Academic Press.

O'Regan, J. K. (1992). Solving the 'real' mysteries of visual perception: the world as an outside memory. *Canadian Journal of Psychology*, **46**, 461–88.

Palmer, J. (1994). Attention and visual search: distinguishing four causes of a set-size effect. *Current Directions in Psychological Science*, **4**, 118–23.

Palmer, J., and McLean, J. (1995). *Imperfect, unlimited-capacity, parallel search yields large set-size effects*. Paper presented at the Society for Mathematical Psychology, Irvine, CA.

Pashler, H. (1987). Detecting conjunctions of colour and form: reassessing the serial search hypothesis. *Perception and Psychophysics*, **41**, 191–201.

Pashler, H. (1998). Familiarity and visual change detection. *Perception and Psychophysics*, **44**, 369–78.

Pashler, H. (1998). *The psychology of attention*. Cambridge, MA: MIT Press

Pashler, H., and Badgio, P. (1985). Visual attention and stimulus identification. *Journal of Experimental Psychology: Human Perception and Performance*, **11**, 105–21.

Pashler, H., and Badgio, P. C. (1987). Attentional issues in the identification of alphanumeric characters. In M. Coltheart (ed.), *Attention and performance XI*, pp. 63–81. Hillsdale, NJ: Erlbaum.

Posner, M. I. (1980). Orienting of attention. *Quarterly Journal of Experimental Psychology*, **32**, 3–25.

Posner, M. I., and Cohen, Y. (1984). Components of visual orienting. In H. Bouma and D. G. Bouwhuis (eds), *Attention and performance X: Control of language processes*, (pp. 531–56). Hillsdale, NJ: Erlbaum.

Posner, M. I., and Presti, D. E. (1987). Selective attention and cognitive control. *Trends in Neuroscience*, **10**, 13–16.

Posner, M. I., Rafal, R. D., Choate, L. S., and Vaughan, J. (1985). Inhibition of return: neural basis and function. *Cognitive Neuropsychology*, **2**, 211–28.

Prinzmetal, W. (1995). Visual feature integration in a world of objects. *Current Directions in Psychological Science*, **4**, 1–5.

Raymond, J. E., Shapiro, K. L., and Arnell, K. M. (1992). Temporary suppression of visual processing in an RSVP task: an attentional blink? *Journal of Experimental Psychology: Human Perception and Performance*, **18**, 849–60.

Reeves, A., and Sperling, G. (1986). Attention gating in short-term visual memory. *Psychological Review*, **9**, 180–206.

Rensink, R. A., O'Regan, J. K., and Clark, J. J. (1997). To see or not to see: the need for attention to perceive changes in scenes. *Psychological Science*, **8**, 368–73.

Shapiro, K. L., Raymond, J. E., and Arnell, K. M. (1994). Attention to visual pattern information produces the attentional blink in rapid serial visual presentation. *Journal of Experimental Psychology: Human Perception and Performance*, **20**, 357–71.

Shiffrin, R. M., and Gardner, G. T. (1972). Visual processing capacity and attentional control. *Journal of Experimental Psychology*, **93**, 72–83.

Simons, D. J., and Levin, D. T. (1997). Change blindness. *Trends in Cognitive Sciences*, **1**, 261–67.

Sternberg, S. (1969). The discovery of processing stages: extensions of Donders' method. *Acta Psychologica*, **30**, 276–315.

Sternberg, S. (1975). Memory scanning: new findings and current controversies. *Quarterly Journal of Experimental Psychology*, **27**, 1–32.

Theeuwes, J., Kramer, A. F., and Atchley, P. (1998). Visual marking of old objects. *Psychonomic Bulletin and Review*, **5**, 130–34.

Tipper, S. P., Weaver, B., and Watson, F. L. (1996). Inhibition of return to successively cued spatial locations: commentary on Pratt and Abrams (1995) *Journal of Experimental Psychology: Human Perception and Performance*, **22**, 1289–93.

Townsend, J. T. (1971). A note on the identification of parallel and serial processes. *Perception and Psychophysics*, **10**, 161–3.

Townsend, J. T. (1976). Serial and within-stage independent parallel model equivalence on the minimum completion time. *Journal of Mathematical Psychology*, **14**, 219–39.

Townsend, J. T. (1990). Serial and parallel processing: sometimes they look like Tweedledum and Tweedledee but they can (and should) be distinguished. *Psychological Science*, **1**, 46–54.

Townsend, J. T., and Roos, R. N. (1973). Search reaction time for single targets in multiletter stimuli with brief visual displays. *Memory and Cognition*, **1**, 319–32.

Townsend, J. T., and Nozawa, G. (1988). *Strong evidence for parallel processing with simple dot stimuli.* Paper presented at the annual meeting of the Psychonomic Society, Chicago, IL, November.

Treisman, A., and Gelade, G. (1980). A feature integration theory of attention. *Cognitive Psychology*, **12**, 97–136.

Treisman, A., and Sato, S. (1990). Conjunction search revisited. *Journal of Experimental Psychology: Human Perception and Performance*, **16**, 459–78.

Treisman, A., and Schmidt, H. (1982). Illusory conjunctions in the perception of objects. *Cognitive Psychology*, **14**, 107–41.

Tsotsos, J. K. (1990). Analyzing vision at the complexity level. *Brain and Behavioral Sciences*, **13**, 423–69.

van der Heijden, A. H. C. (1975). Some evidence for a limited-capacity parallel selfterminating process in simple visual search tasks. *Acta Psychologica*, **39**, 21–41.

Ward, R., and McClelland, J. T. (1989). Conjunction search for one and two identical targets. *Journal of Experimental Psychology: Human Perception & Performance*, **15**, 664–72.

Ward, R., Duncan, J., and Shapiro, K. (1996). The slow time-course of visual attention. *Cognitive Psychology*, **30**, 79–109.

Watson, D. G., and Humphreys, G. W. (1997). Visual marking: prioritizing selection for new objects by top-down attentional inhibition of old objects. *Psychological Review*, **104**, 90–122.

Weichselgartner, E., and Sperling, G. (1987). Dynamics of automatic and controlled visual attention. *Science*, **238**, 778–80.

Wolfe, J. M. (1994). Guided Search 2.0. *Psychonomic Bulletin and Review*, **1**, 202–238.

Wolfe, J. M. (1998). What can 1 million trials tell us about visual search? *Psychological Science*, **9**, 33–9.

Wolfe, J. M. and Alvarez, G. A. (1999). Give me liberty or give me more time! Your visual attention is faster if you don't tell it what to do. *Investigative Ophthalmology & Visual Science*, **40**, 5796.

Wolfe, J. M. (1999). Inattentional amnesia. In V. Coltheart (ed.) *Fleeting memories*, pp. 71–94. Cambridge, MA: MIT Press.

Wolfe, J. M., and Bennett, S. C. (1997). Preattentive object files: shapeless bundles of basic features. *Vision Research*, **37**, 25–43.

Wolfe, J. M., and Pokorny, C. W. (1990). Inhibitory tagging in visual search: a failure to replicate. *Perception and Psychophysics*, **48**, 357–62.

Wolfe, J., Cave, K., and Franzel, S. L. (1989). Guided Search: an alternative to the Feature Integration Model for visual search. *Journal of Experimental Psychology: Human Perception and Performance*, **15**, 419–33.

Wolfe, J. M., Klempen, N., and Dahlen, K. (2000). Post-attentive vision. *Journal of Experimental Psychology: Human Perception and Performance*, **26**, 693–716.

Yantis, S., Meyer, D. E., and Smith, J. K. (1992). Analyses of multinomial mixture distributions: new tests for stochastic models of cognition and action. *Psychological Bulletin*, **110**, 350–72.

Zohary, E., and Hochstein, S. (1989). How serial is serial processing? *Perception*, **18**, 191–200.

10

Visual attention moves no faster than the eyes

Robert Ward

Abstract

How quickly can visual attention move from one object to another? Evidence is reviewed here from behavioural and electrophysiological studies of visual search, from single-cell recordings of visual search, from behavioural and electrophysiological studies of covert and overt orienting, and from studies of attentional dwell time. In none of these studies or paradigms are movements of attention faster than approximately 150 ms identified; in most cases time to move attention appears to be 200 ms or more. Visual attention therefore appears to move no faster than the eyes.

We have only limited capacity to identify visually presented objects. Many experiments have demonstrated costs in accuracy and speed when multiple objects must be identified, as compared to single-object performance (e.g. Broadbent 1958; Duncan 1980; Treisman 1969). Because visual processing capacity is limited, it is best to process selectively the objects most relevant to current goals, and ignore irrelevant or distracting objects. Theories of visual attention describe such selective processing. A long-standing debate in visual attention, triggered in large part by results from visual search tasks, centres on how limited capacity is allocated for visual identification: serially to one item at a time, or in parallel to multiple items simultaneously? I will argue that shifts of visual attention occur no faster than eye movements, and that this simple temporal constraint greatly limits the space of possible attention models and the nature of capacity allocation. I first describe serial and parallel models, and then apply these characterizations to behavioural and neurophysiological data, including measures of cuing, visual search, and attentional dwell-time.

High-speed serial models. These models generally assume two stages of processing: a parallel preattentive stage followed by a serial stage of selective attention (Neisser 1967). Early preattentive processes are meant to segregate, group, and otherwise organize the visual array to facilitate the subsequent allocation of attention. They might even rank and prioritize regions of the visual input that are most promising for the current task (Cave and Wolfe 1990). However, beyond this role as an organizational front-end, preattentive processing is meant to be crucially limited. Only focused attention to an object can deliver the integrated and durable representations needed to support conscious awareness and overt report. For example, in many recent models, attention acts to bind separate feature dimensions, such as colour and shape, into a single integrated object description (Luck *et al.* 1997; Treisman and Gelade 1980; Treisman and Gormican 1988; Wolfe 1994). For such reasons, limited-capacity attentional processing is meant to be allocated serially, so that once processing of one object is complete, attention is reallocated.

Much of the empirical motivation for serial models has come from results of standard visual search tasks, in which subjects look for a single target among a variable number of distractors. In such tasks, search times for the target often increase linearly with the number of display items, and will often demonstrate the signature of a serial self-terminating search, an approximate 2:1 ratio of target-absent to target-present slopes. Slope ratios of roughly 2:1 have a straightforward explanation in models of serial, self-terminating search. When the target is absent, each item in the display must be searched before an 'absent' response can be made with certainty. When the target is present, on average it will be found after searching half the display items; search then terminates, and a 'present' response is made. In a serial self-terminating search, RTs in the target-absent displays should represent an exhaustive search of all display items, and the slope of the target-absent searches is therefore the rate of item processing (Schneider and Shiffrin 1977).

The rate at which attention can be shifted from item to item is an interesting and important feature of serial models. We would expect some variability in this rate based on the similarity between targets and distractors: the more similar, the slower the rate of switching (Sternberg 1969). However, as a rough approximation, this rate is frequently estimated to be around 50 ms per switch (see Moore and Wolfe, this volume, Chapter 9; Duncan *et al.* 1994; Treisman and Gelade 1980). Within a 50 ms cycle, attention is meant to move rapidly and accurately to an item in the display, process the item, compare it to a desired target specification, and if necessary compute the coordinates for the next focus of attention. Since this all this activity is meant to occur within a 50 ms period, we have previously called this the *high-speed model* of attention (Ward *et al.* 1996).

While this 50 ms rate may be something of a convention, it is interesting to consider the history of this estimate. For example, search rates in the range of 5–10 ms/item have been judged to be too fast for serial processing mechanisms, and attributed instead to the efficient operation of parallel preattentive processing. But why should this be so? How did differences of a few dozen ms in search slopes come to be attributed to different search processes, one parallel and one serial? Early experiments by Treisman and Gelade (1980; see also Treisman *et al.* 1977) compared easy searches for a unique feature to difficult searches for a combination of features, and found two things: First, search slopes were much slower in conjunction search (about 60 ms/item for target-absent displays) than in the feature search (between 3 and 25 ms/item). But second, slope ratios approximated to the signature 2:1 ratio of serial self-terminating search only in the slower conjunction search, but not in the fast feature search. Treisman and Gelade (1980) concluded there were two qualitatively different modes of search. In feature search, the presence of a unique feature could be determined in a parallel preattentive operation. In conjunction search, display items were meant to be sequentially scanned by focused attention. These initial indications were therefore consistent with the possibility that a quantitative difference in search rate might be associated with a qualitative difference in the nature of the search.

However, subsequent experiments have not strongly supported this notion of qualitatively different forms of search. In particular, Wolfe's (1998) meta-analysis of his large database of search experiments does not support the idea that differences in search rates might be associated with qualitatively different modes of processing. First, Wolfe's analysis of this database, representing over a million trials of search collected over a 10-year period, shows no evidence of a qualitative split between faster parallel and slower serial searches. Instead, search rates very continuously from rates of less than 1 ms per item to rates of over 400 ms per item. Second, search rates were not good predictors of slope ratios, and both the fastest and slowest searches frequently produced slope ratios of 2:1 and greater (Wolfe 1998, Figure 2); in fact, a median slope ratio between 2 and 2.5:1 appears to hold for all search rates (Wolfe 1998, Figure 4). That is, there is no search rate particularly predictive of the 2:1 slope ratio of serial self-terminating search, and searches with the 2:1 serial signature occur at all different rates.

On the simple basis of standard visual search data, then, there is no reason to believe that search rates averaging 10 ms per item are executed in any qualitatively different way from search rates of 50 ms or more per item. What is curious, then, is that serial models of attention are still associated with a 50 ms per item rate. There is a 'three bears' quality to the issue: search rates of the order of 10 ms per item are generally assumed to be too fast for serial shifts of attention, and have been taken to reflect pre-attentional processes; switching rates of the order of 500 ms (Duncan *et al.* 1994; I will return to this issue later), are assumed to be too slow for serial shifts of attention, and taken to reflect post-attentional target processing; but rates of the order of 50 ms per item are frequently accepted as reasonable for serial attentional mechanisms. The obvious question is, on what basis? Clearly such an assumption would not be based on behavioural evidence from standard search tasks. It may reflect beliefs about speed limits of neural processing; but then again, there is little evidence of controlled modulation at even the 50 ms rate in single cell studies (but see Crick 1984). This may be an interesting issue for future serial models. However, to anticipate the arguments ahead, for the present purposes, any model that assumes that attention can be moved more quickly than the eyes can be treated as a high-speed model.

The Guided Search model (Cave and Wolfe 1990; Wolfe 1994) is currently the most developed of the serial models of attention, and has been applied successfully to a variety of search experiments. Among its attractions, the model suggests how both relatively fast (e.g. 10 ms per item) and relatively slow searches (e.g. 50 ms per item) can be accounted for by a high-speed serial attention mechanism. In visual search for a designated target, preattentive processes are used if possible to prioritize items in the display for subsequent focused processing. In essence, preattentive processes provide a ranked and ordered list of where in the display the target is most likely to be located. When preattentive guidance is effective, search rates will increase, not because individual items are being scanned more quickly, but because fewer items need to be searched. In the event that preattentive guidance is completely ineffective, the Guided Search model approximates to a simple serial self-terminating search. Regardless of the extent of preattentive guidance, serial application of attention from one item to another is meant to occur at a relatively high rate, for present purposes nominally 50 ms per item.

Parallel models. As a class, limited-capacity parallel models of attention show much more variability in their architecture than serial ones. Their most distinguishing characteristic is of course the fact that limited capacity for visual identification can be allocated to multiple objects simultaneously, at least to some degree (Bundesen 1990; Eckstein 1998; Humphreys and Muller 1993; Ward and McClelland 1989). The notion of 'some degree' is unfortunately critical, because, as a group, these models do not make consistent predictions about how much identification can occur in parallel. For example, in some models multiple objects may be simultaneously identified and buffered in a visual short-term memory. Other models allow parallel processing over the display, but not in a way that allows different identities to become simultaneously available for report (Bundesen 1990). Still others are ambiguous. For example, in the diffusion model of search used by Ward and McClelland (1989), evidence accumulates in parallel for each item in the display about its match to the target identity, but the model says nothing about whether non-targets are fully identified or simply determined not to match the target template.

In standard search tasks, the aim is to find a single target among variable numbers of similar distractors. These tasks frequently find increased reaction time (RT) to find the target as the number of display items is increased. Such results are clear evidence of limited capacity for visual identification. But do these data tell us anything more about the mechanisms of capacity allocation? In particular, can they give us any insight into whether attention is allocated serially, or in parallel, to multiple items?

The answer is well known to be: no, definitely not. It is well established that limited-capacity parallel models can mimic the behaviour of serial models in circumstances like visual search for a single target (Broadbent 1987; Townsend and Ashby 1983). Earlier models of search have made simultaneous quantitative fits to mean RT, trends in variance, and error rates (Bundesen 1990; Humphreys and Muller 1993; Ward and McClelland 1989). Again, the diverse range of architectures makes a single generalized explanation difficult; but essentially, as the number of items in the display increases, the amount of capacity available to each decreases. This results in generally longer search times when more items are in the display. The speed of search depends also on similarity between target and distractors: in general, less capacity is needed for easy discriminations than for hard.

The indeterminance of standard visual search as a diagnostic reflects a more general equivalence that can be found in the behaviour of serial and parallel processes. This does not mean that the two forms of processing cannot be distinguished. In fact, while the work of Townsend and colleagues may be most frequently cited in reference to the *ambiguity* of serial and parallel processing, much of their work is devoted to analysing conditions and experiments that *disambiguate* these processes (for example, Townsend 1990) However, in the case of models for visual attention, disentangling serial and parallel allocation may be especially difficult: beyond the central difference of capacity allocation, parallel models tend to share many of the important assumptions of serial models. Like serial models, parallel models assume that there is a problem of limited resources to solve, and attempt to allocate capacity effectively to the visual objects of most importance. Again, as in serial models, many parallel models also assume a role for a relatively capacity-free preattentive processes, which again are meant to organize the display prior to allocating attention.

Given these characterizations of the two models, it is easy to imagine many comparisons that would fail to discriminate them. First, while the allocation of limited capacity is different between the models, both will demonstrate capacity limitations in identification. Second, while the functions and capabilities of preattentive processing in the serial model are not the same as the limited-capacity processing in the parallel model, both classes will demonstrate some form of parallel processing across the display. Consider how these similarities conspire to produce similar predictions about the effects of non-targets on processing, as a function of target–non-target similarity. When there are good cues available to distinguish targets and non-targets, parallel processing in both models may prevent non-targets from engaging attentional resources. When cues for selection cannot easily segregate targets from non-targets, then non-targets may engage limited capacity in a way that is very similar to targets. It has therefore proved extremely difficult, using behavioural measures alone, to investigate the nature of non-target processing: not only are behavioural measures of non-target processing necessarily indirect, but a variety of models will make similar predictions about performance. More generally, it is widely recognized that without carefully designed quantitative predictions, demonstrations of either limited capacity or parallel processing will tend to be indeterminate signatures of both models (Broadbent 1987; Townsend 1990).

Implications of covert and overt shifts of attention

The long-standing distinction between overt and covert shifting of attention is generally not addressed in either high-speed or parallel models, but has important implications. When focusing attention towards a sudden onset, it is natural to orient the eyes, head, and body towards the source of stimulation. These overt movements are a critical component of the selective visual attention system, as they redirect the fovea to allow focused visual processing of potentially interesting regions. These overt shifts can be planned and executed over a period of a few hundred ms. For example, in difficult visual search tasks, the eyes are moved between pauses of 200–800 ms (Bouma 1978). Skilled readers can program and execute eye-movements at a rate of roughly once every 250 ms when reading coherent text (Just and Carpenter 1980). Saccadic reaction times to unpredictable peripheral targets also fall within the range of 200 ms (Carpenter 1981). If the stimulus at fixation is immediately turned off with the onset of the saccadic target, saccadic latencies are reduced 30 ms or more (the so-called 'fixation offset effect': cf. e.g. Kingstone and Klein 1993). Such a reduction (including the effects of 'express saccades', cf. e.g. Fischer and Boch 1983) suggests that the physical mechanics of eye movements—such as contracting the appropriate muscles, overcoming inertia of the eyeball, etc.—are probably not limiting factors for saccadic reaction times.

Obviously, both serial and parallel models must allow in principle for overt sequential shifts, even though they might fall outside the model's domain of application. This would be the case, for instance, if the operation of a model was meant to describe processing within a single attentional 'episode'. Overt shifts of attention would then define the focus and items of a new attentional episode. By such a view, the constraint of overt shifts appears to be broadly compatible with all classes of models.

However, it is also well recognized that attention may be reallocated from one location to another covertly, in the absence of eye or body movements. This is evident both from passing introspection and also from effects in the Posner cuing paradigm (Posner 1980). For high-speed serial models, the distinction between overt and covert attention shifts is reflected in their presumed timecourse. Shifts of attention in high-speed serial models generally refer to covert rather than overt shifts, so that the rate of serial search reflects in part the speed of covert shifts. Because search rate would also vary with the speed of other processes, such as the processing of item identity, the speed of serial search might be considered a conservatively slow estimate of covert shift speed.

High-speed serial models would now be characterized as follows. Eye movements, executed on the order of every 200 ms or so, define the region over which attention is currently operating. Within this region, selective visual attention darts about quickly from one location to another, identifying the objects present. The image that comes to mind is the eyes in the form of something like a behemoth Death Star, moving slowly into position before releasing a series of high-speed probes in the form of covert attention shifts, and then lumbering on to new territory.

For parallel models, the crucial assumption is that the speed of overt shifts places a lower boundary on the speed of covert shifts. That is, if an overt shift can be planned and initiated in 200 ms, then a covert shift can be planned and initiated in approximately the same time-frame. Within the attentional episode defined by the focus of attention, objects are processed in parallel. In such models, shifts of the eyes and of attention could occur over similar timecourses. While some degree of separation might be possible, the two would generally move together, sometimes one leading, sometimes the other, in a coordinated waltz through visual space.

What are the implications of incorporating slow processing shifts into the serial and parallel models? There are at least two. The first is a somewhat negative result. Suppose we were to find direct evidence for the serial reallocation of visual attention. Would this necessarily disconfirm parallel models? Without quantitative considerations, almost certainly not. All models must allow for sequential shifts of attention over the time-scale of eye, head, and body movements. A second implication is more positive, and suggests a way of distinguishing serial and parallel models. According to high-speed serial models, covert shifts should be much faster than overt shifts, at around a rate of 50 ms per item rather than 200 or more. Parallel models are consistent with the notion of very similar, relatively slow timecourses for *both* covert and overt shifts. In this chapter, I will survey some of the evidence available on the speed of attention shifts. Evidence for fast shifts of attention, in particular shifts significantly greater than the speed of overt orienting movements, would lend support to high-speed serial models. Evidence of slow shifts comparable to the speed of eye movements might be taken to support parallel models.

Evidence from cuing

At first thought, attentional cuing paradigms (e.g. Posner and Cohen 1984) may seem the most obvious method for assessing the speed of covert shifts of attention. In simple peripheral cuing, the onset of a peripheral visual cue is used to attract attention to its location, while eye position is maintained centrally. On cued trials, a subsequent visual target appears at the cue location; on uncued trials, the target appears at a different location. Effects of cuing are measured as a function of stimulus onset asynchrony (SOA) between the cue and target. There is ample evidence to suggest that relatively short SOAs of 50 ms are sufficient to produce significant reaction-time advantages for targets on cued compared to uncued trials (e.g. Danziger and Kingstone 1999; Posner and Cohen 1984), suggesting that the cue onset promotes a shift of attention from fixation to the cue location.

   However, the time between cue and target that is necessary to induce a facilitation effect does not correspond to the time required to shift attention. A benefit for cued locations at an SOA of 50 ms clearly does not mean that attention was shifted to the cue 50 ms after cue onset. In fact, 50 ms after cue onset, the visual signal produced by the cue will probably not have had time to reach even subcortical structures like the superior colliculus (Goldberg and Wurtz 1972). Likewise advantages for cued locations at cue-target SOAs of 0 ms (Posner and Cohen 1984) do not imply an immediate attentional shift. Instead, the response to the target is the end-point of a process that lasts several hundred ms, beginning with cue onset and terminating with response execution. Effects from the cue could exert their influence at almost any point over this extended timecourse. By themselves, facilitation to cues at short SOAs would be consistent with either rapid shifts of attention of approximately 50 ms, or slower shifts requiring a much longer time (but obviously less than total RT to the target), for example 200 ms or more.

   To get more detail about the temporal dynamics of covert attention shifts, we can measure cuing effects at multiple SOAs, and encourage observers to make multiple covert shifts. Experiments measuring inhibition of return might be taken to demonstrate this approach: although RTs to cued targets are faster than uncued ones at SOAs of 0–150 ms, the effect frequently reverses at longer SOAs (Posner and Cohen 1984). This advantage for invalid over valid locations at relatively long SOAs has been termed inhibition of return (IOR). Attention-based accounts of IOR assume that attention is initially drawn to the cue location, but is subsequently withdrawn when the target does not appear soon afterwards (Posner *et al.* 1985; there may be motoric influence as well: Kingstone and Pratt 1999). By such accounts we can use SOA manipulations to measure a maximum time required to remove attention from the cued location to another location. This maximum time would be measured by the difference in SOAs necessary to produce a reversal of cuing effects. This difference seems to be of the order of several hundred ms. For example, in Posner and Cohen's study, at a 0 ms SOA (simultaneous presentation of cue and target) an advantage for cued targets was found; at an SOA of 200 ms, there was little or no difference between cued and uncued target; at later SOAs, an advantage for uncued targets was found, which increased to an SOA of 500 ms. One could argue that, at the 'cross-over' point of 200 ms, attention has not yet been fully reallocated away from the cue, and that a 'peak-to-peak' latency difference of 500 ms might be a more appropriate measure of time to reallocate attention. Without making any interpretative commitment at this point, these results suggest that any shift away from the cue required a maximum of between 200 and 500 ms. Maximum estimates computed in this way can be longer, depending on task difficulty (Lupianez *et al.* 1997).

By this logic above, IOR effects can only constrain the maximum time to reallocate attention to an uncontroversial estimate of hundreds of ms. However, it may be worth noting that accounts proposing relatively slow shifts have survived potential falsification. For example, if SOAs for early facilitation and later IOR had been separated by 25–50 ms, that could constitute evidence for fast covert shifts.

Recent electrophysiological studies may offer greater detail on the timecourse of cued shifts of attention, which is otherwise difficult to address with purely behavioural methods. Müller, Teder-Sälejärvi, and Hillyard (1998) tracked the effects of attention over time on the steady-state visual evoked potential (SSVEP) to a visual target. A peripheral visual target flickering at a constant rate, for example at 20 Hz, will induce entrained electrophysiological activity in contralateral recording sites placed over visual cortical areas. This entrained activity is the SSVEP. The amplitude of the SSVEP has been shown to increase when attention is covertly directed to the target (Morgan *et al.* 1996). Müller *et al.* (1998) measured changes in the amplitude of the SSVEP during an endogenous cuing paradigm. They found that the increase in SSVEP amplitude associated with the cued target developed relatively slowly in the 450–650 ms interval following the cue. Further, the development of SSVEP amplitude correlated highly with the timecourse of detection performance measured behaviourally. Although these findings are suggestive of a slow timecourse for cued attention shifts, they are not yet conclusive, as Müller *et al.* (1998) note that this timecourse includes both the time to shift attention and the time needed to process the cue. In addition, there were clear individual differences in shifting times as measured by SSVEP amplitude. Nevertheless, the SSVEP technique offers what seems to be an extremely direct and powerful method for measuring brain activity associated with a target of interest, and tracking that activation through changes in the attentional state. The SSVEP offers great promise for future investigations on the timecourse of the attentional state.

Evidence from visual search

The standard visual search paradigm has been an invaluable tool for investigating many interesting phenomena such as visual similarity and visual categories. For example, asymmetries in search rates have been used as a diagnostic for defining psychologically primitive visual feature maps. However, as has been already discussed, the standard search paradigm is poorly suited to address mechanisms of capacity reallocation. It is difficult or impossible to interpret search rates directly, since rates could reflect either the processing time of single items under the assumptions of serial models, or the time to process multiple items under parallel models.

Careful design may allow estimation of the timecourse of visual processing using variants of standard search tasks. Consider the methods of Egeth and Dagenbach (1991). Their procedure required subjects to make speeded discriminations about whether two objects had the same or different identities. The visual quality of each object (low or high) was manipulated independently, with the assumption that low-quality items would require more time to process than high-quality ones. In other words, if a high-quality item required T ms to process, a low-quality item would require T+ΔT ms to process. If the two objects are processed serially, we would then expect reaction times to increase linearly with the number of low-quality items. On the other hand, if the two objects are processed in parallel, reaction times will be determined by the slowest process to finish. That is, each of two high-quality items should be processed in about the same time, T ms. Each of two low-quality items should also be processed in about the same time, but this time will be greater than that for the high-quality items, T + ΔT ms. When a high and a low-quality stimulus are presented together, the high-quality item will take T ms, but a decision must be delayed until completion of the low-quality item, requiring T+ΔT ms. So, for parallel models, the effects of visual quality are subadditive.

Egeth and Dagenbach (1991) used this procedure on stimulus sets requiring either easy discriminations (for example, distinguishing Xs and Os) or hard (rotated Ts and Ls). With the easy discriminations, the results fit well with a parallel account, and for the difficult discrimination, with a serial account. However, using their assumptions, we can go somewhat further and estimate the actual processing times required for individual items, T, and the extra time-cost for low-quality items, ΔT. We must also allow a constant time, K, to reflect the time needed for response organization and initiation. The time to respond to two high-quality objects in the hard discrimination condition was 586 ms. Under the assumptions of the serial model, this time should be equal to 2T+K. Similar estimates would be made from the results of Kwak, Dagenbach, and Egeth (1991), in which case 2T+K would be about 700 ms. If we are willing to make wild guesses about plausible values for K, we can refine these estimates further. For the processing time per item T to reach 50 ms, we would have to assume K equal to 486 ms. If K is assumed to be about 150 ms, then we would estimate 2T at around 436 ms, or T to be around 218 ms, an estimate consistent with slow reallocation of attention. Although these values of K are strictly speculative at present, it may be possible to develop experiments with additional conditions to allow direct estimation of K values, and thereby direct estimation of the time T to process a single object. Such a method would offer new possibilities for inferring processing time from latency measures.

Recently there have been attempts to move beyond purely behavioural measures by directly measuring brain activity during visual search tasks, using single-cell recordings. As was described earlier, for many behavioural measures, parallel and serial models make similar predictions about the effects of non-targets on responses. Single-cell recordings allow measurements of processing activity related to both targets and non-targets. Chelazzi et al. (1993) recorded from cells in inferotemporal (IT) cortex of alert macaque monkeys. The animals were trained to search for a visual target in the following procedure. First, a central image was presented to the monkeys that indicated the target item for the upcoming search array. After this image was removed, there was a variable delay, followed by the search array, containing the target image and 1 to 4 distractor images. The monkeys were trained to maintain fixation at the screen centre until the target was found within the search array; at that point, a saccade would be made to the target. The stimuli were all pictures of complex objects, like fruits and plants. Behavioural data from this task showed approximately linear increase in reaction time with increasing numbers of distractors, and a slope in target-present trials of approximately 25 ms per item.

During the search task, recordings were made from IT cortex. Individual IT neurons showed different sensitivities to the possible images. So, for example, a cell that responded strongly to an isolated image of a flower (the effective stimulus) might respond weakly to an image of a banana (the ineffective stimulus). The crucial comparison made by Chelazzi *et al.* (1993) was to measure a neuron's firing rate to a display containing two stimuli: the cell's effective and ineffective stimulus, and measure neural firing rate as a function of whether the effective or the ineffective stimulus was specified as the target. From the onset of the search array and continuing for about 200 ms, firing rate increased regardless of whether the effective or ineffective stimulus was the target. In other words, for the first 200 ms, the activity of IT neurons simply reflected whether or not their effective stimuli were in the search array. During this time, the IT neurons were not sensitive to whether the effective stimuli were the target or not. Subsequently, however, activity remained high only when the effective stimulus was the target. That is, when the ineffective stimulus was the target, responses to the effective stimulus were gradually suppressed during an interval between 200 and 300 ms after array onset. The timecourse of this suppression shows no evidence of a rapid timecourse such as might be expected by high-speed models. Although search slopes are in the region of 25–50 ms per item, there is no evidence of such rapid changes in neural activity. Chelazzi (1999) notes that these results are consistent with a parallel and competitive model of capacity allocation, in which neural activity increasingly reflects the properties of the attended stimulus, while responses to non-targets are gradually suppressed.

However, Chelazzi (1999) has argued that serial models might also predict the observed pattern of results. Chelazzi (1999) notes that first, in a serial model, activity of a single IT neuron should reflect the properties of a single item at a time. Second, in a serial model, the response activations reported by Chelazzi *et al.* would be averaged over two scan paths: one in which the target is attended first, and another in which the distractor is first attended, followed by a shift of attention to the target. Under the assumptions of serial processing, consider the trials in which the ineffective stimulus is the target. Neural responses here will be an average of two kinds of trials. In some portion of trials, the ineffective stimulus is attended first. This results in a subsequent saccade to the target, but very little activity from the IT cell. In the rest of trials, the effective stimulus will be attended first. The effective stimulus will generate a burst of activity from the IT cell, but because the attended item is not the target, focused attention will move on to the ineffective stimulus. At this time, the activity of the IT neuron will drop back to baseline levels prior to the saccade to the target. On average then, the neural response on trials in which the target is the ineffective stimulus will show an initial rise in activity followed by subsequent decay.

This reasoning is correct, but it will now be valuable to make some quantitative considerations. Suppose attention moves from item to item at a rate of x ms per item. Again, on trials where the ineffective stimulus is the target, but the effective stimulus is attended first, we should observe an initial burst of activity, reflecting the response of the neuron to the effective item. However, after x ms, activity should return to baseline levels as attention is shifted to the ineffective stimulus. That is, in these circumstances, the speed of switching should be reflected in the speed of activation decay. If attention is moved at a rate of about 200 ms per item, then we would predict results just like those that Chelazzi *et al.* (1993) found: in all cases, firing rates following the onset of the search array would increase for about 200 ms, regardless of whether the effective stimulus was the target. However, if attention shifts between items at a rate of 50 ms per item, then we would have expected this interval of indiscriminate activity to last only for approximately 50 ms. By this argument, the results of Chelazzi *et al.* (1993) are consistent with either parallel processing of the entire display or serial processing at a rate of approximately 200 ms per item. However, they are not consistent with high-speed shifts of attention.

Another approach to measuring brain activity during visual search is through the use of event-related potentials (ERP). Woodman and Luck (1999) measured the N2pc potentials during a visual search task. The N2pc potential has been previously characterized as a negative voltage deflection observed over contralateral occipital sites, associated with finding a target in difficult search tasks (Luck and Hillyard 1994). Like the use of SSVEP (Müller *et al.* 1998), tracking of the N2pc (Woodman and Luck 1999) suggests a potentially powerful method for non-invasively measuring the allocation of attentional capacity across time. In the Woodman and Luck (1999) experiments, subjects searched an array of squares for a left-facing gap while maintaining central fixation. The search array consisted of two critical squares that might contain the target, for example one red and one green, embedded within a larger search array. One of the critical squares was in the left visual field, the other in the right. Woodman and Luck reasoned that, under the assumptions of serial search, as subjects shift their attention from one hemifield to the other in their search for the target, the N2pc should likewise shift from one hemisphere to the other. The speed of covert attention shifts between the critical items would then be reflected in the speed of N2pc shifts. In their experiments, Woodman and Luck (1999) found that shifts in the N2pc required approximately 150 ms. If this time is taken to reflect the speed of serial reallocation of attention, it charitably appears to be more or less on the boundary separating high-speed serial models from parallel models. It is probably faster than many eye movements, though not all, yet much slower than a 50 ms per item rate.

Alternatively, the results of Woodman and Luck (1999) can also be explained without resorting to serial shifts of attention, using a parallel model like that of Ward and McClelland (1989). In this model, based on the diffusion model of Ratcliff (1978), multiple target detectors representing different locations in the visual display, gather evidence in parallel for and against the presence of a target. Target-present responses are made once a single detector crosses its positive criterion; target-absent responses are made once all (or almost all) detectors cross their negative criterion. We might assume that an N2pc component is generated when such a detector crosses either response criterion. In this case, switches in the N2pc would not reflect a single spotlight of attention moving from one item to another, but the independent decision times of multiple detectors examining different parts of the display.

An analogy may help. Imagine a row of potential targets to be examined, each with a bell in front. In the serial search case, a single inspector is responsible for searching all the items. The sequential examinations of the inspector from one item to the next would correspond to the serial allocation of attention. This search could be optimized in a variety of ways: for example, the inspector might move to the most likely target locations first. In any case, once the inspector has examined an item he rings the bell in front of it, so that the movement of the inspector during the search process is reflected in a series of sequential bell rings. The inspector might even ring the bell differently to indicate whether or not the inspected item was the target. In this analogy, the ringing of the bell, like the N2pc, corresponds to a perceptual decision made following examination of an item. The time between rings (or N2pc peaks) would reflect the time required for the inspector to reallocate search from one object to another. However, in the parallel case, we could have multiple inspectors working simultaneously, so that multiple objects are processed in parallel. For simplicity we could assign a separate inspector to each potential target (although to maintain the notion of limited capacity in this system, we could have to assume that the more inspectors we have, the slower they become). Because there will necessarily be some random variation in the rate at which inspectors work, search with multiple inspectors will also produce a pattern of successive bell rings. However, in this case, the sequence of ringing bells (or N2pc peaks) would simply reflect variation in the examination time of the many inspectors, not a single inspector darting from one object to another.

Evidence from studies of attentional dwell-time

Given that visual processing capacity is limited, capacity allocated to one item will be unavailable for another. Another behavioural approach to measuring the speed of capacity reallocation is therefore to measure the timecourse of interference from one attended object on another. When two visual targets are presented at separate times, attention to the first can interfere with processing of the second for relatively long periods, lasting several hundred milliseconds (Broadbent and Broadbent 1987; Chun and Potter 1995; Duncan *et al.* 1994; Moore *et al.* 1996; Raymond *et al.* 1992; Shapiro *et al.* 1994). A typical study presents two targets separated in time by a variable SOA. Attention is engaged on the first target (T1), by requiring its identification or other visual processing. Performance on the second target (T2) is then assessed as a function of the SOA between T1 and T2. Interference on T2 processing may last several hundred ms before performance returns to asymptotic levels. This sustained interference from one attended item on another has been called the 'attentional blink', or AB deficit (Raymond *et al.* 1992), and has been argued to represent the dwell-time of attention once it has been allocated (Duncan *et al.* 1994; Ward *et al.* 1996).

Interference from T1 on T2 is eliminated when T1 can be ignored (Raymond *et al.* 1992), suggesting that the blink is not the result of forward masking from T1, or of other involuntary processing initiated by T1 onset. Interference does not reflect simply the requirement to respond to the T1 target, as the size of the blink is not affected by whether one or two responses are made to a single T1 object (Ward *et al.* 1996). This same finding rules out the possibility that sustained interference is a result of the low short-term memory load needed for report of T1 attributes. Finally, interference can be found using a variety of presentation methods. Early studies used rapid serial visual presentation (RSVP), in which T1 and T2 were embedded within a stream of non-target items, appearing in rapid succession at a single location (e.g. Broadbent and Broadbent 1987; Raymond *et al.* 1992). However, an equivalent timecourse of interference is found if the nontargets are removed and masked T1 and T2 items are presented either at separate locations (Duncan *et al.* 1994), or successively at the same location (Ward *et al.* 1997).

These recent investigations have been motivated and developed in part from earlier studies of Sperling and colleagues (Reeves and Sperling 1986; Sperling *et al.* 1971; Weichselgartner and Sperling 1987) using the RSVP attention shift paradigm to measure the timecourse of attention shifts. In this procedure subjects monitor one RSVP stream for a target that signals them to shift attention (but not the eyes) to a second, concurrent, RSVP stream. Subjects report one or more of the characters from as early as possible in the second stream. Reported items from the second stream tend to occur 300 to 400 ms after presentation of the first target, suggesting a slow reallocation of visual processing from the first target to the second stream (Reeves and Sperling 1986; Weichselgartner and Sperling 1987). Interestingly, items in the second stream appearing simultaneously with the target are also likely to be reported (Weichselgartner and Sperling 1987). A similar sparing is found in the attentional blink, where T2 accuracy is often higher when T2 is temporally adjacent to T1 (within 100 ms of T1 onset), compared to when T2 is further away (200–600 ms from T1 onset: cf. e.g. Raymond *et al.* 1992).

Other findings compatible with long dwell-times of attention measure identification accuracy for multiple items, comparing simultaneous and successive presentation. For example, Pashler and Badgio (1987) measured accuracy for identifying the highest of four digits, as a function of two presentation conditions. In one case, digits were presented simultaneously in a single display lasting 67 ms; in the other, two successive displays containing two digits each presented, each display lasting 67 ms. Pashler and Badgio (1987) reasoned that if attention could be reallocated within 67 ms, there should be an advantage for the successive presentation. In fact performance was equivalent for simultaneous and successive displays at these short exposure durations. When the experiment was run using 500 ms exposures rather than 67 ms exposures, a clear advantage emerged for successive processing. These results are consistent with long dwell-times of 500 ms, not rapid times of 67 ms.

At present, the size and generality of these dwell-time results offer a powerful argument against high-speed serial models. However, a number of arguments might be raised concerning how extended dwell-times relate to the speed of attentional shifts. Moore *et al.* (1996) and Moore and Wolfe (this volume, Chapter 9) raise a number of specific points to be addressed:

1. *Dwell-time experiments tend to use masking, whereas search experiments do not. It may be the case that masking produces catastrophic disruption of otherwise high-speed mechanisms.*

It is true that most search experiments do measure reaction time to unlimited-duration displays. However, search rates can also be estimated by measuring accuracy as a function of variable-exposure duration to briefly presented, masked search arrays (Braun 1994; Eckstein 1998). Estimates of average processing time per item are comparable using either method. For example, a rate of 30 ms per item was estimated for search through a masked array of rotated Ts and Ls, comparable to estimates from unlimited-duration RT experiments (Braun 1994). Search rates for identifying a digit among letters, as calculated from accuracy measures with RSVP techniques, estimate search rates of 10–15 ms per item (Sperling *et al.* 1971). It should be clear that such search rates, whether obtained by latency or accuracy measures, are not in themselves diagnostic of high-speed serial processing. Instead, the point is that masking in itself does not seem severely to disrupt estimates of processing rate.

It is certain that the efficiency of masking would affect the estimated search rate: as masking becomes less effective, apparent search rate will increase. Efficiency of masking has been repeatedly demonstrated to affect estimates of dwell-time. When targets are unmasked, dwell-times can be reduced or eliminated (Giesbrecht and Di Lollo 1998). Such effects may indicate that poor masking allows processing to be moved 'off-line' to avoid the interference that would otherwise result from processing items appearing close together in time (Arnell and Duncan 1997). Such effects would probably be expected regardless of the nature or speed of capacity reallocation.

2. *Dwell-time experiments use successive presentation of stimulus items, while search experiments use simultaneous presentation of items. Simultaneous presentation of items may allow for more efficient planning of attention shifts and scan path, and lower dwell-times.*

It is first important to note that long dwell-times can be generated even when there is no uncertainty about scan path. Extended interference of 300–600 ms is commonly found in RSVP paradigms (Raymond *et al.* 1992), in which all items are presented successively at fixation and no planning of scan path is required. In addition, it is likely that any benefit to search rates arising from the preplanning of scan paths is extremely limited. Pashler and Badgio (1987) tested one variation of this suggestion by successively presenting characters across the breadth of a search array. Single characters would either appear in a predictable left to right order, or in an unpredictable sequence across locations. If preplanning of scan path were a crucial factor in generating rapid attention shifts, we would expect to see better performance in the predictable presentations. However, Pashler and Badgio (1987) found no difference between the two presentation types. More recently, Horowitz and Wolfe (1998) compared visual search performance in two conditions: one in which elements in the search array randomly changed position every 111 ms, and one in which elements remained in a constant position throughout the trial. Preplanning of scan path would be useless or impossible in the random condition, since the position of items is continually changing. However, Horowitz and Wolfe (1998) found that visual search rates were unaffected by this manipulation, suggesting that visual search processes are 'amnesic' and do not register transient changes in the display. The evidence above suggests that the inability to plan a scan path does not catastrophically disrupt search, and probably produces little or no effect on either visual search rates or dwell-time.

3. *Long dwell-times would make it impossible to reject potential targets accurately when presented at the fast rates (about 10 items per second) that are used in RSVP studies.*

There are two issues to be addressed here. We may typically think of parallel visual processing as computing the properties of multiple concurrent objects. However, if visual processes are slow relative to the onset and offset of objects, then processing of one object might not be completed before a second object appears. Is it then possible to process multiple objects across time? That is, are there visual processes that simultaneously compute properties of objects appearing at different times? Studies of visual masking are a clear source of parallel processing across time. Visual masking paradigms routinely demonstrate interactions occurring both forwards and backwards from times of onset (Breitmeyer 1984; Di Lollo 1980), and indicate that items appearing at separate times are being processed in a parallel, interactive manner. Recent accounts of object-masking consider such interactions in detail (Enns and Di Lollo 1997). Studies investigating report of objects from rapid visual streams routinely find conjunction errors in which features of a target item are combined with features of other items appearing nearby in time (Botella and Eriksen 1991, 1992).

But even if we accept concurrent processing in time, there is a second issue. If every object onset attracted limited visual capacity, capacity would quickly drain away to nothing, particularly when the rate of new object onsets exceeds the rate of capacity reallocation, as in RSVP presentations. Parallel models with long dwell-times must assume that this does not happen because visual capacity is allocated selectively and with reasonable efficiency. This appears to be the case, so that when there are strong cues available to distinguish relevant target items from irrelevant non-targets, the non-targets do not produce interference on target processing. In experiments measuring dwell-time, good selectivity is typically demonstrated in control conditions in which the T1 item is ignored and interference on T2 reports is eliminated (Raymond *et al.* 1992). However, when there are only weak cues available to distinguish targets from non-targets, then even non-targets can produce long-lasting interference on subsequent targets (Duncan *et al.* 1994; Ward *et al.* 1997).

A summary of the evidence

Both serial and parallel models must allow for sequential shifts of attention. For example, no one would suggest that this book can be read without such shifts (would they?). However, serial and parallel models can differ in the proposed speed of attention shifts. High-speed serial models, as characterized here, require attention to be moved at a much faster rate than is possible with the eyes or other overt orienting movements. Parallel models are consistent with the possibility of similar timecourses for both overt and covert shifts of attention.

Findings from a variety of experimental paradigms and various behavioural and neurophysiological measures have been surveyed with the aim of identifying clear evidence for fast shifts of attention. Behavioural data from visual search studies have been repeatedly identified as too ambiguous for this purpose. However, evidence from a variety of other methods is consistent with a shift time for attention of 200 ms or more. This time seems in good agreement with measures of saccadic latencies measured under a variety of conditions. In fact, the fastest shifts identified here have been measured in ERP studies of visual search (Woodman and Luck 1999): at about 150 ms, these shift times would live on the fast end of saccadic latencies; but saccadic latencies of this speed have been found before, even excluding 'express saccades'.

In none of the studies or paradigms examined has a shift time approaching 50 ms been identified. In fact, the only time such a small number is found is when computing average per item search rates in visual search tasks. Again, such data can be explained as a result of parallel processes reallocated only at relatively slow rates. In short, no unambiguous evidence for high-speed attention shifts has been found here.

Although this survey has been entirely consistent with parallel models of attention as characterized here, the survey has also been limited only to estimates of how quickly visual attention can be reallocated from one object to another. A variety of other results demonstrate evidence of parallel processing in other ways, for example, redundant target gains (Mordkoff and Yantis 1991), visual search functions unaccountable by serial models but simulated with parallel ones (Ward and McClelland 1989), and context-specific migration errors between concurrently presented words (McCelland and Mozer 1986).

Together these findings constitute a challenge for high-speed serial models. How might such models account for the range of results presented here suggesting relatively slow shifts of attention? Or what if a rate of even 150 ms is accepted as the maximum rate of serial reallocation of attention? If search rates are significantly faster than the rate of attention switching, then the implication is that parallel preattentive processes have reduced the need for serial shifts of attention. However, the large majority of published visual search data report search rates much faster than 150 ms per item, suggesting preattentive segregation of targets and distractors in all kinds of search tasks, including many previously suggested to require focused attention. This is significant because a central theoretical emphasis of many serial models is the distinction between the capabilities of preattentive and attentional processes. The functional distinctions between preattentive and attentional processing would therefore need to re-evaluated under these assumptions.

Some open issues

The variety of results surveyed here, as well as behavioural data from visual search experiments, is consistent with parallel models assuming long dwell-times and relatively slow re-allocation of attention. The general claim is that the timecourse of reallocating attention from one object to another is comparable to the timecourse of eye movements from one object to another. If attention can be serially reallocated only at this very slow rate, this implies substantial parallel processing within each attentional episode. This notion of a similar timecourse for shifts of visual attention and overt orienting raises some interesting open issues.

Independence and coupling of oculomotor and attention systems. The relationship between brain systems for visual attention and for oculomotor behaviour is extremely controversial. Movements of the eyes and shifts in visual attention are tightly coordinated in many everyday behaviours. In the laboratory, visual performance is improved when targets occupy the location of an upcoming eye movement. Furthermore, oculomotor and attention control are supported by overlapping networks of brain regions (Rizzolatti, 1994). The premotor theory of attention makes the strong claim that the neural structures used to program eye movements are the same structures used to program shifts of attention (Rizzolatti *et al.* 1994). Others have argued that although the operations of the oculomotor and attention systems are tightly coordinated, the two systems can be functionally and dissociated (Klein and Pontefract 1994; Posner 1980).

The similar timecourse suggested here for eye movements and shifts of visual attention is compatible with the notion of a tight anatomical and functional coupling of the two systems, but does not demand it. The two systems could have similar timecourses for other reasons. For example, both systems may rely upon some third supervisory component; alternatively, similar temporal dynamics may simply have developed as an efficient way of coordinating the two systems.

Costs and benefits of serial and parallel processing. The argument made here is for a system that processes items in parallel within the current attentional episode, and can only slowly reallocate resources to a new episode. Optimizing the performance of such a system poses some interesting trade-offs. In a parallel distributed processing system, simultaneous processing of a group of objects will inevitably produce cross-talk between representations of the attended items, or uncertainty about the representation of single items within the group (Hinton *et al.* 1986). Cross-talk produced by parallel processing of multiple items would be manifest behaviourally as costs in performance relative to presentations of single items. However, serial reallocation of resources, say from one group of items to another, or from a group of items to a single item, would produce its own cost, in this case in the extra time required for reallocation. Depending on task requirements and the nature of the stimuli, either parallel, (slow) serial, or a mixture of processes will be the most efficient.

Finally, although the survey of studies taken in this chapter suggests a similar timecourse for shifts of attention and the eyes, more satisfying would be the results of experiments directly addressed at the issue. Although studies capable of discriminating parallel and serial models can be difficult to design, the question of whether shifts of visual attention can occur much more quickly than eye movements may be simpler. The characterization given here of high-speed and parallel models suggests possibilities for future research aimed at investigating mechanisms for allocating limited visual capacity.

References

Arnell, K. M., and Duncan, J. (1997). Speeded responses or masking can produce an attentional blink. Talk given at the 38th Annual Meeting of the Psychonomic Society, Philadelphia, PA.

Botella, J., and Eriksen, C. W. (1991). Pattern changes in rapid serial visual presentation tasks without strategic shifts. *Bulletin of the Psychonomic Society*, **29**, 105–8.

Botella, J., and Eriksen, C. W. (1992). Filtering versus parallel processing in RSVP tasks. *Perception and Psychophysics*, **51**, 334–43.

Bouma, H. (1978). Visual search and reading: eye movements and functional visual field: a tutorial review. In J. Requin (ed.), *Attention and performance VII*, pp. 115–47. Hillsdale, NJ: Erlbaum.

Braun, J. (1994). Visual search among items of different salience: removal of visual attention mimics a lesion in extrastriate area V4. *Journal of Neuroscience*, **14**, 554–67.

Breitmeyer, B. G. (1984). *Visual masking: an integrative approach*. Boston, MA: MIT Press.

Broadbent, D. E. (1958). *Perception and communication*. London: Pergamon.

Broadbent, D. E. (1987). Simple models for experimentable situations. In P. Morris (ed.), *Modelling cognition*, pp. 160–85. New York: Wiley.

Broadbent, D. E., and Broadbent, M. H. P. (1987). From detection to identification: response to multiple targets in rapid serial visual presentation. *Perception and Psychophysics*, **42**, 105–13.

Bundesen, C. (1990). A theory of visual attention. *Psychological Review*, **97**, 523–47.

Carpenter, R. H. S. (1981). Oculomotor procrastination. In D. Fisher, R. A. Monty, and J. Senders (eds), *Eye movements: cognition and visual perception*, pp. 237–46. Hillsdale, NJ: Erlbaum.

Cave, K. R., and Wolfe, J. M. (1990). Modelling the role of parallel processing in visual search. *Cognitive Psychology*, **22**, 225–71.

Chelazzi, L. (1999). Serial attention mechanisms in visual search: a critical look at the evidence. *Psychological Research*, **62**, 195–219.

Chelazzi, L., Miller, E. K., Duncan, J., and Desimone, R. (1993). A neural basis for visual search in inferior temporal cortex. *Nature*, **363**, 345–7.

Chun, M. M., and Potter, M. C. (1995). A two-stage model for multiple target detection in rapid serial visual presentation. *Journal of Experimental Psychology: Human Perception and Performance*, **21**, 109–27.

Crick, F. (1984). Function of the thalamic reticular complex: the searchlight hypothesis. *Proceedings of the National Academy of Science USA*, **81**, 4586–90.

Danziger, S., and Kingstone, A. (1999). Unmasking the inhibition of return phenomenon. *Perception and Psychophysics*, **61**, 1024–37.

Di Lollo, V. (1980). Temporal integration in visual memory. *Journal of Experimental Psychology: General*, **109**, 75–97.

Duncan, J. (1980). The locus of interference in the perception of simultaneous stimuli. *Psychological Review*, **87**, 272–300.

Duncan, J., Ward, R., and Shapiro, K. (1994). Direct measurement of attentional dwell time in human vision. *Nature*, **369**, 313–15.

Eckstein, M. P. (1998). The lower visual search efficiency for conjunctions is due to noise and not serial attentional processing. *Psychological Science*, **9**, 111–18.

Egeth, H., and Dagenbach, D. (1991). Parallel versus serial processing in visual search: further evidence from subadditive effects. *Journal of Experimental Psychology: Human Perception and Performance*, **17**, 551–60.

Enns, J. T., and DiLollo, V. (1997). Object substitution: a new form of masking in unattended visual locations. *Psychological Science*, **8**, 135–9.

Fischer, B., and Boch, R. (1983). Saccadic eye movements after extremely short reaction times in the monkey. *Brain Research*, **260**, 21–6.

Giesbrecht, B., and Di Lollo, V. (1998). Beyond the attentional blink: visual masking by object substitution. *Journal of Experimental Psychology: Human Perception and Performance*, **24**, 1454–66.

Goldberg, M. E., and Wurtz, R. H. (1972). Activity of superior colliculus in behaving monkey: I. Visual receptive fields of single neurons. *Journal of Neurophysiology*, **35**, 542–59.

Hinton, G. E., McClelland, J. L., and Rumelhart, D. E. (1986). Distributed representations. In D. E. Rumelhart and J. L. McClelland (eds), *Parallel distributed processing: Vol. 1: Foundations*, pp. 77–109. Cambridge, MA: MIT Press.

Horowitz, T. S., and Wolfe, J. M. (1998). Visual search has no memory [see comments]. *Nature*, **394**, 575–7. (Comment in: *Nature*, **394**, (6 Aug. 1998) (6693): 519).

Humphreys, G. W., and Muller, H. J. (1993). Search via Recursive Rejection (SERR): a connectionist model of visual search. *Cognitive Psychology*, **25**, 43–110.

Just, M. A., and Carpenter, P. A. (1980). A theory of reading: from eye fixations to comprehension. *Psychological Review*, **87**, 329–54.

Kingstone, A., and Klein, R. M. (1993). Visual offsets facilitate saccadic latency: Does predisengagement of visuospatial attention mediate this gap effect? *Journal of Experimental Psychology: Human Perception and Performance*, **19**, 1251–65.

Kingstone, A., and Pratt, J. (1999). Inhibition of return is composed of attentional and oculomotor processes. *Perception and Psychophysics*, **61**, 1046–54.

Klein, R. M., and Pontefract, A. (1994). Does oculmotor readiness mediate cognitive control of visual attention? Revisited! In C. Umilta and M. Moscovitch (eds), *Attention and performance*, Vol. XV, pp. 333–50. Cambridge, MA: MIT Press.

Kwak, H., Dagenbach, D., and Egeth, H. (1991). Further evidence for a time-independent shift of the focus of attention. *Perception and Psychophysics*, **49**, 473–80.

Luck, S. J., and Hillyard, S. A. (1994). Spatial filtering during visual search: evidence from human electrophysiology. *Journal of Experimental Psychology: Human Perception and Performance*, **20**, 1000–14.

Luck, S. J., Girelli, M., McDermott, M. T., and Ford, M. A. (1997). Bridging the gap between monkey neurophysiology and human perception: an ambiguity resolution theory of visual selective attention. *Cognitive Psychology*, **33**, 64–87.

Lupianez, J., Milan, E. G., Tornay, F. J., Madrid, E., and Tudela, P. (1997). Does IOR occur in discrimination tasks? Yes, it does, but later. *Perception and Psychophysics*, **59**, 1241–54.

McClelland, J. L., and Mozer, M. C. (1986). Perceptual interactions in two-word displays: familiarity and similarity effect. *Journal of Experimental Psychology: Human Perception and Performance*, **12**, 18–35.

Moore, C. M., Egeth, H., Berglan, L. R., and Luck, S. J. (1996). Are attentional dwell times inconsistent with serial visual search? *Psychonomic Bulletin and Review*, **3**, 360–5.

Mordkoff, J. T., and Yantis, S. (1991). An interactive race model of divided attention. *Journal of Experimental Psychology: Human Perception and Performance*, **17**, 520–38.

Morgan, S. T., Hansen, J. C., and Hillyard, S. A. (1996). Selective attention to stimulus location modulates the steady-state visual evoked potential. *Proceedings of the National Academy of Science USA*, **93**, 4770–4.

Müller, M. M., Teder-Sälejärvi, W., and Hillyard, S. A. (1998). The time course of cortical facilitation during cued shifts of spatial attention. *Nature: Neuroscience*, **1**, 631–4.

Neisser, U. (1967). *Cognitive psychology*. New York: Appleton-Century-Crofts.

Pashler, H., and Badgio, P. C. (1987). Attentional issues in the identification of alphanumeric characters. In M. Coltheart (ed.), *Attention and performance XII*, pp. 63–81. Hillsdale, NJ: Erlbaum.

Posner, M. I. (1980). Orienting of attention. *Quarterly Journal of Experimental Psychology*, **32**, 3–25.

Posner, M. I., and Cohen, Y. (1984). Components of visual orienting. In H. Bouma and D. G. Bouwhuis (eds), *Attention and performance X*, pp. 531–56. Hillsdale, NJ: Erlbaum.

Posner, M. I., Rafal, R. D., Choate, L. S., and Vaughan, J. (1985). Inhibition of return: neural basis and function. *Cognitive Neuropsychology*, **2**, 211–28.

Ratcliff, R. (1978). A theory of memory retrieval. *Psychological Review*, **85**, 59–108.

Raymond, J. E., Shapiro, K. L., and Arnell, K. M. (1992). Temporary suppression of visual processing in an RSVP task: an attentional blink? *Journal of Experimental Psychology: Human Perception and Performance*, **18**, 849–60.

Reeves, A., and Sperling, G. (1986). Attention gating in short-term visual memory. *Psychological Review*, **93**, 180–206.

Rizzolatti, G., Riggio, L., and Sheliga, B. M. (1994). Space and selective attention. In C. Umilta and M. Moscovitch (eds), *Attention and performance XV*, pp. 232–65. Cambridge, MA: MIT Press.

Schneider, W., and Shiffrin, R. M. (1977). Controlled and automatic human information processing: I. Detection, search, and attention. *Psychological Review*, **84**, 1–66.

Shapiro, K. L., Raymond, J. E., and Arnell, K. M. (1994). Attention to visual pattern information produces the attentional blink in rapid serial visual presentation. *Journal of Experimental Psychology: Human Perception and Performance*, **20**, 357–71.

Sperling, G., Budiansky, J., Spivak, J. G., and Johnson, M. C. (1971). Extremely rapid visual search: the maximum rate of scanning letters for the presence of a numeral. *Science*, **174**, 307–11.

Sternberg, S. (1969). The discovery of processing stages: extensions of Donders' method. In W. G. Koster (ed.), *Attention and performance II*. Amsterdam: North-Holland. (Reprinted from *Acta Psychologica*, **30**, (1969), 276–315.)

Townsend, J. T. (1990). Serial and parallel processing: sometimes they look like Tweedledum and Tweededee but they can (and should) be distinguished. *Psychological Science*, **1**, 46–54.

Townsend, J. T., and H. Ashby, F. G. (1983). *Stochastic modeling of elementary psychological processes.* Cambridge: Cambridge University Press.

Treisman, A., and Gelade, G. (1980). A feature integration theory of attention. *Cognitive Psychology*, **12**, 97–136.

Treisman, A., and Gormican, S. (1988). Feature analysis in early vision: evidence from search asymmetries. *Psychological Review*, **95**, 15–48.

Treisman, A., Sykes, M., and Gelade, G. (1977). Selective attention and stimulus integration. In S. Dornic (ed.), *Attention and performance VI*. Hillsdale, NJ: Erlbaum.

Treisman, A. M. (1969). Strategies and models of selective attention. *Psychological Review*, **76**, 282–99.

Ward, R., and McClelland, J. L. (1989). Conjunctive search for one and two identical targets. *Journal of Experimental Psychology: Human Perception and Performance*, **15**, 664–72.

Ward, R., Duncan, J., and Shapiro, K. (1996). The slow time-course of visual attention. *Cognitive Psychology*, **30**, 79–109.

Ward, R., Duncan, J., and Shapiro, K. (1997). Effects of similarity, difficulty, and nontarget presentation on the time course of visual attention. *Perception and Psychophysics*, **59**, 593–600.

Weichselgartner, E., and Sperling, G. (1987). Dynamics of automatic and controlled visual attention. *Science*, **238**, 778–80.

Wolfe, J. M. (1994). Guided Search 2.0. *Psychonomic Bulletin and Review*, **1**, 202–38.

Wolfe, J. M. (1998). What can 1 million trials tell us about visual search? *Psychological Science*, **9**, 33–9.

Woodman, G. F., and Luck, S. J. (1999). Electrophysiological measurement of rapid shifts of attention during visual search. *Nature*, **400**, 867–9.

11

Perceptual links and attentional blinks

Jane E. Raymond

Abstract

A functional visual system needs mechanisms to integrate, or link, information when it pertains to a single object and other mechanisms to segment information when it describes different objects. Parsing processes like these must operate both in space and in time. The attentional blink (AB) effect, previously studied with mutiple series of different objects, reveals costs for segmenting object-level information in time. If this effect is studied using a series of images of the same object presented in multiple views and successive tasks require detection of its transiently presented features, then AB effects disappear. This suggests that perceptual linking mechanisms allow feature information to bypass the temporal constraints that apply to rapid selection of successive objects.

Our visual environment, coupled with our own active motor behaviour, provides a dynamic visual array that is replete with frequent, abrupt changes in the retinal image. Our perceptual experience, however, is characterized by continuity and stability rather than visual chaos. To achieve this, the brain uses numerous low- and high-level mechanisms to ensure that our conscious visual world does not include the jittery, rapid temporal transients that characterize the retinal image. The processing of visual motion may play an important role in maintaining this coherence in perceptual experience. For example, motion-sensitive mechanisms use retinal image slip to initiate compensatory smooth eye movements when there is self-motion or object motion, so that the image of a fixated object can remain stationary on the retina. Other motion-based mechanisms mediating saccadic suppression (Volkman 1976; Burr *et al.* 1994) operate to 'edit out' visual episodes of extremely rapid retinal image motion (resulting from saccades) or to render conscious experience insensitive to retinal motion blur, producing perceived clarity in rapidly moving objects (Burr *et al.* 1986).

In addition to these lower-level visual motion mechanisms, higher-order perceptual processes contribute to our sense of environmental coherence and stability. Evidence of this is that our awareness seems largely structured from stable object representations that remain constant in spite of our changing viewpoints, the natural plasticity of many objects, and temporal events such as temporary occlusion resulting from object or self-movement. If a person makes their hand into a fist and then lays it open once again, we do not see the hand in its different postures as two different objects but rather as a single, plastic, and animated object. Visual motion processing may mediate 'perceptual links' between representations of the hand in each posture, enabling object constancy and stability in the presence of scene dynamics. Perceptual linking of this type seems able to unify perception of surprisingly different images, as the many uses of 'morphing' motion in cinematic contexts have shown. Central to the idea of object-based perceptual linking is the notion that the visual system parses the scene into sets of objects and then uses these perceptual units to construct awareness. Analogies and names for these object-based units of awareness are abundant in the cognitive literature and include 'object files' (Kahneman and Treisman 1983), 'FINSTs' (Pylyshyn 1989), and 'object tokens' (Kanwisher and Driver 1992). These models assume that perceptual linking mechanisms maintain continuity in object representations in the face of retinal image change.

While linking related visual information into unified object representations is necessary in many situations, there are other circumstances where perceptual linking of similar images would be disadvantageous. Thus mechanisms to limit perceptual linking are needed so that we can distinguish similar but different objects from one another and recognize *bona fide* new objects when they appear. If you are talking to someone and another person passes between you and your interlocutor, you do not perceptually link the two faces into a single, albeit changing, face. Although the two faces may have much in common visually, two distinct people are recognized. This requires a perceptual mechanism that is capable of determining when too much change in an image has occurred. When sufficient change is detected, a 'new' object representation must be generated. An important issue is how much change is necessary for the generation of a new object representation and what the costs associated with it are.

In this chapter I report two experiments that explore this issue. The purpose was to examine the presence of temporal bottlenecks associated with the construction of 'new' object representations as opposed to updating the features of an 'old' but changing object. If awareness trades in object units, then there should be an 'on-line' mechanism for rapidly updating an 'old' object's representation, causing little or no temporal bottleneck in perception; whereas processing new objects should reveal a measurable bottleneck. If however, awareness primarily deals with novelty in the retinal image, then the perceptual history of the object to which the novel information is attached should largely be irrelevant. Before describing the experiments, a brief discussion of mechanisms of scene parsing in space and time may be useful to set the context for these studies.

Scene parsing

Our sensory systems evolved to enable the identification and location of physical objects so that we can react to and interact with them. Since we are only able to interact with a small number of objects at a time, it can be argued that the primary goal of the visual system should be to render an accurate representation of only currently *actionable* objects, i.e. those relevant to the task at hand. A first step in meeting this goal would be to parse the visual array into different objects. Scene parsing must occur across space (as in static images) as well as time. Thinking about scene parsing as occurring within spatio-temporal dimensions provides a framework for understanding how the perceptual system might represent real objects in a sensory world characterized by changing visual inputs.

There is a large literature on the mechanisms of static scene parsing, most of it suggesting that a number of automatic, 'data-driven' sensory mechanisms work to define image discontinuities (in luminance, colour, texture, motion depth, etc.) that are needed to define discrete surfaces (e.g. Marr 1982). Similarly, in the temporal domain, obligatory mechanisms responsive to luminance change or to abrupt onsets and offsets (Alais *et al.* 1998) may mediate data-driven parsing into 'events' (which may be considered as the temporal analogues of surfaces or objects).

Such surface-defining and event-defining processes, however, cannot proceed unchecked in the natural world if actionable object information is to be gained. Images involving textures (e.g. wallpaper, foliage, etc.) or scenes with temporary occlusions (e.g. watching a butterfly through a picket fence) would yield too many discrete objects or events to guide action sensibly. This problem is illustrated in Fig. 11.1, wherein only two objects are depicted, but, because they are both textured, many separate small objects or surfaces are presented to the visual

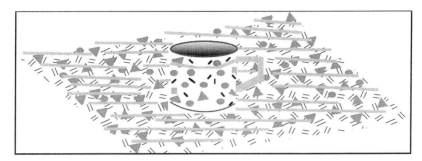

Fig. 11.1 A complex scene in which parsing on the basis of simple image discontinuities would yield too many 'objects' for sensible analysis. Integration of information is needed to distil the image into two simple objects.

system. To deal with this problem, the visual system must deploy image processing mechanisms to 'smooth', or integrate, across behaviourally irrelevant image discontinuities, for example across shadows or within textured regions (Dakin and Watt 1997; Glass 1969) so that coherent object surface areas can be seen.

In the temporal domain, different successive images are linked, resulting in motion perception. Different edges with different motion vectors can be 'smoothed' to create the global perception of a unified moving object. A natural example would be a horse galloping across a field, with tail and feet moving in different relative directions. A clear global perception of the horse's movement direction is experienced in spite of local variations of movement direction in the object. Examples used in the laboratory are partially coherent random dot kinematograms (Williams and Sekuler 1984), and other motion stimuli resulting in motion capture phenomena (e.g. Ramachandran 1987).

Smoothing across local variations in texture, depth, motion, and colour acts to reduce the information available in the visual array. However, these operations acting in isolation would still yield too large a number of object representations to guide action. An additional task-sensitive selection mechanism is needed to detect behaviourally relevant objects, *targets*, in the presence of a background of behaviourally irrelevant objects, *distractors*. This mechanism has been called *selective attention*, and can be described as a set of neural mechanisms acting on perception that facilitate processing of task-relevant objects and inhibit processing of task-irrelevant objects. Representations of attended objects are more likely to be made available to awareness, to direct responding, and to be used later (i.e., contribute to memory and learning) than non-attended distractors.

An important aspect of the boosted processing of attended objects may involve creating or enhancing coherency in the object's mental representation. Some theories of attention go so far as to propose that selective attention is required to bind the representations of discrete object features into a coherent object representation (Luck and Beach 1998; Treisman and Gelade 1980). If attention plays a role in promoting coherency of object representations, then attention might be expected to enhance perceptual linking of visual information, for example via motion processing, during transient changes to an object's appearance. Indeed, psychophysical and neurophysiological evidence of attentional modulation of motion processing been reported (see Raymond 2000 for a review). These ideas suggest that as long as an object is attended, its representation may be rapidly modified to incorporate small image changes, such as the addition or deletion of a feature or a shift in orientation or size, without the need to reinitiate object recognition routines.

Successive attention to objects

Most current concepts of attention allow that only a very small number of objects can be attended at any given time, and this is consistent with the idea that human action systems can only interact with a small number of objects at a time. This suggests that, if many objects must be considered in a task, attention must be allocated to them successively. This immediately raises the question of how fast one object can be selected and processed before the next one can be selected. Research on selective attention indicates that successive selection of different objects from complex scenes can be quite slow, even when the information needed from an object is very simple. This can be demonstrated using rapid serial visual presentation (RSVP) stimuli and asking observers to select two objects in succession from the stimulus series. Although a single object can be easily selected when the interval between the onsets of the successive stimuli is only about 100 ms, the task of selecting a second object becomes very difficult if the interval between the two target objects is less than about half a second. This difficulty in selecting the second target has been called the attentional blink (AB), because it depends on attention to the first target (Raymond *et al.* 1992). It is generally thought to reflect a bottleneck in the processes needed to bring a representation of a target to a point where it can support accurate report (Shapiro *et al.* 1997). Occupation with processing the first target leaves the representation of the second target in a less than stable state, making accurate reporting of it deficient. In all the studies leading up to this general conclusion, the first and second targets were *different* objects.

Does the AB reflect the perceptual processing system's difficulty in dealing successively with different *objects* or with different *stimuli*? If attention operates to speed perceptual processing of target *objects*, then AB effects should not be observed when successive targets involve the same object but different images. If, however, attention is needed to process novelty *per se* in an image, then AB effects should be found with different images even when the same object is portrayed. I did two experiments to test these possibilities. Using conventional AB methods, I presented the same object in RSVP, with each frame displaying a different viewpoint. This created an apparent rotation of the object around its midpoint, and allowed visual motion mechanisms to link successive views into a unified and continuous object percept. In this RSVP task, observers were asked to report on the abrupt acquisition of two different features by the twisting object. I found that if only a single object was used throughout the RSVP stream, AB effects were not found. If however the target features appeared suddenly as part of a 'new' object, then AB effects were found. These data suggest that updating of information relevant to an attended object is rapid and does not impose the temporal bottlenecks produced when the same information is relevant to a new object.

Attentional blink experiments with 'old' and 'new' objects

Unlike previous AB studies in which different objects are presented serially, I presented (in a standard RSVP stream of 11 items/sec) the same object (a trident, see Fig. 11.2) appearing in different randomly chosen orientations. Choppy apparent rotation of the object in 3D space was

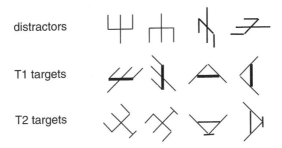

distractors

T1 targets

T2 targets

Fig. 11.2   An illustration of the stimuli used. Filler stimuli could appear in one of 18 different orientations. T1 and T2 stimuli were those shown here.

produced. The first task was to detect a long, thick bar (T1) and the second task was to detect a short thin bar (T2) that appeared on the rotating object. However, on half the trials, the object bearing the defining T1 feature was a different object (an arrowhead-like object). In the first experiment, the object bearing the T2 defining feature was always an 'old' object (trident). In the second experiment, the object bearing the T2 defining feature was the 'new' object (the arrowhead). The results suggest that the AB occurs as a result of processes operating on object-level representations, and that changes in an object's features do not exact the same cognitive 'cost' as a change of object.

Methods. Ten young adults participated in Experiment 1 and eight in Experiment 2. Participants viewed the display on a monochrome computer monitor from a distance of 40 cm with their head position stabilized using a chin rest.

Five different types of black letter-like stimuli, shown in Fig. 11.2, were presented in RSVP. The size of each symbol was approximately 1.0 deg square. All the distractor items were simple tridents (see Fig. 11.2A) presented in one of 18 different orientations. T1 items were either 'tridents' or 'arrowheads' with a thickened central line (Fig. 11.2B and C). When T1 was a trident, it had one of two orientations not used for the filler stimuli. T2 items were either 'tridents' (Experiment 1) or 'arrowheads' (Experiment 2), with a 'foot' (a short thin line) attached at right angles to the axis of symmetry of the figure at the figure's base (Fig. 11.2D and E). The orientation of T2 items was never identical to that of any T1 item or any filler.

Each shape was presented singly for 75 ms at the same location in the centre of a uniform grey field. An interstimulus interval of 15 ms was used, producing a RSVP presentation rate of 11.11 items/second. Although this presentation rate when used with letter stimuli (Raymond *et al.* 1992) produces the perception of discrete changing letters, using the trident stimuli it produced the perception of a single trident twisting somewhat jerkily into different orientations in three-dimensional space whilst maintaining the same centre anchor position.

Procedure. Each subject participated in one session of 240 RSVP trials. Trials were self-initiated and began with a 180 ms presentation of a small, white fixation dot. This was followed immediately by a rapid serial presentation of between 16 and 24 items. Trials could contain both T1 and T2, just one target (either T1 or T2), or neither target. Distractor items were always tridents.

The participant had two tasks. The first was to identify whether T1 (defined by a thickened central line) was present and, if so, to name it as 'trident' or 'arrowhead'. The target was absent on one-third of trials, was a trident in another third of trials, and an arrowhead on the remaining third of trials. Trial type was randomized within the session. The number of items appearing prior to T1 was randomly chosen on each trial to be between 7 and 15. When no T1 item appeared, a distractor trident not previously shown in the trial was used instead. Eight items always followed T1.

The second task was to report whether T2 (defined by a short line appearing at the base of one of the items) was present or absent. A T2 was present on half the trials. It could be presented at post-T1 serial positions 1, 3, 5 or 7. T2 was presented 10 times in each position, yielding 120 T2-present and 120 T2-absent trials, divided equally among the three T1 conditions. The order of T1–T2 trial type was randomized within the session. In Experiment 1 T2 was always a trident and in Experiment 2 the T2 item was always an arrowhead.

Results: Performance on the T1 task. Detection and identification of T1 was generally very high, with participants in Experiments 1 and 2 being correct on 93.8 per cent and 90.9 per cent of trials, respectively. The group mean percentages of trials on which participants were correct on the first task (detecting a thick central bar and naming the object in which it was imbedded) are shown in Table 11.1 for both experiments and for each T1 condition. The most interesting aspect of these data is the decrement in performance when T1 was a trident compared to an arrowhead. When T1 was a trident and therefore similar to all the distractor items, including the immediately trailing item (+1 item) or 'mask', performance was poorer by 9 per cent than when T1 was an arrowhead. A mixed design ANOVA with Experiment and T1 type as factors showed a non-significant difference between experiments and a significant effect of target type, $F(2,32) = 15.2$, $p < 0.001$. If the size of an AB effect is determined by T1 task difficulty (Chun and Potter 1995; but see also Shapiro *et al.* 1994; McLaughlin *et al.* in press) related perhaps to effectiveness of post-target masks, then these data predict larger AB effects in the trident T1 conditions. As we will see, this was not the case.

Table 11.1 Group mean percentage correct for each T1 condition

T1	Trident	Arrowhead	None
Experiment 1 95.2%	88.5%	98.1%	
Experiment 2 95.3%	84.5%	93.1%	

Results: Performance on the T2 task. Performance on the T2 task was generally high in both experiments. Group mean percentage correct was 82.2 per cent in Experiment 1 and 91.6 per cent in Experiment 2. However, these data were considerably variable across conditions, and ranged from 10 per cent to 100 per cent. This variability was partly due to differences among participants in false alarm rates. Although generally low, group mean false alarm rates were higher in Experiment 1 (8.2 per cent) than in Experiment 2 (2.6 per cent). To deal with these characteristics of the data, an a' signal detection measure (a non-parametric measure of the area that falls under the ROC curve) was used to reflect sensitivity to the T2 targets (Creelman 1991; Donaldson 1992). This measure was used instead of d' because of its greater sensitivity when data contain extreme values.

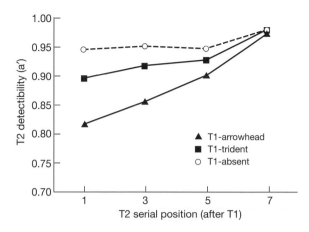

Fig. 11.3 The group mean a' score for T2 detection as a function of the post-T1 serial position of T2 for Experiment 1 (T2 = trident). Serial positions 1, 3, 5, and 7 correspond to target onset asynchronies of 90 ms, 270 ms, 450 ms and 630 ms, respectively. Open circles represent data from conditions with no T1 target. Data from conditions with a T1-trident (squares) and T1-arrowhead (triangles) are represented by filled symbols.

Experiment 1 (T2-trident). Fig. 11.3 plots the group mean a' values for T2 detection as a function of post-T1 serial position of T2 for each of the three T1 conditions. A repeated measures ANOVA (with T1 and serial position as factors) on these data showed a significant interaction effect of T1 type and serial position, $F(6,54) = 3.12$, $p = 0.01$. As can be seen in the figure, except for serial position 7, there was no serial position effect in the T1-absent or T1-trident conditions, whereas there is an obvious serial position effect over the range of values tested for the T1-arrowhead condition. Performance was highest for T2s presented at serial position 7 for all T1 conditions, suggesting that detection of targets presented at this time was enhanced by factors unaffected by the attentional demands of T1.

The presence of an AB effect was tested by comparing means for the T1-present and T1-absent conditions. For the T1-arrowhead condition, significant ($p < 0.01$) 'blink' effects (i.e., decrements for T1 present) for serial positions 1 and 3 were found. However, no such evidence for AB effects for the T1-trident condition for any serial positions was obtained.

Experiment 2 (T2-arrowhead). Fig 11.4 plots the group mean a' values for T2 detection as a function of post-T1 serial position of T2 for each of the three T1 conditions for Experiment 2. A repeated measures ANOVA (with T1 and serial position as factors) on these data showed a significant interaction effect of T1 type and serial position, $F(6,36) = 4.87$, $p < 0.01$. As in Experiment 1, performance was best for all three T1 conditions when T2 was presented at serial position 7. Although overall performance was somewhat higher than in Experiment 1, a similar clear lack of serial position effect was seen in the T1-absent and the T1-trident condition. The serial position effect for the T1-arrowhead condition is small but evident.

As before, the presence of AB effects was tested by comparing means in the T1-present and T1-absent conditions. For the T1-arrowhead condition, a significant ($p < 0.01$) 'blink' effect for serial position 1 was found, providing evidence of a small, short-lived AB. For the T1-trident condition, no such evidence for AB effects for any serial position was found.

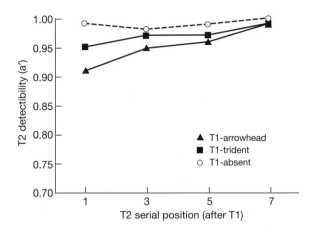

Fig. 11.4 The group mean *a′* score for T2 detection as a function of the post-T1 serial position of T2 for Experiment 2 (T2 = arrowhead). Serial positions 1, 3, 5, and 7 correspond to target onset asynchronies of 90 ms, 270 ms, 450 ms and 630 ms, respectively. Open circles represent data from conditions with no T1 target. Data from conditions with a T1-trident (squares) and T1-arrowhead (triangles) are represented by filled symbols.

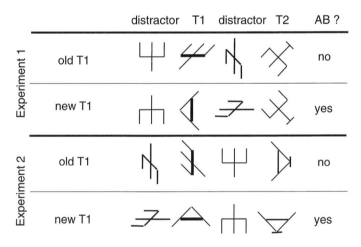

Fig. 11.5 A summary of experimental conditions and results.

Discussion of experimental results

A summary of the stimuli used in each experiment and the result obtained can be seen in Fig. 11.5. In both experiments the distractor items were 'twisting' tridents and the T1 target was a thick bar imbedded in either a trident or an arrowhead. The Experiments differed in the T2 task (hence T2 performance data from each experiment were not directly compared). In Experiment 1, the T2 target was always a trident with a small foot. On half the trials in this experiment, participants observed only twisting tridents sprouting and then losing features. No AB effect was found for this condition. In the rest of the trials, a 'new' object appeared briefly.

When it did, it was always a T1 target, and this provoked an AB effect. In Experiment 2 there was at least one appearance of a 'new' item on half the trials, because the T2 target when present was always an arrowhead. When its only appearance was as the T2 item, no AB effect was observed. However, when the arrowhead stimulus was seen first as a T1 item and then as a T2 item, a modest, short-lived but significant AB effect was found. The results can be summarized by noting that AB effects were only observed for those conditions in which T1 was an 'new' object (i.e., different from that depicted by the preceding distractors).

In Experiment 1, the presence of an AB effect when T1 and T2 were different objects (arrowhead and trident, respectively), coupled with the absence of an AB effect when they were the same object, suggests at first glance that AB occurs whenever different objects are successively judged. However, such a conclusion is unwarranted because a similar condition in Experiment 2 (T1 as trident, T2 as arrowhead) provides no evidence of an AB. In fact, in this experiment, when the same object is used for both T1 and T2, a blink is found. The critical difference between these two experiments is that in Experiment 1 the target in the 'same object' condition is the same as all the distractor items in the RSVP series, whereas in Experiment 2, the target in the 'same object' condition was 'new' compared with the distractor objects. Thus it appears that, when T1 is a 'new' object, an AB results.

These data support the idea that the AB reflects limitations of processes operating on object level-representations, as opposed to part- or feature-based representations. Clearly, since an AB effect was not observed in all the conditions studied here, it is not an automatic consequence of directed attention in a dual target RSVP task. This means that simple accounts of the AB effect involving object-substitution notions (Seiffert and DiLollo 1997) or perceptual bottlenecks (Chun and Potter 1995) are inadequate unless they are elaborated to take object-level representations into account in their formulations. It is not difficult to envision how this could be done, however, and these results should not be taken as strong evidence for or against the different theories of the AB. They do however suggest that the AB operates on object-level representations, not on part- or feature-based representations.

Before putting forward a possible theoretical account for these data, the contribution of differential masking must be considered. Since AB effects are known to be absent without a mask presented immediately after T1 (Raymond *et al.* 1992), it is possible that the trident masks used were less effective for trident T1s than for arrowhead T1s. For all T2 serial positions except the first post-T1 position, the masking item was always a trident, and it might be expected that such a stimulus would fail to mask another trident effectively, thus eliminating an AB effect. However, the T1 error data do not support this. T1 errors were significantly higher, not lower, in the T1-trident conditions than the T1-arrowhead conditions in both experiments, suggesting that tridents may have been slightly more effectively masked than were the arrowheads.

Object files and the attentional blink

The pattern of data seen in the two experiments described here supports the general idea that the AB reflects the activity of mental operations that deal in the currency of objects rather than visual features or even object parts. This possibility is consistent with numerous earlier studies that have indicated that visual selection may operate in an object-oriented manner (e.g. Duncan, 1984; Egly *et al.* 1994). The general finding of these studies is that once an object has been selected for attention, the perceptual benefits of that selection are observed for judgements involving all of the object's parts and features. The data presented here extend these ideas to suggest that the perceptual benefits of selective attention extend in time to allow 'real-time' modifications of represented objects to be processed with minimal cost.

Various analogies have been developed to conceptualize how the brain might operate using an object-oriented system of representations (Kahneman and Treisman 1983; Pylyshyn 1989; Kanwisher and Driver 1992). One of these, the 'object file' (Kahneman and Treisman 1984) may be particularly useful for the present purpose. In this analogy, when a new object is attended, a 'file' is opened, and 'tags' representing all the various sensory features (e.g., colour, depth plane, etc.) of the image occupying the attended spatio-temporal location are included in the file. Inclusion of feature 'tags' in this set may be guided by stimulus borders, perceptual organization rules, and top-down expectations. Since awareness of an object is thought to require accessing the object file, it is assumed that, once available to awareness, the entire content of the file is readily accessible. The 'object file' is thus a metaphor for feature binding and for object-based perceptual awareness. The present data provide insight into the time required for the different operations an object file system might use, and thus may inform theories of feature binding and awareness.

Suppose that in all the experimental conditions tested here an object file for a trident is opened when the first trident in the RSVP series is presented. Activation of motion-sensitive mechanisms by pre-T1 items (all tridents) acts to maintain the stability of the object file in spite of significant changes to the retinal image in successive items. The absence of an AB effect when T1 and T2 are also tridents suggests that the process of updating or adding information to an active object file either takes very little time or can be done for different features in parallel. When T1 is a trident but T2 is not, a new object file for T2 must be created. Lack of an AB effect in this condition suggests that this can be initiated readily after updating an already opened file. However, when T1 is *not* a trident and thus requires a new object file to be created when encountered, AB effects are always seen regardless of what T2 is. This suggests that the creation of a new object file may be the basis for the AB effect. An important point to note is that in Experiment 2, when T1 and T2 were the same (i.e., both arrowheads), the AB effect was smaller than when T1 was also new but different from T2 (arrowhead T1 in Experiment 1). This finding suggests that switching between active files may incur some cost. It further suggests that updating features 'on-line' in an active object file cannot be conducted immediately after the opening of that file. The data indicate a 'refractory' period of about 300 ms after the object is first seen.

Since we obviously do not have object files in our brain, what neural mechanisms might operate to enable awareness of coherent object representations? One possibility is coordination of diverse neural mechanisms processing a visual object's different features (Duncan 1996; Luck and Beach 1998). It has been suggested that frontal lobe activation using re-entrant pathways to the visual system may modulate activity in an object-specific manner, thus implementing such coordination. The current data suggest that this coordination requires significant time (of the order of 500 ms) to set up, but once established remains flexible, updating awareness in an 'on-line' manner. However, when current visual information is task-relevant and activates significantly different visual mechanisms than those currently active in the coordinated network, the network is 'dissolved' and reorganized as rapidly as possible. Here we postulate that this reorganization would be required for an 'arrowhead' T1 preceded by a series of tridents. Such reorganizing of the active networks might not be necessary when T1 is also a trident. Why? Perhaps this is where motion-processing plays an important role. When some significant amount of motion information can be processed, linking at least some elements of the successive images together, the motion signal can be used to prevent collapse of the coordinating network, thus maintaining the object file in spite of substantial image change. Motion in this way links the processing of similar images across time, so that rapid on-line updates of changes in an object's features or parts can be readily available to awareness. Such a mechanism makes efficient use of the brain's current pattern of activity. The linking functions of motion are limited, however, and when the correspondence in successive images exceeds some threshold, the links are terminated and the activity pattern disassembles. This allows a new interpretation of the current visual scene and permits the recognition of a new object. The attentional blink appears to reflect the breaking of perceptual links, and thus mirrors the temporal parsing of the visual scene.

References

Alais, D., Blake, R. and Lee, S-H. (1998). Visual features that vary together over time group together over space. *Nature Neuroscience*, **1**, 160–4.

Burr, D. C., Ross, J. and Morrone, M. C. (1986). Seeing objects in motion. *Proceedings of the Royal Society of London*, **B227**, 249–65.

Burr, D. C., Morrone, M. C. and Ross, J. (1994). Selective suppression of the magnocellular visual pathway during saccadic eye movements. *Nature*, **371**, 511–13.

Chun, M. M., and Potter, M. C. (1995). A two-stage model for multiple target detection in rapid serial visual presentation. *Journal of Experimental Psychology: Human Perception and Performance*, **21**, 109–27.

Creelman, C. D. (1991). *Detection theory: a user's guide*. Cambridge University Press: Cambridge, England.

Dakin, S. C. and Watt, R. J. (1997). The computation of orientation statistics from visual texture. *Vision Research*, **37**, 2227–59.

Donaldson, W. (1992). Measuring recognition memory. *Journal of Experimental Psychology: General*, **121** (3), 275–7.

Duncan, J. (1984). Selective attention and the organisation of visual information. *Journal of Experimental Psychology: General*, **113**, 501–17.

Duncan, J. (1996). Co-ordinated brain systems in selective perception and action. In *Attention and performance XVI*, ed. T. Inui and J. L. McClelland, pp. 549–78. Cambridge, MA: MIT Press.

Egly, R. Driver, J. and Rafal, R. (1994). Shifting visual attention between objects and locations: evidence from normal and parietal lesion subjects. *Journal of Experimental Psychology: General*, **123**, 161–77.

Glass, L. (1969). Moiré effect from random dots. *Nature*, **223**, 578–80.

Kahneman, D. and Treisman, A. (1983). Changing views of attention and automaticity. In R. Parasuraman, R. Davies, and J. Beatty (eds), *Varieties of attention*, New York: Academic Press.

Kanwisher N. G. and Driver, J. (1992). Objects, attributes, and visual attention: which, what, and where. *Current Direction in Psychological Science*, **29**, 303–37.

Luck, S. J. and Beach, N. J. (1998). Visual attention and the binding problem: a neurophysiological perspective. In R. D. Wright (ed.), *Visual attention*. New York: Oxford University Press.

McLaughlin, E., Shore, D. I. and Klein, R. M. (2001) The attentional blink is immune to masking-induced data limits. *Quarterly Journal of Experimental Psychology*, **57A**, 169–96.

Marr, D. (1982). *Vision: a computational investigation into the human representation and processing of visual information*. San Francisco: W. H. Freeman.

Pylyshyn, Z. (1989). The role of location indexes in spatial perception: a sketch of the FINST spatial-index model. *Cognition*, **32**, 65–97.

Ramachandran, V. S. (1987). Interaction between colour and motion in human vision. *Nature*, **328**, 645–7.

Raymond, J. E. (2000). Attentional modulation of visual motion perception. *Trends in Cognitive Sciences*, **4**, 42–50.

Raymond, J. E., Shapiro, K. L. and Arnell, K. A. (1992). Temporary suppression of visual processing in an RSVP task: an attentional blink? *Journal of Experimental Psychology: Human Performance and Perception*, **18**, 849–60.

Seiffert, A. E. and DiLollo, V. (1997). Low-level masking in the attentional blink. *Journal of Experimental Psychology: Human Performance and Perception*, **23**, 1061–73.

Shapiro, K. L., Raymond, J. E. and Arnell, K. M. (1994). Attention to visual pattern information produces the Attentional Blink in RSVP. *Journal of Experimental Psychology: Human Performance and Perception*, **20**, 357–71.

Shapiro, K. L., Arnell, K. and Raymond, J. E. (1997). The attentional blink: a view on attention and a glimpse on consciousness. *Trends in Cognitive Science*, **1**, 291–5.

Treisman, A. and Gelade, G. (1980) A feature integration theory of attention. *Cognitive Psychology*, **12**, 97–136.

Volkman, F. (1976). Saccadic suppression: a brief review. In R. A. Monty and J. W. Senders (eds), *Eye Movements and psychologicial processes*, pp. 73–84. Hillsdale, NJ: Erlbaum.

Williams, D. and Sekuler, R. (1984). Coherent global motion percepts from stochastic local motions. *Vision Research*, **24**, 55–62.

12

A spatiotemporal framework for disorders of visual attention

Masud Husain

Abstract

Disorders of visual attention are common consequences of damage to certain brain regions, particularly the parietal lobes. Visual extinction is the failure to report a contralesional stimulus in the presence of a competing ipsilesional stimulus, whereas visual neglect refers to unawareness of contralesional stimuli regardless of the presence of a competing ipsilesional stimulus. Until recently, both these deficits have been considered to represent impairments in the ability to disengage and/or direct *spatial* attention. New evidence demonstrates an impairment of the *temporal* dynamics of attention in these conditions. A similar deficit may also exist in dorsal simultanagnosia, a disorder that occurs after bilateral parietal lesions, which is characterized by the ability to see only one object at a time, even when overlapping figures occupy the same spatial location. Here, I discuss the evidence for both spatial and temporal deficits, and consider extinction, neglect and dorsal simultanagnosia within the context of a unifying spatiotemporal framework for understanding disorders of visual attention.

Introduction

This chapter is concerned with three disorders of visual perception—extinction, neglect and dorsal simultanagnosia—that have traditionally been considered to be critically dependent upon damage to the parietal lobes (De Renzi 1982). The purpose of this discussion is not to present an exhaustive review of these conditions. Rather, my aim here is to formulate a useful framework in which to embed our current understanding of these complex disorders of high-level visual processing. I begin with a few definitions and descriptions.

Patients with *extinction* acknowledge the presence of single visual stimulus when it is presented in either left or right visual hemifield. However, when two stimuli are simultaneously presented bilaterally—one in each hemifield—they report seeing only the ipsilesional one, i.e. the one presented on the same side as their lesion. In other words, they fail to acknowledge the presence of the contralesional stimulus *when there is a competing stimulus* in the ipsilesional hemifield.

Neglect is a failure to acknowledge a stimulus presented in space opposite the side of the brain lesion, regardless of the presence or absence of a competing stimulus in ipsilesional space. The same definition could apply to hemianopia, which is generally considered to be a primary sensory field defect due to lesions in the visual pathway leading from the lateral geniculate bodies to the primary visual area in striate cortex. If neglect is very dense it can be difficult to distinguish from hemianopia. Indeed, some patients with large lesions suffer from both a hemianopia and neglect.

However, the clinician is often alerted to the presence of neglect by the patient's persistent turning of eyes and head towards the ipsilesional side; by finding that unawareness of contralesional stimuli can vary and is not absolute; by observing that the apparent visual field loss does not obey the vertical meridian; and by the patient's failure to orient fully into contrale-sional space on simple pen-and-paper tasks such as line bisection and cancellation (Fig. 12.1). A patient with hemianopia (without neglect) may be slow in performing these tasks, but will usually explore contralesional space. Finally, an important clinical clue comes from the patient's history. Most patients with neglect are not aware they have a problem, whereas those with a hemianopia (without neglect) complain bitterly that they have difficulty seeing on one side of space.

That the focus of most investigations of extinction and neglect has been directed at the spatial deficit is not surprising, since both these disorders are defined by a spatially defined failure to report ipsilesional stimuli. *Simultanagnosia* is, at least overtly, rather different. In general, this term refers to a disorder of vision in which the individual parts of a display are better perceived that than the entire scene. The classical demonstration is that of getting patients to describe what they see in a complex picture. They may describe some of the details meticulously, but still not appreciate what is happening overall in the picture.

Farah (1990) has provided a critical review of this subject, and proposed that there may well be two types of simultanagnosia associated respectively with lesions of the dorsal (parietal) and ventral (temporo-occipital) visual systems. In this discussion, I am concerned with the disorder she has termed *dorsal simultanagnosia*, which is characterized by an inability to perceive more than one object at a time, regardless of its spatial extent. In this regard, the condition appears to be strikingly different from either extinction or neglect, where the clinical presentation is dominated by what appears to be a spatial deficit.

All three disorders—extinction, neglect and simultanagnosia—are traditionally considered to be associated with parietal lobe damage (De Renzi 1982). However, it has become apparent that

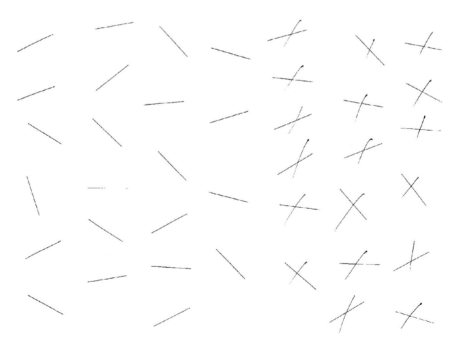

Fig. 12.1 Cancellation of targets by a patient with left-sided visual neglect. Targets on the left of the page have been ignored.

they may also follow damage to other regions of the brain. For example, visual neglect may be associated with lesions of the frontal lobe, and even subcortical structures such as the basal ganglia and thalamus (Heilman *et al.* 1994; Mesulam 1981). However, neglect following sub-cortical lesions may be associated with underactivity of overlying cortex (Perani *et al.* 1987). Thus, the critical lesion site may nevertheless be cortical. Finally, for neglect, there appears to be an important hemispheric asymmetry: it is more common and severe following right-hemsiphere lesions than left (De Renzi 1982).

Representation of space in neglect and extinction

Given the spatial nature of the deficits in neglect and extinction, it is natural to find that many explanations for these conditions have been framed in terms of spatial representations. For neglect, the argument reads like this: there is a disruption of the representation of contralesional space that leads to unawareness of stimuli located there and a failure to explore that region. Thus the primary disorder is one of representing space, and all the rest follows as a consequence.

Some consider the representation that is disrupted to be that of contralesional *egocentric* (body-centred) hemispace, with rotation of the perceived mid-sagittal plane away from the neglected side (Ventre *et al.* 1984; Karnath 1997). Others have considered the possibility that there is an ipsilesional translation of an egocentric reference frame (Vallar *et al.* 1995), while others still have argued that the representation of contralesional space is compressed (Hallignan and Marshall 1991; Milner and Harvey 1995; Bisiach *et al.* 1996).

For extinction, the arguments about the nature of the spatial structure of the representation that is disrupted have not been so elaborate. Rather, it is generally considered that the strength of the representation of a contralesional stimulus is weaker than that of an ipsilesional one. Furthermore, it has been shown that in a patient with left-sided extinction, if two stimuli are presented in the left visual field, the most leftward one can still be extinguished by a stimulus to its right (di Pellegrino and De Renzi 1995). Thus relative spatial location, rather than absolute spatial position, may be an important factor in determining which stimulus is extinguished.

However, a recent investigation of cross-modal extinction in a right-hemisphere patient suggests that the spatial structure of representations underlying extinction deserves further consideration. In this case, when the patient's gaze was directed straight ahead, a tactile stimulus delivered to the left hand could be extinguished by a visual stimulus in the right visual field. The critical finding was that when the patient's gaze was directed to the left and the right visual event now occurred at the same location in space as the left tactile stimulus, there was no longer any extinction (Kennett, Rorden, Husain and Driver, unpublished observations). In both conditions (gaze straight ahead and gaze left), the visual stimulus occupied the same retinotopic location in the right visual field and the tactile stimulus was delivered to the same location in somatotopic space. Nevertheless, the degree of extinction was modulated by where the stimuli were with respect to the body. Thus, cross-modal extinction may operate in an egocentric frame of reference.

For visual neglect, it has been argued that evidence in favour of disrupted egocentric representations comes from demonstrations of amelioration of neglect by manoeuvres that orient gaze leftwards, such as caloric vestibular stimulation (Karnath 1994; Cappa *et al.* 1987; Vallar *et al.* 1993), optokinetic (Pizzamiglio *et al.* 1990; Karnath 1996) or neck-muscle proprioceptive stimulation (Karnath 1994; Karnath *et al.* 1993; Karnath 1995) and leftward trunk rotation relative to the head (Karnath *et al.* 1991). In my opinion, such improvements in unawareness of contralesional stimuli need not have anything to do with an underlying disruption of egocentric spatial representations. In neurologically normal individuals, these postural manoeuvres simply orient the eyes towards one side of space. And there is no reason to believe that this is *due to* some effect on egocentric representations: low-level reflexes may be sufficient. In patients with neglect, I would argue, these manoeuvres have the very same effect, directing the eyes towards one side of space and thus allowing previously neglected objects to fall into view. In other words, all these manoeuvres are simply physically directing the eyes toward objects in previously neglected space. The improvement in report of contralesional stimuli cannot therefore be taken to be a demonstration of the nature of the neural representation that is disrupted.

There are three further difficulties for the egocentric model. The first is the failure by some groups to find a correlation between direction of deviation of subjective midline and neglect (Bartolomeo and Chokron 1999; Farnè *et al.* 1998; Chokron and Bartolomeo 1997). This suggests that rotation of the mid-sagittal plane is not the primary problem in neglect syndrome.

The second difficulty for egocentric models is the claim that neglect can be object-based rather than space-based (for a review see Driver (1999)). Thus some right-hemisphere patients may consistently neglect the left side of objects regardless of whether they are located in the left or the right hemispace. This may not be such a problem for proponents of space-based models, because it is quite possible that there may be different types of neglect. Furthermore, it has been argued that some of these demonstrations of object-based neglect may not be truly 'object-based', but rather due to the size of the spatial aperture over which visual processing operates. Robertson and her colleagues have presented evidence to suggest that lesions of the right temporo-parietal junction lead to focusing of attention to local forms at the expense of global ones (Robertson *et al.* 1988). If the spatial window over which attention operates is small, there may be local neglect of features wherever attention is directed, and this may lead to apparently object-based effects (Driver 1999).

The third difficulty encountered by egocentric models is that, in and of themselves, they offer no explanation for the modulation of the severity of neglect by manipulations that alter the stimulus characteristics of a scene. For example, reducing the density of targets or target: distractor ratio or increasing the salience of targets with respect to distractors may also have the same effect (Eglin, Robertson, and Knight 1989; Humphreys and Riddoch 1993*a,b*; Rapcsak *et al.* 1989; Kaplan *et al.* 1991; Mark *et al.* 1988; Kartsounis and Findley 1994; Husain and Kennard 1997). Strictly spatial models do not accommodate these effects, although the addition of a selection mechanism may do so (Pouget and Sejnowski 1997).

Clearly, explanations of neglect based *only* on egocentric representations have difficulty in accounting for all the behaviour seen in neglect patients. However, my main concern over space-based accounts is their lack of explanatory power. For neglect, space-based accounts argue that the contralesional representation of space (whether it is body-centred, head-centred, retinotopic) is degraded. But such an argument avoids the real issue: What computational processes do those representations perform? In my opinion, there is little further to be gained by answering that 'they represent space'.

The same reply could be given to someone who was curious about what hemianopia told us about the function of primary visual cortex (V1). We would surely not claim to understand the function of striate cortex by learning simply that hemianopia is loss of the retinotopic representation of contralesional space? Furthermore, appreciating that hemianopia is a retinotopic loss does not really help in understanding the function of V1. Of course, striate cortex does indeed hold a retinotopic representation of visual space, but so also do the retina and other visual areas in the occipital lobe. Presumably, all these regions perform quite different operations on the visual input. Knowing that their loss leads to retinotopic deficits is unhelpful in unravelling their functions.

It might be argued that whilst hemianopia is retinotopic, neglect may be body-centric. The representation destroyed in this condition is likely therefore to require an association of signals from the retina with eye-position and head-position signals such as that which is known to occur in monkey parietal cortex (Andersen *et al.* 1985; Andersen 1997). Thus if one could prove that neglect was an impairment of representing body-centred space, we would understand the properties of the representation that has been destroyed. But such an argument also misses the point.

The visual pathways do remap (or re-present) the world in a spatiotopic manner, whether it be retinotopic or not. But simply understanding the *mapping* should not be confused with understanding the operations performed within the maps that exist in visual areas. So, just as there are multiple retinotopic representations of space along the visual pathway, it has now become clear that there are also many areas in monkey brain, other than parietal cortex, in which retinal and eye-position signals are combined (for a review, see Battaglini *et al.* 1997). Understanding the spatial structure of these representations does not permit a complete understanding of their functions. Similarly, I would argue, even if we knew that a body-centred representation of space is disrupted in neglect—or extinction—we would not have a full understanding of these syndromes from such knowledge.

Spatial attention in neglect and extinction

An alternative view to the representational account has been to consider that the primary disorder in neglect and extinction is one of *spatial* attention. In other words, the primary problem is one of directing attention into contralesional space—and all the rest follows. A landmark study that has been most influential in promoting this view is the one performed by Posner and his colleagues (Posner *et al.* 1984). They used a paradigm which had been developed to probe the covert orienting of attention in neurologically normal subjects (Posner 1980).

Subjects viewed a visual display consisting of three boxes in a horizontal row, each separated by a visual angle of eight degrees. They fixated the central box and waited for a cue stimulus (transient illumination of the left or right box), to which they were requested not to make a response. At a varying interval after the onset of the cue, a target stimulus (an asterisk) would appear in the left or right box, to which subjects were requested to make a speeded manual response. Throughout a trial, subjects were required to maintain central fixation, and this was monitored by the investigators. In 80 per cent of trials the cue was 'valid', appearing at the location of the subsequent target. In the remaining 20 per cent of trials the cue was 'invalid', appearing in the hemifield opposite that in which the target subsequently appeared. In neurologically normal subjects reaction times to targets appearing where valid cues had appeared were significantly shorter than when invalid cues were presented.

For patients with right parietal lesions, Posner and his colleagues found that reaction times to valid targets in the right hemifield were only modestly faster than they were to those occurring in the left hemifield. Furthermore, there was no significant interaction between the hemifield in which the target appeared and the time-interval between cue and target. Thus they concluded that attention could be summoned equally well to left or right targets. However, there was a significant impairment in these patients when an invalid cue appeared in the right hemifield and the subsequent target appeared on the left. Under these circumstances, parietal patients were extremely slow to detect targets on the left.

The investigators' interpretation of this finding was that once spatial attention had been directed to the right by the cue, right parietal patients had difficulty in disengaging *and* shifting attention to a target in left hemispace. The authors called this an 'extinction-like' reaction time pattern, because ipsilesional stimuli had an effect on the perception of contralesional ones. A similar effect, but for invalid cues appearing in the left hemifield, was also found in patients with left parietal lesions.

The patients in this study were chosen on the basis of the site of their lesion and not on the basis of their clinical syndrome. Thus five out of thirteen of them had no demonstrable signs of visual extinction or neglect, while most of the rest suffered from extinction alone, which varied in its severity and consistency. Only one patient suffered from moderate neglect. However, subsequent investigations using the Posner paradigm have demonstrated the presence of a similar 'extinction-like' reaction-time pattern in patients with neglect (Morrow and Ratcliff 1988). In their initial study, Posner and his associates found that the lesion site that correlated best with the 'extinction-like' pattern was the superior parietal lobe (Posner *et al.* 1984). In a subsequent study, the best anatomical correlate was found to be the temporo-parietal junction (Friedrich *et al.* 1998), the area that is classically associated with neglect (Vallar and Perani 1986).

Posner and his colleagues have investigated the presumed disengage and shift deficit further. They found that the extinction-like reaction-time pattern was not dependent upon the invalid cue occurring in the ipsilesional field and the target in the contralesional one. Rather, it appeared to be directional effect: for right-parietal patients, for example, even if the invalid cue was *presented within the left hemifield* there was a delay in responding to targets that subsequently appeared to the left of the cue position (Posner *et al.* 1987). These investigators concluded that the impairment is therefore one of disengaging attention and shifting it in a contralesional *direction*.

On the other hand, Baynes and her colleagues have presented data using a variant of the Posner paradigm that suggest that this may not be an adequate explanation and that hemispace, as well as direction of attentional shift, plays a part (Baynes *et al.* 1986). They presented their subjects with cues—valid or invalid—and targets that within a block of testing would *all* occur either to the left or right of fixation. But the variation in their task was that cues and targets appeared in boxes that were aligned vertically, so that any shifts of attention would need to be made up or down within a hemifield, rather than left or right. They found that in their right-hemisphere lesion patients (all of whom had involvement of the parietal lobe) with clinically demonstrable left-sided visual extinction there was still a demonstrable cost on trials with invalid cues when they were presented in the left hemifield. Thus, they concluded that once attention had been deployed to the left hemifield, subsequent orienting of attention within that hemifield was impaired even if the shift of attention was vertical.

Regardless of whether the impairment is directional or hemifield-specific, problems for the disengage-and-shift deficit theory have been presented by at least two recent reports of patients with left-sided extinction. Di Pellegrino and his colleagues demonstrated that a left-sided stimulus can be extinguished by a right-sided one, even if the left-sided stimulus is presented several hundred milliseconds before the right one (di Pellegrino *et al.* 1997). Similarly, Rorden and co-workers showed that left-sided extinction patients judged a right stimulus to have come first, even when the left one had been presented 200 ms beforehand (Rorden *et al.* 1997). Thus, these experiments demonstrate a left-sided disadvantage even when there is no initial right-sided stimulus for attention to engage upon. Furthermore, Rorden and his colleagues have pointed out that a major difficulty for the Posner hypothesis is that even if there was a disengage-and-shift deficit in left-sided extinction or neglect, this would not explain why attention would always first be engaged on a stimulus on the right (Rorden *et al.* 1997).

The findings of di Pellegrino *et al.* and Rorden *et al.* are better explained by a combination of two factors. First, there appears to be a *spatial bias* to direct attention (or a limited-capacity visual processing system) to the right in both left-sided extinction and neglect, presumably because of unopposed contralesional hemisphere activity (Kinsbourne 1987). Second, there is an impairment of processing information on the left that disadvantages perception of left-sided stimuli compared to those on the right. Thus, even if a stimulus is briefly presented on the left just before one on the right, it loses in the competition to be perceived through a limited-capacity processing system. Viewed from this perspective, extinction and neglect may be exaggerated versions of normal phenomena. Thus, when neurologically normal subjects are asked to report letters presented in very brief displays, they too demonstrate a limited capacity (cf. Duncan 1980). Patients with extinction and neglect may have a spatial bias and also a very limited processing capacity.

It remains to be seen whether both extinction and neglect lie on a continuum of disorders, or whether the neglect syndrome is qualitatively different from extinction. For example, there is evidence to suggest that parietal neglect is associated with a disorder of initiating contralesional movements to targets in the contralesional hemifield (Mattingley *et al.* 1998). This impairment may be quite separate from the attentional disorder of visual perceptual processing in neglect. However, from the perspective of the argument I am attempting to develop here, the next important issue is to consider what the visual processing defect in attentional disorders consists of.

Dorsal simultanagnosia

In my opinion, some important insights into unilateral 'spatial' neglect and extinction are to be obtained from a disorder that is most often seen in its severest form following bilateral parietal damage. The term simultanagnosia (or simultaneous agnosia) was first introduced by Wolpert (1924) to describe a condition in which patients with brain damage are able to report individual elements (objects or parts of objects) in a complex picture, but are unable to integrate them to form a coherent understanding of the entire display. It is as if they see fragments of the scene but are unable to piece the whole together. Although Wolpert coined the term in 1924, it is clear that similar and indeed more severe problems had earlier been noted in the patients described by Bálint and Holmes.

Bálint (1909) first described a triad of symptoms that may follow bilateral parietal lobe damage: 'psychic paralysis of gaze', neglect of objects in the visual surround and misreaching to visual targets (so-called optic ataxia). Here, we shall not be concerned with the reaching disorder of Bálint's syndrome, but concentrate on the visual perceptual impairments. For a fuller discussion of Bálint's original case see Husain and Stein (1988).

Bálint noted that his patient experienced great difficulty in looking at objects other than the one he was fixating. On cursory examination one might have thought he had paralysis of eye movements. More careful observation showed that the patient could in fact move his eyes with a full range of movement, his apparent paralysis of gaze being due to an inability to notice spontaneously visual objects other than the one fixated. Bálint had to direct him verbally to seeing other objects in the visual surround, even though on careful testing the patient's visual fields appeared to be intact. Bálint therefore described the fixation of gaze as due to a constriction of the 'psychic field-of-view'. It is as if the patient's vision were locked into the object at fixation, because he was simply not aware of other objects in the visual surround. Interestingly, Bálint observed that his patient could attend to only one object at a time, *regardless* of the size of the object or the number of objects placed in front of him.

Holmes (1918) subsequently described in great detail further cases of bilateral posterior parietal damage. His emphasis was on what he called their 'visual disorientation' (Holmes 1918; Holmes and Horrax 1919). Typically, when asked to touch an object in front of him, a patient would reach in the wrong direction and grope hopelessly until his hand or arm came into contact with it. Through careful observation, Holmes came to the conclusion that his patients' difficulties were in judging visual location. But they also had difficulty in counting small objects such as coins placed in front of them. Furthermore, like Bálint's patient, they had difficulty in seeing more than one object at a time. One of his patients even remarked: 'I can only look at one at a time.' Holmes concluded that in addition to visual mislocalization there was also a constriction or limitation of visual attention in these patients.

From the perspective of the spatiotemporal framework I am attempting to develop here, the most important observation in the study of simultanagnosia following bilateral posterior parietal lobe damage came forty years later, from Luria (1959). He studied a young army officer who had been shot in the head. Luria assessed his patient's visual perception using controlled tachistoscopic displays. One of the displays he showed consisted of two overlapping triangles in the configuration of a Star of David. The patient incurred no difficulty in reporting a star. However, when one of the triangles was drawn in red and the other in blue ink, only one or other of the triangles was reported, and the patient never said he saw two triangles or a star. Thus, when two forms (triangles of different colours) overlapped in space, only one was perceived. Simultaneous perception of two forms was not possible at the same location, even though the patient could apparently see both these forms when they were drawn in the same ink to form one object—the star.

It is difficult to argue that this problem in perceiving overlapping figures is primarily a spatial disorder, since the objects displayed occupied essentially the same location in space. Even if one wanted to argue that a spatial disorder may not be manifest because bilateral parietal damage effectively nulls the spatial bias seen following unilateral parietal damage, this would still not offer a clear explanation of the residual non-spatial deficit.

More recently, Humphreys and colleagues have extended the observation made by Luria. They examined two patients with Bálint's syndrome, one who had clearly suffered bilateral strokes involving the parietal lobes, and the other who had suffered cerebral anoxia with evidence of at least a left parietal lesion on imaging (Humphreys *et al.* 1994). (Anoxia is usually associated with bilateral brain damage, although this may not be easily demonstrable on imaging.) When either single words or pictures of objects were briefly presented, the patients had no difficulty in reporting them. However, when overlapping displays consisting of pictures and words were presented, the patients most often reported either the picture alone or both the word and the picture. Words alone were very rarely reported. These patients were further tested on a number of different paradigms assessing their simultaneous form perception. However, the key conceptual point developed by the investigators was that these Bálint's syndrome patients suffered from a *non-spatial form of extinction*. In the experiment I have described here, pictures tended to extinguish words in the same location in space.

The deficit in simultanagnosia need not occur only with bilateral lesions. Unilateral brain injury can lead to a milder form of the disorder. Furthermore, a defect that resembles simultanagnosia can occur in the context of unilateral visual neglect. Patterson and Zangwill (1944) gave a particularly clear account of a young patient with a very focal lesion (their Case 1). He had been unfortunate enough to be present at an explosion. A steel nut penetrated his skull and lodged in the underlying brain. Subsequent surgery demonstrated injury to the right supramarginal and angular gyri of the parietal lobe, the regions most commonly associated with visual neglect (Vallar 1993). In addition to left-sided visual extinction and neglect, this patient suffered from 'a complex disorder affecting perception, appreciation and reproduction of spatial relationships in the central field of vision' (p. 337).

Patterson and Zangwill (1944) observed 'that the patient always drew complex objects or scenes item by item and detail by detail. He appeared to lack any real grasp of the object as a whole' (p. 342). They characterized this problem as a 'piecemeal approach' that 'could be defined as a fragmentation of the visual contents with deficient synthesis' (p. 356). They went on to remark: 'it might be argued that the piecemeal effect arises from undue restriction of attention to one aspect of the drawing with consequent neglect of the rest. This implies neglect both of what has already been drawn and of what still remains to be added' (p. 356). They noted a similar piecemeal approach in their Case 2, a man with a traumatic right parieto-occipital injury, who also had mild left-sided neglect.

Luria (1959) appreciated that Paterson and Zangwill's descriptions of a 'piecemeal approach' might be a defect of visual synthesis, that is qualitatively similar to simultanagnosia following bilateral parietal injury. From the perspective of the argument I am developing here, the important point is that there may be a common mechanism involved in simultanagnosia following bilateral parietal injury and visual neglect following unilateral (most frequently right) parietal damage. Thus the key to understanding that mechanism may therefore not be the spatial aspect of the disorder that is so impressively evident following unilateral (most commonly right) parietal damage.

Bilateral parietal injury, as Humphreys *et al.* (1994) have observed, may lead to non-spatial extinction, restricting visual perception to seeing one thing at a time—even though (overlapping) objects may occupy the same location in space. Is there any evidence that there is a non-spatial disorder of attention in spatial neglect?

Non-spatial attentional deficit in spatial neglect and extinction

To answer this question we have used a rapid stimulus visual presentation (RSVP) paradigm that presents a sequence of stimuli rapidly at only one location in space (Broadbent and Broadbent 1987; Raymond *et al.* 1992). Just as pen-and-paper cancellation tasks involve searching for targets over space, this task involves searching for targets in stimuli presented sequentially over time *at one location in space*. Subjects view a fixation point at the centre of a computer monitor placed directly in front of them. When they are ready, they press a start key and a sequence of letters is presented, one at a time (at a rate of 5.5 letters per second) at the same location in the centre of a grey screen. Each letter subtends only 0.7 degrees of visual angle, and is presented for 131 ms with an interval of 49 ms between letters.

Figure 12.2 depicts only a few of the stimuli in a trial, but is nevertheless useful to demonstrate some of the important features of the paradigm. First, all the letters in a sequence are

Display parameters

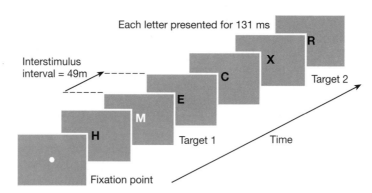

Fig. 12.2 Some of the stimuli in an RSVP stream. All the letters are black except target 1 (T1), which is white. Here it is an 'M', but it can be any letter in the alphabet except X. In each trial the number of letters before T1 varies randomly from 7 to 15. None of the letters that precede T1 is an X. In each trial, there are always 10 black letters following T1. On *half* the trials, an X is presented at some point in the ten-letter sequence that follows T1. X is referred to as target 2 (T2).

black except one, which is white. We call this the first target, or T1. In Fig. 12.2 it is an M, but it can be any letter in the alphabet except X. In each trial the number of black letters before T1 varies randomly from 7 to 15. None of the letters that precede T1 is an X. In each trial, there are always 10 black letters following T1. On half the trials, an X is presented at some point in the ten-letter sequence that follows T1. We call the X the second target, or T2. So there are 7–15 letters before T1, there is always a T1, and there are always ten letters after T1. On half the trials one of the ten black letters that follow T1 is an X (T2). It is the time between T1 and the ten letters that follow it that is the focus of our interest.

Subjects are tested under two conditions. In the *dual-report condition* we ask them to identify what the white letter (T1) was, and also say whether they saw an X (T2) at any stage in the 10-letter sequence following T1. So they are asked to do two things: identify T1 was and also report if they saw an X (T2). In the *single-report condition* subjects see the same sequences of stimuli but we simply ask them to say whether they saw an X (T2). They can ignore the white letter (T1). Single and dual report tasks are run in different counterbalanced blocks.

The results of ten neurologically normal control subjects aged between 60 and 80 on the single and dual report tasks are shown in Fig. 12.3. Their ability to detect T2 (the X) is plotted against the time between when T1 (the white letter) and when T2 (the X) was presented. Thus, this is a plot of how accurately subjects can say whether they saw T2, depending upon *when* it occurred after T1. In the single-report task, where subjects ignore T1, their ability to detect T2 (the X) does not vary according to when it occurred relative to T1. Their ability to see T2 is the same regardless of whether it is presented straight after T1, or it is the 10th letter after T1.

However, in the dual-report task, when they have to identify T1 as well as detect if T2 is present, there is a dip in their ability to detect T2 if it is presented within 360 ms of T1. So when normal subjects identify T1, if T2 occurs within 360 ms they are significantly impaired in detecting it. If they see the same sequences and ignore T1 (single-report task), they have no difficulty in detecting T2 during this period. This dip—which has been demonstrated in many experiments in normal individuals of varying ages—is referred to as the attentional blink or dwell-time

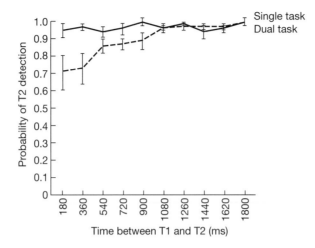

Fig. 12.3   Performance of neurologically normal subjects (mean age = 73). The probability of detecting T2 was significantly lower in the dual task compared to the single task for T1–T2 intervals of up to 360 ms. This dip in performance on the dual task is referred to as the attentional blink.

(Raymond *et al.* 1992; Duncan *et al.* 1994; Shapiro *et al.* 1997). In effect, it measures the temporal dynamics of visual processing: the time it takes for the visual system to engage and identify T1 before it is free to detect a subsequent stimulus—T2. This is the time during which the visual system fails to be aware of stimuli it can readily detect in the single-target task. And because this time is so short (only 300–500 ms) its been called 'the attentional blink'.

We used the same paradigm to study eight stroke patients with left-sided neglect. Three had lesions of the right inferior parietal lobe, the area that is classically associated with neglect (Vallar 1993). Four had lesions of the right inferior frontal lobe, the critical lateral frontal site associated with neglect (Binder *et al.* 1992; Husain and Kennard 1996). One patient had a basal ganglia lesion. The mean age of patients was 64, and they were tested on average 34 days after stroke. The performance of the neglect patients on the single- and dual-report tasks is shown in Fig. 12.4. On the single-report task, like normal subjects, they had little difficulty in detecting T2 regardless of when it occurred after T1. However, on the dual task their performance was very different. Like normal subjects, they encountered difficulty in identifying T2 if it came straight after T1, but the impairment was significantly worse, and it improved only gradually afterwards. Performance on the dual task was significantly different from that on the single task for up to 1.4 seconds. So when patients identified T1 they had an impairment in detecting T2 for up to 1.4 seconds afterwards. If they saw the same sequences but did not have to identify T1, they detected T2 very accurately, regardless of when it was presented. There was no difference between patients with parietal and frontal lesions.

Eight patients with right-hemisphere strokes without neglect were also tested. Their mean age, lesion volume and time from stroke onset were all not significantly different from those of the neglect patients. We found that their pattern of performance is not significantly different from that of neurologically normal individuals either on the single- or the dual-report tasks. So their attentional blink is normal, lasting only 360 ms. In summary, the patients with neglect we have tested have a significantly more severe and more protracted attentional blink than normal subjects or control right-hemisphere stroke patients without neglect. Thus there is evidence of a non-spatial aspect in spatial neglect (Husain *et al.* 1997).

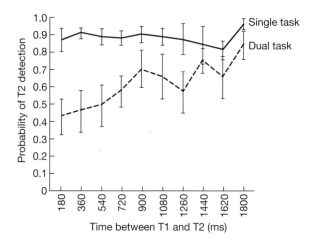

Fig. 12.4 Performance of eight right-hemisphere patients with left visual neglect. T2 detection was significantly impaired compared to T1 performance for T1–T2 intervals less than 1440 ms. Furthermore, the magnitude of visual unawareness during the attentional blink was significantly greater than in age-matched neurologically normal subjects (Fig. 12.3) or stroke patients without neglect.

Di Pellegrino and his colleagues have also used an RSVP paradigm to investigate the time-course of visual processing in the ipsi- and contralesional visual field in a patient with left-sided visual extinction (di Pellegrino *et al.* 1998). Their patient had suffered a right temporo-parietal infarct, and initially demonstrated left visual neglect. By the time he was tested, 5–7 months after stroke-onset, he had no evidence of visual neglect or visual field defect, but consistently demonstrated left-sided extinction on double simultaneous stimulation.

The patient was asked to keep his gaze fixed at a central cross throughout the experiment. Within a block of trials, two letters (which I shall refer to as T1 and T2) were presented rapidly (100 ms) in quick succession in *either* the left or the right field. Note that both T1 and T2 occurred in the same visual field and at the same location. The stimulus onset interval between T1 and T2 was varied between 100 and 1000 ms. There were no masks or distractor items. Before each block of trials, the patient was told in which field the stimuli would appear. His task was to report either both T1 and T2 (dual-report task) or simply report T2, ignoring T1 (single-report task).

In the single task, report of T2 was good in both visual fields (93 per cent accuracy on the left vs. 94 per cent on the right). Furthermore, it did not vary as a function of T1–T2 interval. In the dual task, regardless of the side of presentation, accuracy of T2 report was impaired if T2 appeared immediately after presentation of the first stimulus and improved with increasing T1–T2 interval—i.e., this was the attentional blink phenomenon. The important finding was that the attentional blink was significantly more severe (T2 report was worse) and more pro-tracted when stimuli appeared in the left visual field compared to the right. It took 600 ms on the left for T2 report on the dual task to reach T2 performance on the single task, compared to only 300 ms on the right.

Thus, once the visual system was engaged in identifying T1, it took longer to detect a subse-quent stimulus (T2) in the left field than the right. Therefore, in this patient with spatial extinc-tion, visual processing appeared therefore to be slower on the contralesional side even when there was no competing ipsilesional stimulus.

These two RSVP studies of patients with neglect and extinction (Husain *et al.* 1997; di Pellegrino *et al.* 1998) demonstrate abnormalities in visual processing over time at the same location in space. For neglect patients, there may well be a gradient of impaired processing extending from the contralesional to the ipsilesional side. For the extinction patient studied by di Pellegrino and his colleagues there is clear evidence for such a proposal. Indeed, the *temporal* phenomenon of slower visual processing on the left that they observed may contribute to the bias against a left-sided stimulus when a right-sided stimulus is concurrently presented (*spatial* extinction). Thus, objects on the left, which are not reported (extinguished) in the presence of concurrent stimuli on the right, require an abnormally protracted time to be selected by attention even when occurring alone. If a competing stimulus is presented on the right, it wins in the competition to be processed by the limited-capacity system. So the spatial bias may be tightly coupled to the temporal processing impairment, and need not be due to some primary bias to orient attention to the ipsilesional side (Kinsbourne 1987).

The concept that emerges is that the spatial bias in extinction and neglect *results from* the processing impairment, which is profound on the contralesional side and gradually improves on the ipsilesional side. That evidence for such a proposition has come from investigations of the temporal dynamics of visual processing does not mean that the key deficit is 'temporal'. It simply means that such paradigms have allowed identification of a deficit that is non-spatial. The important issue is to examine why visual processing may be slowed.

What is the non-spatial disorder of visual processing?

To answer this question, perhaps the best place to start is to consider what causes the attentional blink in neurologically normal individuals. Studies to date have raised the possibility of at least two types of underlying mechanism. First, devoting resources to processing a visual stimulus (T1) may cause the visual system to cease processing new stimuli for the duration of the attentional blink. So a subsequent stimulus (T2) may be only partially processed if it occurs within a few hundred milliseconds of T1 (early selection theory). Alternatively, there may be competition between T1 and T2 for selection *after* features belonging to each item have been correctly conjoined, i.e., after both are fully represented in the visual system (late selection theory).

Early selection and feature binding

One possible mechanism underlying an early selection process is that the visual system takes time correctly to conjoin or bind features (e.g., colour, luminance, orientation, etc.) of a stimulus before features belonging to a subsequent item can be processed. Studies of (spatial) visual search have suggested to some investigators that attention is the means by which the features of a particular object are correctly bound. Furthermore, according to Treisman and Gelade (1980), features belonging to an object are correctly assigned to it because they occupy the same location in space. Is there any evidence that feature binding is impaired in the three disorders of attention considered in this review?

Cohen and Rafal (1991) studied a left-hemisphere lesion patient with right-sided visual extinction. The patient's task was to identify a central digit at fixation, and then report the colour and identity of a concurrent peripheral target letter that was presented with a distractor letter O. The distractor could appear in one of several different colours. The patient made disproportionately more conjunction errors (for example, reporting that the target had the colour of the distractor) than feature errors (for example, reporting that the target had a colour that was not presented) in the right hemifield than in the left.

On (spatial) visual search tasks, neglect patients have been shown to be slower to detect a target when conjunction of features is required to discriminate it from distractors than in searches where simple feature discrimination will suffice—for example detecting a red circle amongst green ones (Riddoch and Humphreys 1987; Eglin, Robertson and Knight, 1989). However, searches for features alone were not normal, and it is possible that poorer performance on conjunction searches occurred simply because they are more difficult. Evidence to suggest that this may indeed be the case comes from a search study that varied target discriminability with respect to distractors using feature-based manipulations that are known to make search harder in neurologically normal subjects. Such manipulations were sufficient to impair search disproportionately on the contralesional side compared to the ipsilesional one (Humphreys and Riddoch 1993*b*).

Finally, what about feature binding in simultanagnosia? A patient with bilateral parietal damage with Bálint's syndrome has recently been investigated in great detail (Friedman-Hill *et al.* 1995; Robertson *et al.* 1997). He was found to make many conjunction errors when reporting the colour and shape of objects presented to him. The investigators, who included Treisman, suggested that an impairment in spatial localization (cf. the visual disorientation emphasized by Holmes) leads to the defect of feature binding in this condition. Further supportive evidence for a specific role of parietal cortex in feature binding comes from a functional imaging study that demonstrated increased bilateral parietal activity when feature conjunction is required (Corbetta *et al.* 1995).

Thus there is some evidence to suggest that feature binding may be impaired in extinction, neglect and simultanagnosia. This may therefore be a candidate mechanism for the non-spatial processing defect seen in all three of these disorders.

Late selection and competition

On the other hand, there is evidence from neurologically normal subjects to suggest that the attentional blink is due to limitations in late selection. Both behavioural and electrophysiological studies (reviewed by Shapiro *et al.* 1997) have provided evidence to support the view that, although subjects often fail to report T2, this target is nevertheless implicitly processed further, to the level of semantic (meaning) analysis. It has therefore been proposed that T1, T2, and stimuli temporally adjacent to them enter a visual short-term memory buffer *after* the features of each of these stimuli are correctly conjoined. However, it is the subsequent competition between T1 and T2 that leads to 'loss' of T2 during selection for report.

To the best of my knowledge, there is no evidence from attentional blink studies of the significance of such a late selection mechanism in extinction, neglect and simultanagnosia. However, there is a growing body of evidence from other paradigms that stimuli that are neglected by patients may nevertheless undergo considerable unconscious processing (see, for example, McGlinchey-Berroth *et al.* 1993).

Object files

Although debates about early and late selection mechanisms continue, it is likely that the complexity of visual processing is not entirely captured by either of these polarized viewpoints. In contrast to Treisman's original formulation of importance of spatial location in attentional processing, some have argued that attention in neurologically normal subjects is object-based. For example, it has been shown that concurrent, attentionally demanding operations may be performed on a single object without cost, whereas the same operations performed on different objects reveal extra attentional costs (Duncan 1984). To account for these results, Treisman and her colleagues have argued that visual processing is conducted on an object representation, which they refer to as an 'object file' (Kahneman *et al.* 1992). So attentional processing occurs on all the features of an object simultaneously. Thus the visual system has to open and keep track of a single object file across space and time.

If this model is applied to the attentional blink, when subjects view an RSVP stream, presentation of the second target (T2) initiates a review to determine if it is consistent with the object file created by the first (T1). Once the visual system has started to analyse an object it does not direct full resources to a stimulus that is considered to be another object for several hundred milliseconds, and this is the cause of the attentional blink. Such a model poses difficult questions. For example, what is an 'object' to the visual system? And on what basis are separate objects distinguished across space and time? From the perspective of the current discussion, the model also raises the possibility that attentional disorders may represent an impairment in the management of object files and their ongoing review and updating with time.

Conclusions

In this review, I have argued that analyses of the spatial impairments in visual neglect and extinction fail to capture all the problems associated with these two parietal syndromes. I have suggested instead that examination of the *non-spatial* impairment in simultanagnosia (following bilateral parietal injury) provides an important clue to understanding the unilateral visual processing disorders in extinction and neglect.

In simultanagnosia, there is evidence that the object currently being processed 'extinguishes'—or impairs processing of—other objects. Recent investigations in neglect and extinction have also demonstrated abnormal temporal dynamics of visual processing at one location in space. It has been shown that the time required to discriminate one object and release processing capacity for another is prolonged in these conditions, even when directional shifts of attention are not required. It is possible that the spatial bias in extinction and neglect *results from* asymmetries in this processing impairment, which is profound on the contralesional side and gradually improves on the ipsilesional side. In simultanagnosia, there may be no spatial bias because bilateral hemispheric damage effectively makes the processing impairment symmetric. Nevertheless, a profound non-spatial impairment of visual processing remains.

A central problem is to understand why visual processing ('attention') is engaged abnormally in extinction, neglect and simultanagnosia. An impairment of feature binding (early selection), or competition between targets and non-targets after feature binding (late selection), or management of 'object files' may be responsible. A second issue concerns the differences between the two spatial disorders—extinction and neglect. Are they caused by the same underlying disorder of visual processing? Or are there additional components of the neglect syndrome that are not explained by one single visual processing disorder? Attempts to answer these questions will, I hope, improve our understanding both of these disorders of attention and of the mechanisms underlying vision in neurologically normal individuals.

References

Andersen, R. A. (1997). Multimodal integration for the representation of space in the posterior parietal cortex. *Philosophical Transactions of the Royal Society of London B*, **352**, 1421–8.

Andersen, R. A., Essick, G. K., and Siegel, R. M. (1985). The encoding of spatial location by posterior parietal neurons. *Science*, **230**, 456–8.

Bálint, R. (1909). Seelenlähmung des 'Schauens', optische Ataxie, raümliche Störung der Aufmerksamkeit. *Monatschrift Psychiatrisches Neurologie*, **25**, 51–81.

Bartolomeo, P. and Chokron, S. (1999). Egocentric frame of reference: its role in spatial bias after right hemisphere lesions. *Neuropsychologia*, **37**, 881–94.

Battaglini, P. P., Galletti, C. and Fattoni, P. (1997). Neuronal coding of visual space in the posterior parietal cortex. In P. Thier and H-O. Karnath (eds.), *Parietal lobe contributions to orientation in 3D space*, pp. 539–53. Berlin: Springer.

Baynes, K., Holtzman, J. D., and Volpe, B. T. (1986). Components of visual attention. Alterations in response pattern to visual stimuli following parietal lobe infarction. *Brain*, **109**, 99–114.

Binder, J., Marshall, R., Lazar, R., Benjamin, J., and Mohr, J. P. (1992). Distinct syndromes of hemineglect. *Archives of Neurology*, **49**, 1187–94.

Bisiach, E., Pizzamiglio, L., Nico, D., and Antonucci, G. (1996). Beyond unilateral neglect. *Brain*, **119**, 851–7.

Broadbent, D. E., and Broadbent, M. H. P. (1987). From detection to identification: response to multiple targts in rapid serial visual presentation. *Perception and Psychophysics*, **42**, 105–13.

Cappa, S., Sterzi, R., Vallar, G., and Bisiach, E. (1987). Remission of hemineglect and anosognosia after vestibular stimulation. *Neuropsychologia*, **25**, 775–82.

Chokron, S. and Bartolomeo, P. (1997). Patterns of dissociation between left hemineglect and deviation of the egocentric reference frame. *Neuropsychologia*, **35**, 1503–8.

Cohen, J. D. and Rafal, R. D. (1991). Attention and visual feature integration in a patient with a parietal lobe lesion. *Psychological Science*, **2**, 106–10.

Corbetta, M., Shulman, G., Miezin, F., and Petersen, S. E. (1995). Superior parietal cortex activation during spatial attention shifts and visual feature conjunction. *Science*, **270**, 802–5.

De Renzi, E. (1982). *Disorders of space exploration and cognition*. New York: Wiley.

di Pellegrino, G., and De Renzi, E. (1995). An experimental investigation on the nature of extinction. *Neuropsychologia*, **33**, 153–70.

di Pellegrino, G., Basso, G., and Frassinetti, F. (1997). Spatial extinction on double asynchronous stimulation. *Neuropsychologia*, **9**, 1215–23.

di Pellegrino, G., Basso, G., and Frassinetti, F. (1998). Visual extinction as a spatio-temporal disorder of selective attention. *NeuroReport*, **9**, 835–9.

Driver, J. (1999). Egocentric and object-based visual neglect. In N. Burgess, K. J. Jeffery and J. O. O'Keefe (eds), *The hippocampal and parietal foundations of spatial cognition*, pp. 66–89. Oxford: Oxford University Press.

Duncan, J. (1980). The locus of interference in the perception of simultaneous stimuli. *Psychological Review*, **87**, 272–300.

Duncan, J. (1984). Selective attention and the organisation of visual information. *Journal of Experimental Psychology: General*, **113**, 501–17.

Duncan, J., Ward, R., and Shapiro, K. L. (1994). Direct measurement of attentional dwell time in human vision. *Nature*, **369**, 313–15.

Eglin, M., Robertson, L. C. and Knight, R. T. (1989). Visual search performance in the neglect syndrome. *Journal of Cognitive Neuroscience*, **1**, 372–85.

Farah, M. (1990). *Visual agnosia: disorders of object recognition and what they tell us about normal vision*. Cambridge, MA: MIT Press.

Farnè, A., Ponti, F., and Làdavas, E. (1998). In search for biased egocentric reference frames in neglect. *Neuropsychologia*, **36**, 611–23.

Friedman-Hill, S. R., Robertson, L. C., and Treisman, A. (1995). Parietal contributions to visual feature binding: evidence from a patient with bilateral lesions. *Science*, **269**, 853–5.

Friedrich, F. J., Egly, R., Rafal, R. D., and Beck, D. (1998). Spatial attention deficits in humans: a comparison of superior parietal and temporo-parietal junction lesions. *Neuropsychology*, **12**, 193–207.

Halligan, P. W., and Marshall, J. C. (1991). Spatial compression in visual neglect: a case study. *Cortex*, **27**, 623–9.

Heilman, K. M., Valenstein, E., and Watson, R. T. (1994). Localization of lesions in neglect and related disorders. In A. Kertesz (ed.), *Localization and neuroimaging in neuropsychology*, pp. 495–524. San Diego: Academic Press.

Holmes, G. (1918). Disturbances of visual orientation. *British Journal of Ophthalmology*, **2**, 449–506.

Holmes, G., and Horrax, G. (1919). Disturbances of spatial orientation and visual attention with loss of stereoscopic vision. *Archives of Neurological Psychiatry*, **1**, 385–407.

Humphreys, G. W., and Riddoch, M. J. (1993*a*). Interactive attentional systems and unilateral visual neglect. In I. H. Robertson and J. C. Mrashall (eds.), *Unilateral neglect: clinical and experimental studies*, pp. 139–67. Hove, UK: Erlbaum.

Humphreys, G. W., and Riddoch, M. J. (1993*b*). Interactions between object- and space-processing systems revealed through neuropsychology. In D. E. Meyer and S. Kornblum (eds), *Attention and performance XIV*, pp. 84–110. Cambridge, MA: MIT Press.

Humphreys, G. W., Romani, C., Olson, A., Riddoch, M. J., and Duncan, J. (1994). Non-spatial extinction following lesions of the parietal lobe in humans. *Nature*, **372**, 357–9.

Husain, M., and Kennard, C. (1996). Visual neglect associated with frontal lobe infarction. *Journal of Neurology*, **243**, 652–7.

Husain, M., and Kennard, C. (1997). Distractor-dependent frontal neglect. *Journal of Neurology, Neurosurgery and Psychiatry*, **30**, 468–74.

Husain, M., and Stein, J. (1988). Rezsö Bálint and his most celebrated case. *Archives of Neurology*, **45**, 89–93.

Husain, M., Shapiro, K., Martin, J., and Kennard, C. (1997). Abnormal temporal dynamics of visual attention in spatial neglect patients. *Nature*, **385**, 154–6.

Kahneman, D., Treisman, A., and Gibbs, B. J. (1992). The reviewing of object files: object-specific integration of information. *British Journal of Ophthalmology*, **2**, 449–506.

Kaplan, R. F., Verfaellie, M., Meadows, M-E., Caplan, L., Pessin, M. S., and DeWitt, D. (1991). Changing attentional demands in left hemispatial neglect. *Archives of Neurology*, **48**, 1263–6.

Karnath, H-O. (1994). Subjective body orientation in neglect and the interactive contribution of neck muscle proprioception and vestibular stimulation. *Brain*, **117**, 1001–12.

Karnath, H-O. (1995). Transcutaneous electrical stimulation and vibration of neck muscles in neglect. *Experimental Brain Research*, **105**, 321–4.

Karnath, H-O. (1996). Optokinetic stimulation influences the disturbed perception of body orientation in spatial neglect. *Journal of Neurology, Neurosurgery and Psychiatry*, **60**, 217–20.

Karnath, H-O. (1997). Spatial orientation and the representation of space with parietal lobe lesions. *Philosophical Transactions of the Royal Society of London B*, **352**, 1411–19.

Karnath, H-O., Schenkel, P., and Fischer, B. (1991). Trunk orientation as the determining factor of the contralateral deficit in the neglect syndrome and as the physical anchor of the internal representation of body orientation in space. *Brain*, **114**, 1997–2014.

Karnath, H-O., Christ, K., and Hartje, W. (1993). Decrease of contralateral neglect by neck muscle vibration and spatial orientation of trunk midline. *Brain*, **116**, 383–96.

Kartsounis, L. D., and Findley, L. J. (1994). Task specific visuospatial neglect related to density and salience of stimuli. *Cortex*, **30**, 647–59.

Kinsbourne, M. (1987). Mechanisms of unilateral neglect. In M. Jeannerod (ed.), *Neurophysiological and neuropsychological aspects of spatial neglect*, 69th edn, pp. 69–86. Amsterdam: Elsevier.

Luria, A. R. (1959). Disorders of 'simultaneous perception' in a case of bilateral occipito-parietal brain injury. *Brain*, **83**, 437–49.

McGlinchey-Berroth, R., Milberg, W. P., Verfaellie, M., Alexander, M., and Kilduff, P. T. (1993). Semantic processing in the neglected visual field: evidence from a lexical decision task. *Cognitive Neuropsychology*, **10**, 79–108.

Mark, V. W., Kooistra, C. A., and Heilman, K. M. (1988). Hemispatial neglect affected by non-neglected stimuli. *Neurology*, **38**, 1207–11.

Mattingley, J., Husain, M., Rorden, C., Kennard, C., and Driver, J. (1998). Motor role of human inferior parietal lobe revealed in unilateral neglect patients. *Nature*, **392**, 179–82.

Mesulam, M-M. (1981). A cortical network for directed attention and unilateral neglect. *Annals of Neurology*, **10**, 309–25.

Milner, A. D., and Harvey, M. (1995). Distortion of size perception in visuospatial neglect. *Current Biology*, **5**, 85–9.

Morrow, L. A., and Ratcliff, G. (1988). The disengagement of covert attention and the neglect syndrome. *Psychobiology*, **16**, 261–9.

Patterson, A., and Zangwill, O. L. (1944). Disorders of visual space perception associated with lesions of the right cerebral hemisphere. *Brain*, **67**, 331–58.

Perani, D., Vallar, G., Cappa, S., Messa, C., and Fazio, F. (1987). Aphasia and neglect after subcortical stroke. A clinical/cerebral perfusion correlation study. *Brain*, **110**, 1211–29.

Pizzamiglio, L., Frasca, R., Guariglia, C., Incaccia, R., and Antonucci, G. (1990). Effect of optokinetic stimulation in patients with visual neglect. *Cortex*, **26**, 534–40.

Posner, M. I. (1980). Orienting of attention. *Quarterly Journal of Experimental Psychology*, **32**, 3–26.

Posner, M. I., Walker, J. A., Friedrich, F. J., and Rafal, R. (1984). Effects of parietal injury on covert orienting of attention. *Journal of Neuroscience*, **4**, 1863–74.

Posner, M. I., Walker, J. A., Friedrich, F. J., and Rafal, R. D. (1987). How do the parietal lobes direct covert attention? *Neuropsychologia*, **25**(1A), 135–45.

Pouget, A., and Sejnowski, T. J. (1997). A new view of hemineglect based on the response properties of parietal neurones. *Philosophical Transactions of the Royal Society of London B*, **352**, 1449–59.

Rapcsak, S. Z., Verfaellie, M., Fleet, S., and Heilman, K. M. (1989). Selective attention in hemispatial neglect. *Archives of Neurology*, **46**, 178–82.

Raymond, J. E., Shapiro, K. L., and Arnell, K. M. (1992). Temporary suppression of visual processing in an RSVP task: an attentional blink? *Journal of Experimental Psychology. Human Perception and Performance*, **18**, 849–60.

Riddoch, M. J., and Humphreys, G. W. (1987). Perceptual and action systems in unilateral visual neglect. In M. Jeannerod (ed.), *Neurophysiological and neuropsychological aspects of spatial neglect*, pp. 151–81. Amsterdam: Elsevier.

Robertson, L. C., Lamb, M. R., and Knight, R. T. (1988). Effects of lesions of temporo-parietal junction on perceptual and attentional processing in humans. *The Journal of Neuroscience*, **8**, 3757–69.

Robertson, L., Treisman, A., Friedman-Hill, S., and Grabowecky, M. (1997). The interaction of spatial and object pathways: evidence from Balint's syndrome. *Journal of Cognitive Neuroscience*, **9**, 295–317.

Rorden, C., Mattingley, J. B., Karnath, H-O., and Driver, J. (1997). Visual extinction and prior entry: impaired perception of temporal order with intact motion perception after parietal injury. *Neuropsychologia*, **35**, 421–33.

Shapiro, K. L., Arnell, K. M., and Raymond, J. E. (1997). The attentional blink. *Trends in Cognitive Sciences*, **1**, 291–6.

Treisman, A. M., and Gelade, G. (1980). A feature-integration theory of attention. *Cognitive Psychology*, **12**, 97–136.

Vallar, G. (1993). The anatomical basis of spatial hemineglect in humans. In I. H. Robertson and J. C. Marshall (eds), *Unilateral neglect: clinical and experimental studies*, pp. 27–59. Hove: Erlbaum.

Vallar, G., and Perani, D. (1986). The anatomy of unilateral neglect after right hemisphere stroke lesions: a clinical CT correlation study in man. *Neuropsychologia*, **24**, 609–22.

Vallar, G., Bottini, G., Rusconi, M. L., and Sterzi, R. (1993). Exploring somatosensory neglect by vestibular stimulation. *Brain*, **116**, 71–86.

Vallar, G., Guariglia, C., Nico, D., and Bisiach, E. (1995). Spatial hemineglect in back space. *Brain*, **118**, 467–72.

Ventre, J., Flandrin, J. M., and Jeannerod, M. (1984). In search for the egocentric reference. *Neuropsychologia*, **22**, 797–806.

Walker, R. (1995). Spatial and object-based neglect. *Neurocase*, **1**, 189–207.

Wolpert, I. (1924). Die Simultanagnosie. Störung der Gesamtauffassung. *Zeitschrift der Gesellsishafts Neurologisches Psychiatrie*, **93**, 397–415.

Index